B. Dennis Banks
Department of Animal Science
Michigan State University

Improving Cattle by the Millions

Leading Bess to the bull

Improving Cattle by the Millions

NAAB and the Development and Worldwide Application of Artificial Insemination

Harry A. Herman

University of Missouri Press
Columbia & London
1981

Library of Congress Cataloging in Publication Data

Herman, Harry August, 1905–
   Improving Cattle by the Millions.

   Bibliography: p. 365
   Includes index.
   1. Cattle—Artificial insemination—History.
2. Cattle—Breeding—History.   3. National Association
of Animal Breeders—History.   4. Artificial insemination—
History.   I. Title.
SF201.H47   636.2′0824   80–25899
ISBN 0–8262–0320–5

To my wife, Lucille, and my sons,
Harry A. Herman, Jr., and Donald L. Herman,
in appreciation of their many years
of encouragement and understanding.

Preface

This book presents an account of man's efforts to improve domestic farm animals, particularly cattle. It deals principally with the worldwide development and application of artificial insemination (AI) as a means of speeding up genetic improvement.

The development of the artificial program for cattle in the United States, Canada, and many other countries is covered in considerable detail. The growth and development of the artificial insemination program in the United States is closely associated with the work of the National Association of Animal Breeders, Inc. (NAAB), Columbia, Missouri. Therefore, the history of NAAB from its early beginning in 1946 is discussed as the story of cattle improvement is related.

Thousands of people at the local, state, national, and international level have played a key role in the development of the artificial insemination program, which in America is a free-enterprise, self-regulated industry. In the early days of the program in America, and likewise the newly founded NAAB, there were few established guidelines to follow. Through research and improved technology as well as field experience, a vibrant and progressive program, adding millions of dollars every year to the income of dairymen and cattle producers, has developed. Much credit is due to agricultural leaders in each state and nation whose faith and perseverance made the present-day program of cattle improvement possible. We have reviewed this history as it is an important adjunct to our agricultural attainments of the twentieth century.

Progress in improving cattle usually means improvement in the standard of living for their owners. All people benefit because cattle are converters of forages into edible nutritious products for the human diet. Livestock improvement comes, under present-day selection methods, from ideas originating in the human mind. Man is the master. When artificial insemination became an applied tool, it gave man the means to shape cattle in mass numbers for the future, largely because superior sires, once identified, can father thousands of offspring. The basic aim in breeding for improvement is to *allow the best to multiply the fastest*. In a span of thirty-five years, the average milk production per cow in the United States has more than doubled; improvement in beef production is making headway; new and sophisticated methods of sire evaluation and selection have been developed; and the spread of venereal disease transmitted by bulls has been reduced. These attainments have, in a large measure, been due to widespread use of artificial insemination. The broad economic impact of this program on cattle improvement methods in developing countries is yet to be fully realized. However, improved genetic potential and greater production per animal unit are fundamental in efforts to feed the expanding population of the world.

Preparation of the material appearing on the pages to follow has involved much research and gathering of information. It would be impossible to list all the persons I am indebted to for assistance. However, I want to express appreciation to members of the National Association of Animal Breeders; the NAAB Board of Directors; Executive Vice-President William M. Durfey and his staff; George Nichols, Assistant Secretary; and all who made this compilation possible and who provided material. I am especially indebted to the extension dairymen in the fifty states; Roy G. Snyder, Ontario Association of Animal Breeders; and workers in the field of animal reproduction in many countries who supplied information.

Tribute should be paid to the late Sam B. Rose, Findlay, Ohio, the first NAAB historian, who compiled and recorded facts in the beginning years of the association.

My thanks are expressed to Robert E. Walton, Madison, Wisconsin; Max Drake, Tiffin, Ohio; and Richard Kellogg, Columbus, Ohio, who reviewed the original manuscript on the NAAB history; and to the staff of the University of Missouri Press for seeing the manuscript to its completion.

I am grateful to Joan Stauffer, Jefferson City, Missouri, for her excellent editorial assistance and for typing the manuscript.

H. A. H.
Columbia, Missouri
September 1980

Contents

Laboratory with adjoining corrals and breeding chutes at the San Carlos Indian Reservation, 1939. (Photograph courtesy of John F. Lasley, University of Missouri–Columbia.)

Jay L. Lush, professor of animal breeding at Iowa State University in Ames, was the first to apply modern concepts in animal breeding. (Photograph courtesy of NAAB.)

Founders of Performance Registry International at Amarillo, Texas, in 1955. *Standing from left:* George Coffee, J. P. Smith, Ralph Harbin, Frank Sims, L. A. Maddox, Jr., M. R. Calliham. *Seated from left:* Clyde Bradford, Colby Conkwright, Percy Powers, Max Blau, and Lee Richardson. (Photograph courtesy of J. P. Smith.)

Original founders, officers, and directors of the Beef Improvement Federation. *Seated from left:* Mack Patton, Bob Purdy, J. David Nichols, M. E. Hammond, Clarence Burch (president), Bill Pope, and John Sutton, Jr. *Standing from left:* Lloyd Schmidt (vice-president), A. F. Flint, Harry A. Herman, Harold G. Thompson, Henry Matthiesen, Stanley E. Anderson (treasurer), Ray Meyer, Dixon Hubbard (ex officio), and John H. Wilson. (Photograph courtesy of American Hereford Association.)

NAAB Board of Directors, 1948–1949. *From left to right:* Fred Hatler (secretary-treasurer), William F. Schaefer, Jr., Max Drake (president), J. W. Pirie, Kenneth Wallin (vice-president), and Maurice Johnson. (Photograph courtesy of NAAB.)

Harry A. Herman, first NAAB executive secretary from July 1953 until his retirement in September 1972.

George Nichols, NAAB assistant
secretary from 1961 to 1976.
(Photograph courtesy of NAAB.)

Dairy Breed secretaries and NAAB board, 1965. *Seated from left:* Marvin Kruse, J. F.
Cavanaugh, Robert Stewart, Robert Rumler, and David Gibson. *Standing from left:* Reuben
Peterson, Kenneth Baushke, Norman Allen, Harry A. Herman, Vincent Jessup, Melvin L.
Kenley, W. H. Armstrong, and David Yoder. (Photograph courtesy of NAAB.)

Sharing the spotlight with master of ceremonies, B. W. Kagy, are *(from left to right)* Ralph
Treet, Harold Rosa, John Schwenk, Max Drake, Stanley Wiggins, D. Master, J. W. Pirie,
Richard Kellogg, D. C. Reid, William F. Schaefer, Jr., David Yoder, Maurice Johnson, and
B. W. Kagy. (Photograph courtesy of NAAB.)

The plaque, commemorating the founding of NAAB, in place at Noba, Inc., in Tiffin, Ohio. (Photograph courtesy of NAAB.)

William M. Durfey, Jr., current NAAB executive secretary. (Photograph courtesy of NAAB.)

Gordon A. Doak, director of Certified Semen Services, Inc. (Photograph courtesy of NAAB.)

NAAB Board of Directors and officers, 1978–1979. *Seated from left:* William M. Durfey, Jr. (executive vice-president), Bryce Weiker (member of executive committee), Wallace E. Erickson (president), Thomas Lyon (vice-president). *Standing from left:* Albert Fox, Brainard Palmer-Ball, Harry Roth, Charles Michaels, Robert Boese, and Kenneth Young. (Photograph courtesy of NAAB.)

Dorothy R. Schnering, 1893–1973, first woman president of Curtiss Breeding Service. Photograph courtesy of Curtiss Breeding Service.

A. Falmer Larsen (Denmark [*left*]) and Enos J. Perry (New Jersey) exchange greetings at the celebration of the start of the first AI cooperative in 1939, which was held in September 1979.

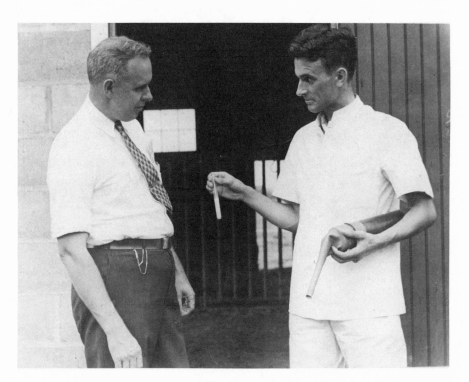

James Henderson (*right*) and Dwight Babbitt check the first collection of semen for AI on 16 May 1938. (Photograph courtesy of Rutgers University.)

J. Rockefeller Prentice
December 17, 1902–June 13, 1972

J. Rockefeller Prentice, 1902–1972, founded, in addition to other AI organizations, the American Dairy Guernsey Associates of Northern Illinois in 1941, which later became the American Breeders Service. (Photograph courtesy of American Breeders Service.)

Former officers and directors of the Ontario Association of Animal Breeders. *Seated from left:* Ward A. Shantz, Charles Campbell, William J. Snowden, Roy G. Snyder (secretary-manager), and Fred Marsh. *Standing from left:* R. J. McDonald, J. M. McLean, Victor Small, M. G. Freeman, Leo Baker, Wilbur J. Shantz, and C. R. Reeds. (Photograph courtesy of Ontario Association of Amimal Breeders.)

Roy G. Snyder, secretary-treasurer-manager of the Ontario Association of Animal Breeders, 1957–1977. (Photograph courtesy of Roy G. Snyder.)

Max Drake, 1946–1951

Kenneth Wallin, 1951–1952

Winthrop Amidon, 1952–1954

J. Stanley Earl, 1954–1958

Richard Kellogg, 1958–1962

Lantz Womack, 1962–1963

Laurel Behnke, 1963–1965

Melvin L. Kenley, 1965–1967

Reuben Peterson, 1967–1968

Kenneth Baushke, 1968–1970

W. L. Campbell, 1970–1972

Robert E. Walton, 1972–1974

David Yoder, 1974–1976

Robert Boese, 1976–1978

Wallace E. Erickson, 1978–1979

Thomas Lyon, 1979–

PART I:
THE DEVELOPMENT OF CATTLE
IMPROVEMENT PROGRAMS

Chapter 1: History and Development of Artificial Insemination

1.1 Introduction

Increased productivity of farm animals adds to the wealth of any nation and helps to provide a supply of nutritious food for its people. Through the ages, man has attempted, through various means, to improve the economic and aesthetic traits of cattle. While progress has been made, no effective method for the large-scale improvement of farm livestock was developed until the field use of artificial insemination was introduced in the early 1930s.

The term *artificial insemination,* commonly called AI, implies the deposition of spermatozoa in the female reproductive tract by artificial (instruments) rather than by natural service involving the male (Figure 1.1).

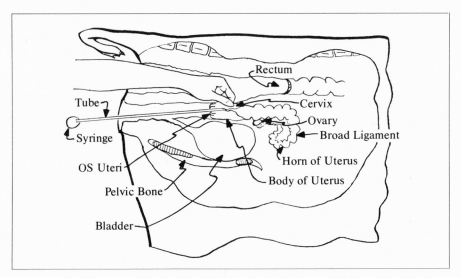

Figure 1.1.   Diagrammatic drawing showing the technique used in making the deep cervical insemination. Source: Louisiana Agricultural Experiment Station Bulletin No. 541.

While the term *artificial breeding* is often used in America, it is not considered correct English. Artificial breeding by definition denotes "not genuine or natural" and in the past was a term repugnant to livestock people because it suggested something unnatural—even to the extent of abnormal inheritance. This interpretation, since the AI program has now been widely adopted, is not considered so taboo, and the two terms have come to mean the same in livestock circles, though artificial insemination is preferred.

By the use of artificial insemination, an outstanding progeny-tested sire can beget thousands of offspring in his lifetime. Several long-lived high production proved dairy bulls used for artificial insemination in the United States have fathered well over fifty thousand offspring. Since such sires usually replace bulls of lesser merit, the gains in production of a given animal population are tremendous. With advancing technology, the future looks even more promising for the mass improvement of domestic animals.

This book deals with man's early efforts to improve cattle and the development of the artificial insemination program to its present level.

1.2 Early Investigations

The use of AI for domestic animals is not a recent development. Arab horsemen are reported to have used the technique for breeding mares in the early fourteenth century,[1] before Anton Van Leeuwenhoek identified sperm by use of his early microscope.

Lazzaro Spallanzani, the Italian physiologist, in 1780 carried out the first-reported scientific studies involving AI. He inseminated a confined dog in heat, using fresh semen, and introduced it directly into the uterus with a syringe. Sixty-two days later the bitch gave birth to three pups that closely resembled the mother and the dog from which semen was procured. This experiment was confirmed by other investigators and helped to demonstrate the practicability of initiating pregnancy, with normal offspring produced, by means of AI. In the course of his work, Spallanzani discovered that the fertilizing power of semen was due to the spermatozoa carried in the seminal fluids. He also found that freezing stallion semen did not kill the sperm but held them in a dormant state until exposed to heat, after which they resumed motility.

In 1799, John Hunter produced pregnancy in the human by this method.[2]

It was nearly a hundred years after Spallanzani's work that efforts were made to use AI in a practical manner. English dog breeders used the technique between 1884 and 1896, finding that several bitches could be bred with one ejaculate and that it is a convenient way to cross dog breeds. In 1890, the French veterinarian Repiquet used AI in horses as a means of improving fertility. Thereafter, horse-breeding centers in several European countries, as a means of overcoming sterility, collected semen from the vagina of mares after they were covered by the stallion. The semen was diluted with cow's milk and by means of a syringe was introduced into the mare's uterus to provide supplementary artificial insemination.[3] It was not until about 1902 that Danish veterinarians found the most important feature of this technique is to breed several mares from one ejaculate. They also viewed AI as a means of extending the influence of a valuable stallion.

1.3 Early Use of Artificial Insemination for Farm Animals

The Russian scientists were the first to utilize seriously artificial insemination as a means of improving reproduction. The leading investigator and pioneer was E. I. Ivanov. In 1899, under his direction, artificial insemination was practiced, with varying results, on numerous government horse-breeding farms. Under his guidance, AI produced a better conception rate than natural service.

Ivanov extended his investigations to cattle and sheep with good results. In 1909, a laboratory was established by the Russian Ministry of Agriculture to study the physiology of reproduction and to train veterinarians in the techniques of artificial insemination. Prior to 1914, about four hundred AI technicians had been trained. From 1914 to 1918, the Bolshevik Revolution interrupted the program. Following the war, Ivanov became director of a livestock breeding station at Moscow. From this laboratory came improved methods for the collection and dilution of semen and practical inseminating techniques for cattle, sheep, and horses. As a result of improved techniques, including the cooling of semen for storage and transport, the number of animals artificially inseminated greatly increased during the 1930s. It is reported that in 1938 some 40,000 mares, 1.2 million cows, and 15 million sheep in Soviet Russia were serviced by means of AI.

In 1914, Prof. G. Amantea, at the University of Rome, devised an artificial vagina to collect semen from the dog.[4] It is believed to be the first. Shortly thereafter, the Russian scientist V. R. Milovanov and others constructed artificial vaginas suitable for collecting semen from bulls, rams, and stallions. The development of the artificial vagina for large animals is credited with making it possible for wide-scale use of AI in the field.

In England, Arthur Walton (1933) of Cambridge pioneered in studies on the properties of semen and its storage and handling. An artificial vagina, known as the Cambridge model, was developed sometime between 1934 and 1935. Walton shipped cooled ram semen

Table 1.1. Leading Countries in Percent of Total Cattle Population Artificially Inseminated, 1973

Country	Estimated Percent of Cattle AI
Israel	99.+
Finland	99.+
Japan	96.0
Czechoslovakia	94.0
Bulgaria	92.0
Denmark	90.0
Norway	90.0
Poland	87.0
France	75.0
West Germany	70.0
Soviet Russia	70.0
Netherlands	69.0
Great Britain	65.0
Canada	53.0 Dairy
	11.0 Beef
United States	55.0 Dairy
	3.0 Beef

Source: Tabulated from 1973 survey by Telesforo Bonadonna, *Zootecnica E. Veterinaria La Fecóndazione Artificiàle,* via Monte Ortigara 35, Milano, Italy, 30, No. 1 and 2, 1975: and from reports compiled by the author during 1972–1979.

(10°C) in a vacuum flask filled with cracked ice to Poland in 1936. Five ewes were inseminated with semen fifty-one hours old and two of them conceived.

Most of the early investigations of AI were chiefly concerned with obtaining satisfactory reproduction. The Russian investigators, along with others, recognized the great potential that AI offered for using the best sires extensively. Practical application of AI could not be undertaken until it was amply demonstrated that the method would result in reproduction equal to natural service, with normal offspring delivered.

Edward Sorensen, of the Royal Agriculture and Veterinary College, in Copenhagen, and others in 1936 organized the first cooperative artificial breeding association in Denmark. The techniques used were largely adopted from the Russians. The association had 220 members, and 1,070 cows were inseminated the first year. The pregnancy rate was 1.68 and slightly better than natural service. In 1937, the Danish veterinarians developed the rectovaginal technique for the artificial insemination of cows (Figure 1.1). This method makes efficient use of semen and improves fertility about 10 percent. The program in Denmark grew steadily, and by 1960 nearly 100 percent of the cattle were bred artificially.

The AI program expanded in most European countries, but was badly hindered during World War II. After the war, it became a valuable tool in propagating depleted herds in Germany, Poland, France, and the Netherlands. With the adoption of the latest techniques, the use of AI has increased steadily, and about 70 percent of European cattle are now artificially inseminated (Table 1.1).

1.4 Early Developments: AI in the United States

Horses. The development and use of AI for farm animals in the United States generally went through the same early stages as it did in Europe. If we accept the definition of artificial insemination as commonly employed, the technique was used by horse breeders in America many years before 1937.

In 1906, L. L. Lewis, of the Oklahoma Agricultural Experiment Station, described the use of the impregnator and in 1911 published a bulletin on artificial insemination in horses.[5] He said, "The use of the impregnator, or other artificial means of introducing the semen, in

Figure 1.2.   A "whiteface" calf born in 1907 as a result of artificial insemination in the R. L. Hughey herd, Alva, Oklahoma. (Photograph courtesy of Graham School, in Garnett, Kansas.)

horse breeding has passed beyond the experimental stage." According to "old-timers" in horse breeding, AI was in use long before Lewis described the use of the impregnator. Horse breeders collected semen in a pan after the stallion dismounted, placed it in a flask, kept it warm (some placed the flask under a sitting hen), transferred the semen to a capsule, and introduced it into the uterus of mares to be bred.

Cattle. Years before the invention of the artificial vagina, some cattlemen were using AI by employing the techniques utilized by horse breeders. In 1907, a "whiteface" calf was born in the herd of R. L. Hughey, Alva, Oklahoma, as the result of the dam being artificially inseminated.[6] The method used in this venture involved collecting semen from the vagina of another cow that had just been bred naturally, putting the semen into a gelatin capsule, and inserting it into another cow in heat. It is also reported that Hughey, in 1910, using the capsule method, bred eight hundred mares artificially to one stallion.

Authentic information indicates that Thomas C. Webster, of Fort Steilacoom, Washington, while in charge of the fort's dairy herd, began using AI in that herd in 1926.[7] In 1930, he became herd manager at Winnebago State Hospital in Wisconsin. Working in cooperation with Glenn Householder, who was then in charge of Wisconsin State Institution herds, Webster artificially inseminated the herds throughout the state.

Howard Clapp, manager of the famous Pabst Farms herd, Oconomowoc, Wisconsin, began using AI in that herd in the mid-1930s. He reported that between 1934 and 1938 a total of 130 calves resulting from AI were born.

In 1937, the practice of artificial insemination was started in the University of Missouri–Columbia dairy herd.[8] Semen was obtained by the ampulla massage method,[9] and the first calf resulting from AI was born 18 April 1938. About the same time, many other agricultural colleges—Nebraska, Minnesota, Wisconsin, Tennessee, and Cornell—began using AI in their station herds. This led to the initiation of research on spermatozoa, semen extension (dilution), storage, shipping, and insemination techniques that were greatly expanded as the organized AI program began to develop.

1.5 Early AI Organizations in the U.S.

While AI had been practiced earlier in private herds, the organized program was started in the United States in 1937 and 1938. Beginning in 1937, C. L. Cole,[10] at the University of Minnesota, North Central

Station, set up an experiment to test the practicability of AI in dairy herds under ordinary farm conditions. The results were encouraging and conception was equal to the results from natural breeding.

The First Cooperative AI Organization. In May 1938, the first farmer-owned cooperative artificial breeding association in the United States began operation in New Jersey. Enos J. Perry, extension dairyman, of Rutgers University in New Brunswick, New Jersey, was the leader. Perry had visited Denmark and had observed the AI program underway in that country. He successfully prevailed upon a group of New Jersey dairymen to form a cooperative, to be organized along the same lines as the Denmark cooperatives. The organization started with 102 members, and 1,050 cows were enrolled by the owners.

In June 1938, the second such association began operations at Hughesville, Missouri, with the Missouri College of Agriculture and the Farm Security Administration cooperating. Harry A. Herman, professor of dairy husbandry, and M. J. Regan, extension dairyman, both of the University of Missouri–Columbia, were the leaders. They were assisted by F. F. McKenzie, a pioneer worker in animal breeding and artificial insemination at the Missouri station. Service, using fresh, undiluted, semen collected every other day from university-owned bulls was provided for about five hundred dairy cows kept on two large cooperative farms and thirty individual herds in the Hughesville area. This project was quite successful but was discontinued in 1944 when the government sold the farms.

Other farmer-owned AI cooperatives were formed in New York, Tennessee, Wisconsin, and other states between 1939 and 1945 (see Appendix A).

By 1946, there were 84 such associations (bull studs) listed in the Dairy Herd Improvement Letter (U.S. Department of Agriculture [USDA]). It was customary at this time to list county cooperative organizations of dairymen whose herds were enrolled on the AI program. There were 608 "Artificial Breeding Associations" located in thirty-four states listed as active during the year. As of 1 January 1947, there were 140,571 herds containing 1,125,040 cows enrolled in AI associations, an increase of 545,563 cows, or 94.1 percent, compared to 1 January 1946.

Cooperative Bull Associations Replaced by AI. The farmer-owned AI cooperatives were responsible for gradually replacing the "cooperatives were responsible for gradually replacing the "cooperative bull associations" in the United States. Beginning in 1906, with the first cooperative bull association formed in Michigan, this program had grown steadily with over four hundred associations active in 1936. The bull association, often a county-wide program, was made up of

Table 1.2. Number of Cows Bred in U.S. by Bull Associations and by AI, 1939–1943

	Bull Associations		AI Organizations	
Year	Bulls	Cows	Bulls	Cows
1939	1,079	35,333	33	7,539
1940	1,266	41,678	138	23,977
1941	1,292	41,478	237	70,751
1942	1,291	43,251	412	112,788
1943	1,230	41,487	574	182,524

Source: Agricultural Research Administration, USDA, January 1944.

cooperating farmers who jointly owned bulls. As a rule, there were four or more "blocks" in each association. A block comprised four herd owners who shared the use of a registered bull kept at one member's farm. The "block bulls" were shifted every two years to avoid inbreeding. While this program helped to provide better genetic potential, it had the disadvantage of having to lead animals some distance for service, did little to control the spread of disease, and only a few proved bulls were in use at one time.

The purebred bull program, strongly backed by agricultural extension workers and the Bureau of Dairy Industry, USDA, helped to improve the genetic potential of dairy herds. There were "swat-the-scrub-bull" programs, "registered bull" campaigns, and similar movements from 1906 forward. It is little wonder that many dairymen, already sold on the value of good herd sires, realized the advantages of AI rather quickly.

It is interesting to note (Table 1.2) that five years after its field application, AI accounted for breeding over four times as many cows to registered bulls as did the bull associations in over thirty-five years of operation.

There were 318 cows bred per AI bull and 33.73 per bull association bull during 1943. This gives an idea of the effectiveness AI provides in using "bull power." By 1970, over 3,600 cows were serviced per AI bull, over one thousand times as many as were serviced by the association's average bull. The number of bull associations declined steadily after 1942. They must be given much credit, however, for many years of promoting improved breeding and progeny testing throughout the dairy industry.

The question in the minds of many in those beginning years of the AI program in America was, "Will it work?" There were many objections raised about this new program that had yet to demonstrate that cows could be settled on time and could deliver normal calves.

1.6 Early Objections Voiced Against AI

The arguments raised by farmers, the agricultural press, religious groups, some college workers, and some state and federal agencies against the use of AI in its beginning, as an organized program in America, were numerous. A few were:

1. The term *artificial* suggested something unreal and even repugnant. Some claimed it was against the laws of God and nature and should be abolished.
2. There were fears that calves resulting from AI would be abnormal, weak, perhaps have two heads, six legs, and be undesirable in general. Many farmers traveled miles to observe the first calves born in the neighborhood as the result of AI to see if they were normal.
3. Artificial breeding would produce more bull than heifer calves.
4. A cow bred artificially once would cease to come into heat and have no further calves after the first AI calf.
5. Bulls forced to serve the artificial vagina continually would become sterile and in a short time be of no use.
6. Cows wouldn't settle on time.
7. It was a nuisance to have to keep a cow in the barn until the AI technician arrived.
8. Disease would be spread from herd to herd.
9. There would be "mix-ups" in semen and one didn't know what bull was used.
10. No one could breed a good herd of cattle unless he had his own bull and bloodlines.
11. AI took too much time due to telephoning someone to breed a cow each time.
12. AI costs too much, and it doesn't cost much to keep a bull.
13. There will be no market for bulls.

All of the above, and other objections, were found invalid, overcome, or lived with as the program demonstrated its practicability.

1.7 Advantages of Artificial Insemination

No major development in modern agriculture, with the exception of hybrid corn, has been accepted and has advanced more rapidly than AI. Economic and management practices made the method readily adaptable for dairy cattle where, among the larger farm animals, the greatest expansion has occurred. Artificial insemination of beef cattle is gaining. It is used for turkeys, swine, dogs, goats, sheep, horses, laboratory animals, bees, and, in some cases, humans.

The chief advantages of artificial insemination, particularly for cattle, are:

1. The use of outstanding sires can be multiplied thousands of times. In natural use, dairy bulls sire about fifty to seventy-five calves a year; in beef, probably twenty-five to thirty as an average. Proved bulls in AI use easily sire five to ten thousand calves in a single year and some have over one hundred thousand progeny in a lifetime.

2. Cattlemen can obtain services from an outstanding bull that would be prohibitive in price for them to own.

3. The danger of spreading disease through natural service, particularly vibriosis, trichomoniasis, leptospirosis, vaginitis, brucellosis, and tuberculosis, is materially reduced with disease-free bulls used in AI.

4. The use of frozen semen permits selective matings to be made anywhere transportation is available. Herd owners now have a wide choice of outstanding bulls and can shape breeding programs accordingly.

5. Young sires can be sampled by use on many cows in many herds quickly so as to obtain adequate proof. Proving young sires is an important part of the cattle AI program.

6. Valuable sires may be kept in service by freezing and storing semen for future use. This is a form of "insurance" against injury or death of a bull that many cattle breeders employ.

7. The cost, danger, and nuisance of keeping a bull on the premises is removed. Prior to the development of the AI program, it was estimated that one farm worker, in the United States, was killed every four days by a bull. Many more were injured.

8. Animals may be mated, irrespective of size, without danger of injury.

9. Infertile bulls are detected early and losses due to delayed calving reduced.

10. As a rule, properly carried out, AI will equal and in some herds exceed the results of natural service at less cost per calf. The gain in genetic potential for performance is always an important added benefit and is the primary purpose for using AI.

11. Crossbreeding can be readily utilized.

Disadvantages of AI. There are fewer arguments against the use of AI today than some forty years ago when the objections (section 1.6) were voiced. The sale of bulls for breeding purposes is reduced and remains a complaint in some quarters. Particularly in beef cattle, AI requires additional labor and equipment. More breeding and calving records as well as cattle identification are necessary and, in general, a more complete herd-disease control is initiated. For the operator who strives to improve his herd income, there are few disadvantages of AI that cannot be overcome. In modern milk and meat production, the path of the "easy going" operator becomes increasingly difficult.

1.8 First Field Use of AI for Beef Cattle

Beef cattle breeders and their respective registry organizations took a dim view of AI in the early days. Regulations were established to prohibit the registration of a calf unless the owner of the dam and the sire were the same. It was argued that since beef cattle ranged over wide areas and were not handled like dairy cows, AI would be of little value to the beef man. Some beef producers that turned to AI in the earlier years did so as a means of controlling the spread of venereal disease. Consequently, until the past twenty years, little effort was made to demonstrate the place of AI for purebred or commercial beef herds.

However, from 1939 to 1942, research workers from the University of Missouri with the cooperation of the Bureau of Animal Industry, USDA, and the Indian Service, U.S. Department of the Interior, demonstrated that the field use of AI for beef cattle can be successful.[11]

John F. Lasley, Ralph Bogart, and F. F. McKenzie, over a three-year period—covering four breeding seasons—inseminated cows in the one thousand-cow herd of the San Carlos Apache Tribe, San Carlos, Arizona. The semen was collected from bulls at the reservation and was used fresh or diluted with the egg yolk-phosphate buffer. The necessary holding pens and breeding chutes were constructed, a laboratory was established, and the cows were checked for heat by visual observation.

Artificial insemination proved to be a practical and efficient method for the breeding of range cows. Over a three-year period, 79 percent of the cows inseminated dropped calves, requiring 1.63 inseminations per calf. As many as five hundred different cows were inseminated with the semen from a single bull over a five-month breeding season. It was concluded that twice as many cows could have been inseminated if the greatest use of the semen had been made.

1.9 Early Regulations Governing AI

The AI program in America began as one of free enterprise and continues so to this day. The first AI organizations, nearly all farmer-owned cooperative, were fostered and guided by the dairy extension specialists, teaching and research personnel at the various agricultural colleges, and dairy farmers.

There was a dearth of information on the organization and administration of such a program. The question of registration of calves resulting from artificial insemination using bulls owned by the AI cooperative, or by another breeder, was often raised. Beginning in 1938, the extension section and the production section of the American Dairy Science Association (ADSA) set up committees to draft guidelines, based on experience, for the conduct of an AI program. From 1940 to 1943, rather concise and complete recommendations were issued for organizing a "cooperative artificial breeding association" and its incorporation; finances and setting of breeding fees; housing and office facilities; selection of personnel including the training of technicians; record keeping, collection, and processing of semen; and other items. These proved of immense value in soundly guiding the program.

In 1938, the secretaries of several dairy breed registry organizations and members of the breeds relations committee of the American Dairy Science Association met in the American Jersey Cattle Club offices in New York City and agreed on the "Between Herds Breeding Certificate." This became the first official form, agreed upon by the respective dairy breed organizations, relative to artificial insemination of purebred dairy cattle.

The Purebred Dairy Cattle Association (PDCA), an organization comprising the six principal dairy breed registry organizations in the United States, was formed in 1939. The PDCA and the various committees of the American Dairy Science Association joined forces in formulating uniform requirements for the artificial insemination of purebred dairy cattle. The first draft of these requirements appeared in 1943. Perry and Herman, representing the ADSA, helped throughout the years in establishing the requirements. The requirements included approval by the PDCA for AI organizations; AI technicians; a uniform breeding receipt for use by technicians; the Between Herds Breeding Certificate; policy statements concerning the ownership of bulls used in AI; and keeping proper breeding records by the approved AI organization and its technicians. The policy of having a representative of the PDCA check each approved AI business annually was also initiated.

In 1954, the PDCA adopted requirements governing the use of frozen semen. The requirement that all bulls producing semen for AI use be blood typed was added. Revisions of the "Requirements Governing Artificial Insemination of Purebred Dairy Cattle" were made as the AI program developed. Revised "requirements" were made effective in 1956, 1960, 1966, and the latest in 1974. The National Association of Animal (Artificial) Breeders (NAAB) collaborated in the revisions through 1974. The "PDCA Requirements," as they became known throughout the cattle-breeding industry, served a useful purpose in bringing uniformity, integrity, high standards, and accuracy in the AI program. Although their specific authority was limited to purebred cattle, these requirements set the pattern for conduct of the program for all dairy cattle.

In late 1976, due to growing differences in policy regarding the conduct of the AI program among the breeds as well as between the PDCA and NAAB, action was taken that each breed would formulate its own policy regarding artificial insemination effective 1 January

1977. At the same time, NAAB adopted a policy of "self-regulation" of the AI program.

Free Enterprise. The AI program in America developed strictly as a free-enterprise venture. The necessary rules and regulations were established by the people conducting the program, with the assistance of dairymen, college workers, breed organization executives, and state and federal officials. While AI programs in other countries were being fostered, financed, and regulated by government decree, the program in America, by a "self-policing" policy, developed without the burden of state and federal regulations. This policy no doubt was influenced in part by the fact that the early AI cooperatives were looked on as being in the same category as the cooperative dairy herd improvement associations and the cooperative bull associations.

The workers in the Division of Dairy Cattle Breeding, Feeding and Management, Bureau of Dairy Industry, USDA, in 1936 and 1940 were not impressed with the possibilities voiced for AI as a tool for dairy cattle improvement. While there was no opposition from the bureau, busy with its cooperative bull association and dairy herd improvement association programs, a "wait and see" attitude was taken. In the 1936 Yearbook of Agriculture, published by the USDA, and largely devoted to "the creative development of new forms of life through plant and animal breeding" and "superior germ plasm," AI is mentioned only once in connection with animal breeding. As the AI program grew and began to demonstrate its potential, the workers in the USDA's dairy and animal research divisions became strong and faithful contributors to its welfare and continue to be so. Without this support of the programs for testing for production, sire evaluation, and proving methods developed by our government workers, the present AI program could not have attained its present heights.

That the AI program in America developed and operates today free of federal and state regulations, except as sire health is involved, is well summarized by this statement:

> This (development of the AI program) has been accomplished without government edict, by farmer cooperation, with competition between organizations, and with assistance and sympathy from appropriate government agencies, dairy and breed groups and the veterinary profession.[12]

It is characteristically American that the persons involved in the conduct of the AI program, in all its phases, would band together in an organization mutually to further their objectives. Such an organization—the National Association of Animal (Artificial) Breeders—had its beginning in December 1946.

Chapter 2: The Revolution in Cattle-Breeding Methods

2.1 Introduction

In terms of years, as well as cattle generations, the wide-scale use of artificial insemination is relatively recent. Its impact on cattle-breeding methods to date, however, is dramatic and far-reaching. Never in the recorded history of the cattle industry have concepts and ideas involving improvement of economic traits been so drastically changed in so short a space of time. These developments are aptly described by I. M. Lerner and H. P. Donald,[1] who state that "artificial insemination has been both a result and a cause of the revolution in large animal breeding."

The achievements of the AI program in the United States in elevating the average milk production per cow is shown in Table 2.1. The average annual milk yield per cow increased from 4,622 pounds in 1940 to 11,240 pounds in 1978. Comparable results have been indicated for other countries where AI has been used extensively. By the use of sires capable of increasing the genetic potential of their progeny for performance—be it milk or meat—the AI program makes a substantial economic input. The savings resulting in terms of national economy are in terms of hundreds of millions of dollars annually. The number and value of all cattle in the United States from 1930 to 1978 is indicated in Table 2.2. The same information for milk cows and heifers from 1935 to 1978 appears in Table 2.3. As is well summarized

Table 2.1. Production of U.S. Dairy Cows, 1940–1978

Year	Milk Cows Number	Avg. Prod. Per Cow		Total U.S. Milk Production
		Milk	Milkfat	
	(1,000)	(lbs.)	(lbs.)	(Billion lbs.)
1940	23,671	4,622	183	109.4
1950	22,000	5,314	210	116.6
1960	17,515	7,029	264	123.1
1970	12,000	9,751	357	117.0
1975	11,143	10,350	380	115.3
1976	11,055	10,879	399	120.3
1977	10,974	11,181	409	122.7
1978	10,848	11,240	412	121.9

Source: National Milk Producers Federation, "1979 Report"; and *Dairy Herd Improvement Letter* 55:2 (1979), USDA.

Table 2.2. Estimated Number and Value of All U.S. Farm Cattle, 1930–1979

| Year | Number | Farm Value | |
		Per Head	Total
	(Thous.)	(Dollars)[1]	(1,000 Dollars)
1930	61,003	56.36	3,438,129
1934	74,369	17.78	1,322,280
1939	66,029	38.44	2,538,154
1945	85,334	68.40	5,836,845
1949	76,830	135.00	10,372,050
1954	95,679	92.00	8,802,468
1959	93,322	153.00	14,278,266
1964	107,903	127.00	13,703,681
1969	110,015	158.00	17,382,370
1974	127,788	293.00	37,441,884
1975	132,028	159.00	20,992,452
1976	127,980	190.00	24,316,200
1977	122,810	206.00	25,298,860
1978	116,375	232.00	26,999,000
1979	110,864	403.00	44,678,192

Source: Agricultural statistics (1965–1979), USDA, Washington, D.C.

[1]The reader's attention is called to the fluctuation in price per head by periods as related to the total number of cattle and its effect on total value. The economic impact of beef price cycles is often traumatic for beef producers and severely depresses the use of AI.

Table 2.3. Estimated Number of Milk Cows and Heifers in U.S., 1935–1979

Year	Cows and Heifers over 2 yrs. Old	Heifers 1–2 yrs.	Heifers under 1 yr.	Total
	(Thous.)	(Thous.)	(Thous.)	(Thous.)
1935	26,082	4,995	5,280	36,357
1939*	24,600	5,122	5,904	35,626
1944	27,704	6,352	7,201	41,257
1949	23,862	5,327	6,081	35,270
1954	23,896	5,873	6,392	36,161
1959	20,132	5,050	5,526	30,708
1964	17,647	4,395	4,692	26,734
1969	14,152[1]	3,990[2]	3,939[3]	22,081
1974	11,286	3,942	3,950	19,178
1975	11,217	4,095	3,399	18,711
1976	11,087	3,958	3,336	18,381
1977	11,035	3,888	3,345	18,268
1978	10,930	3,896	3,312	18,138
1979	10,853	3,936	3,500	18,289

Source: Agricultural statistics (1979), USDA; Economic Research Service reports, USDA; and "1979 Dairy Producer Highlights," National Milk Producers Federation, Washington, D.C.

[1]From 1965 forward listed as cows that have calved.

[2]From 1965 forward listed as heifers 500 pounds and over kept for milk.

[3]Heifers under 1 year kept for milk, estimated on basis of 33 calves per 100 cows.

*AI program began in 1938. In 1945, total dairy cows equaled 27,704,000 head, the highest in U.S. history.

by studies on milk yields in dairy cattle, about 25 percent of the phenotypic variance is genetic and 75 percent environmental.[2] The increased production per cow in the United States, during the past thirty-five years, can be attributed to both improved inheritance and changes in herd management. Herds are larger; dairying is much more specialized than in the days of the small family-size herd; concentrate feeding is much heavier and culling pressure is greater. There is little question, however, about the fact that profitable dairy cows and, likewise, beef cattle must have the necessary genetic potential if they are to respond to good feeding and management. Animal husbandry at the present and in increasing tempo for the future depends upon AI to supply the necessary genetic potential. The application of population genetic principles in the selection of breeding stock, as has been well demonstrated during the past twenty-five years, offers assurance than an increasing number of farm animals can be endowed with a high genetic potential for production.

2.2 Genetic Progress: Early Days of AI

During the first fifteen to twenty years of the AI program, major emphasis was on (a) methods of preserving and extending semen; (b) improving inseminating techniques and maintaining a satisfactory conception rate; (c) control of venereal diseases spread by the bull in natural service; and (d) perfecting the establishment, administration, and promotional programs of AI organizations.[3]

During the first decade of the organized AI service program (1938–1948), the caliber of dairy bulls used was not disregarded. The AI organization eagerly sought sires with the highest verified production capabilities through the program of the National Cooperative Dairy Herd Improvement Association (DHIA). Bulls that proved in breeders' herds through the DHIA testing program and through the herd-testing programs of the breed association—often both programs—were procured locally and nationwide. Promising young sires were also put in use. Each AI organization usually had a "sire committee" composed of dairy farmers. The sire committee members, along with the bull stud management, spent much time and money inspecting bulls, their ancestors, progeny, the bull owner's herd, and his management practices. Often sire committees were assisted by the extension dairymen from the state agricultural college in evaluating bulls.

There were many problems in the selection of bulls. Proved bulls were seven to eight years of age and usually had few tested daughters in only one herd when purchased. Some failed to produce highly fertile semen, and some would not work very well on the artificial vagina. Others had to be bypassed because of disease, injuries, price, and unpopular pedigree. Too, there were pressures brought by local breeders, some of them directors or patrons of the AI cooperative, to have one of their bulls in the stud. Usually it was these breeders who had been selling registered bulls in the area, and they enjoyed a following among local dairymen. Also, in many cases, they had bulls as good as some of those hundreds of miles away, with a lot of "printers ink" on the pedigree. As the AI organizations learned, as soon as the daughters of their bulls began finishing lactations, they were not doing as good a job, in many cases, as purebred breeders in raising milk-production levels.

The main weakness was that AI organizations were using the same selection methods followed by breeders in choosing sires. The breeder had a better-than-average cow herd and provided better-than-average feeding and management. When the bull that proved in one herd, under good conditions, was used through AI on a cross section of the area cow population, his daughters did not equal his daughter average in the original herd. There was criticism of the AI program by purebred breeders because "the AI bulls weren't good enough." Some of this criticism came about as breeders saw the bull market beginning to dwindle, and the old idea that "every registered bull calf could make someone a herd sire" faded as the AI program grew. A comparison of the production of naturally sired and AB (artificial breeding) progeny, however, indicated that the *repeatability* of the production proof of the naturally proved bulls in AI was low. The AI organizations, combining efforts through the National Association of Artificial (Animal) Breeders, were quick to recognize the problem and to resolve it.

In 1947, Raymond Albrectsen, Cornell University extension dairyman, presented a paper at the annual meeting of the American Dairy Science Association.[4] He indicated that both the level of production and the amount of increase over dams were much lower in the AI-sired cows than those of the daughters of the same sires that were naturally serviced. This raised the question—already uppermost in the minds of AI management—"Will natural service proofs be repeated in AI?" The New York workers reported at this time that they had found the proof on young sires selected and proved in AI to be more reliable in predicting future production than the "one herd natural-proof sires." About this same time, several college research workers all reported the repeatability of natural-service proofs in AI to be low.

In 1953, William F. Schaefer, Jr.,[5] Tunkhannock, Pennsylvania, manager of Northeastern Pennsylvania Artificial Breeding Cooperative and chairman of the NAAB Sire Recognition Committee, compared the original proof with AB daughters of two hundred sires used in Pennsylvania and New York. His study, involving 34,178 production records on AI-sired daughters, indicated little correlation between a bull's natural proof and the results obtained through wide usage of artificial insemination. He found the transmitting ability estimated by pedigree analysis, and weighting of the ancestor's production, predicted future production of a sire better than the original natural proof.

2.3 Sire Evaluation Methods Assessed

With the experience gained in evaluating the performance of naturally proved bulls in AI, there became a need to revise the sire evaluation methods. Before proceeding, we will briefly review sire selection methods prior to the AI program.

Early Sire Selection Methods. Until the early 1920s, there was little scientific information available. The first Cow Testing Association (DHIA) started in Michigan in 1905, and shortly thereafter breed registry associations adopted some form of testing to measure milk and butterfat. The use of production information took hold slowly. Phenotypic selection was the vogue. Cattle breeding was considered an art and the theory that "like begets like" prevailed.

As production information accumulated, methods for compiling sire proof, based on the progeny test, came into the picture. The local, state, and national diary shows, as did beef shows, played an important role in establishing the type and ideals for body conformation. Herd sires were selected largely on the basis of pedigree and bloodlines with emphasis on the sire and the dam. Type and show ring winnings also received strong consideration. Production was stressed with the breed-testing programs and later the DHIA records were used. Various schemes for indexing and comparing sires, based on progeny performance, evolved. Among these were the *Daughters Production Average,* computed to a mature equivalent (six to seven years of age) basis; the *Equal-Parent Index,* based on the premise that each parent contributes equally to the producing ability of the offspring; and the *Daughter-Dam Difference,* whereby a sire was indexed on the basis of the plus or minus difference between the average milk production (mature equivalent) of the daughters and dams. The weakness of the daughter-dam comparison in evaluating bulls is that the records of the daughter and dam are made in different years, and their environment may vary. Also, the level of production of the dams must be considered. A rather mediocre bull can have a large plus difference if mated with low-producing cows. There were other evaluation indexes proposed, but the "Daughter-Dam Difference" and the "Equal-Parent Index" were the chief measures utilized when the AI program began in America.

Breed registry associations supported production testing, and various forms of recognition for superior performing sires were established (usually based on rather broad and sometimes not very significant standards). With the tools available, dairy breed associations today greatly encourage better breeding and keep pace with progress as new concepts develop. In the earlier days, it was often said by college professors that "a man who picks three good bulls in a row is a successful breeder." This feat was a difficult one, as most breeders found through experience. Some breeders achieved acclaim by being "lucky" in choosing a bull that proved highly outstanding. There was also much emphasis on "nicking" (epistasis), with experience indicating that some bloodlines when crossed produced better offspring than other crosses. There was also the theory of avoiding "cold blood" and the keeping of family lines intact.

Breeders of cattle through the years have made progress and have laid the foundation for cattle improvement today. Their efforts should not be taken lightly, as the information and scientific aids we have today are of recent application. It must be acknowledged, however, that the earlier methods of sire evaluation and the selection of breeding stock did not result in the continuous building of a genetic makeup conducive to high performance.

Taking stock of the situation, animal geneticists proceeded to study methods for evaluating AI sires and to develop plans for producing sires whose transmitting ability, based on the original proof, could be predicted with considerable accuracy for future offspring.

2.4 Improving Repeatability of AI Proofs

Numerous investigators[6] found the regression of the average production of the AI daughter on natural-serviced daughter-dam difference to be about .20 to .30. Geneticists then turned their attention to the repeatability of the proof on the first group of production-tested AI daughters on future daughters by the same sire. This trend started the development of concepts and their application that revolutionized evaluation and selection methods in cattle breeding.

2.4.1 Population Genetics Concepts Introduced

Following the rediscovery in 1900 of Mendel's classic work, which was first reported in 1865, the knowledge and theory of modern genetics began to be applied in the AI program. As a result, mathematical procedures are now used for predicting the results and for explaining the variations in inheritance among farm animals. Beginning about

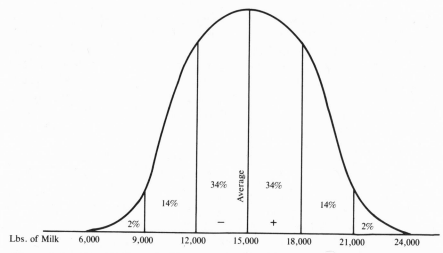

Figure 2.1. Distribution of milk-production records of 1,000 daughters of a sire.

1918, the quantitative approach in large animal breeding was initiated. The application of biometrics resulted in adopting principles of population genetics. Using the basic laws of heredity, geneticists are able to portray in mathematical terms the transmission of genes from generation to generation and to predict, on the average, the expected characteristics of succeeding generations. The population geneticists work with significant numbers (data) in terms of the entire population (Figure 2.1).

2.4.2 The AI Program Paves the Way

As is true of most worthwhile accomplishments, the concepts of population genetics were established over a period of many years. The pioneer work of Francis Galton (1822–1911), an English statistician who is responsible for "the law of filial regression," laid the foundation for the development of biometry. Ronald A. Fisher, an English biometrician (1890–1962), and Sewall Wright (1889–), former USDA animal husbandman, led the way in the early studies of population genetics with animals. Wright was one of the early leaders in applying the concepts of population genetics in large animal improvement programs. The greatest advocate of Wright's early theories in America is Jay L. Lush (1896–), a professor in animal breeding at Iowa State University, Ames. Beginning some fifty years ago, Lush and his many eminent students have continually developed concepts leading to the application of scientific principles in livestock breeding.

For years, these concepts were taken lightly by cattlemen, but were adopted and found highly applicable by poultry and swine producers. To demonstrate effectively the results of applying scientific principles to inheritance, the population geneticist requires adequate numbers. Artificial insemination, with dairy sires having hundreds, in some cases thousands, of production-tested daughters in many herds provided the answer. Large-scale breeding operations and mass improvement of cattle became a reality. Like most evolutionary changes, the application of sound concepts in animal selection had to wait until the time was right. The AI program added a new dimension—the sampling and proving of sires for its use.

2.5 Innovations in Sire Selection

As the AI program expanded and progeny of the early sires in use came into production, it became apparent that (a) despite the most careful selection, bulls vary widely in their transmitting ability for milk production and other traits; and (b) that the only reliable test of a bull's value in AI is his proof in AI use.

Investigations beginning in the early fifties, particularly at Cornell University, by Charles R. Henderson and his associates[7] indicated that the accuracy of AI proofs depended upon several factors: (a) the number of tested progeny available for a given bull; (b) the number of herds in which daughters were tested; (c) the use of herd-mate or contemporary comparisons in evaluating the milk production of the progeny of AI bulls; and (d) corrections of the production records for age, year, season of freshening, and the herd in which the record was made.

Henderson and his associates[8] developed a plan of contemporary comparisons as a means of evaluating AI sires. This plan, now in general use, but with some modifications as experience was gained, involved making adjustments for age differences for all lactations in a herd and for comparing the daughters of a sire with other cows in the same herd freshening during the same season of the year. It was also found that by using the contemporary comparison method and by adjusting herd differences, the regression for the number of daughters (N) could be expressed as $N/N+15$. The accuracy of natural-service proofs as compared to AI proofs with varying numbers of daughters is shown in Table 2.4.

The new concepts in breeding resulted in creating more accurate methods in sire evaluation. These included: herd-mate or contemporary comparisons; Predicted Difference, as a means of estimating a bull's transmitting ability; and use of repeatability or confidence measures to express the probability of a sire proof being repeated on future daughters. Early in these developments, R. D. Plowman, Frank N. Dickinson, Ben McDaniel, E. L. Corley, and others, all in the Animal Husbandry Division (USDA) in Beltsville, Maryland,

Table 2.4. Accuracy Natural-Service Proofs and AI Proofs with Varying Numbers of Daughters

Number of daughters	Regression of future AI proofs on various information
Natural service—avg. 10 daughters	.25
10—AI daughters	.40
25—AI daughters	.62
50—AI daughters	.77
100—AI daughters	.87
500—AI daughters	.97

Source: H. Wilmot Carter, "Effectiveness of Artificial Insemination in Dairy Cattle Improvement," *Journal of Dairy Science* 45: (1962) 277–81.

conducted research on DHIA records. These studies were made, often in cooperation, with the research workers in animal breeding at various universities. These efforts led to the establishment of a computerized program and of the modern sire and cow evaluation concepts utilized for the USDA "Sire and Cow Summaries" now in use. The AI organizations joined forces in these efforts and put them into use. The sampling and proving of young sires gained momentum and became a major project for most bull studs.

The application of modern breeding concepts in AI programs in Europe, New Zealand, and other countries paralleled that in the United States and in some cases was earlier. New Zealand began using herd-mate comparisons in 1950. Great Britain and the United States (New York State) began about 1954.[9]

2.6 Modern Methods for Estimating Breeding Value

Most estimates of breeding value are directed at sire evaluation. The sire and the dam of an individual each *contribute a sample half of the genes inherited*. The sire, however, has many offspring. In AI use, this may range from a few hundred to tens of thousands. Some of the well-proved, more popular bulls in AI account for thirty to fifty thousand services per year. In four generations of cattle, about 93 percent of the genes (inheritance) in a herd is accounted for by the sires used in those four generations.

2.6.1 The USDA-DHIA Daughter–Herd-mate Comparison

In Daughter–Herd-mate Comparison, the milk production of a bull's daughters is compared with the production of other cows (herd-mates) sired by different bulls in the same herd, and at the same time. This method largely rules out preferential treatment or environmental effects and represents a most forward step in evaluating sire proofs. All records are calculated to the standard 305-day lactation, 2× milking, Mature Equivalent (ME) basis (six to seven years of age). Adjustments are also made for season of freshening, year, herd production level, and number of herd-mates.

The Daughter–Herd-mate Comparison replaced the Daughter-Dam Comparison as the official USDA Sire Summary method in 1962. Among other advantages, it permitted a more accurate comparison of bulls and their ranking used in different herds.

Contemporary Comparisons. In a few countries, true contemporary comparisons, in which only first lactation records of progeny and herd-mates in milk at the same time, are used. It has the advantage of reducing errors in sire evaluations due to lactation records beginning at different times, heavy culling of two year olds, and genetic trends. It has the disadvantage of reducing the number of records available for comparison. Most countries, including the United States, use all lactation records of herd-mates in sire summaries with proper weighting. In 1974, the USDA adopted the *Modified Contemporary Comparison* (MCC) for calculating sire summaries. In this comparison, herd-mates are divided into two groups: first lactation and succeeding lactations. The record of each bull's daughter is compared with the modified contemporary average for the herd. The average of noncontemporary herd-mates is figured as a single contemporary in computing the average of contemporaries.

2.6.2 USDA-DHIA Modified Contemporary Comparison

The theories utilized in developing the first *Predicted Difference* (PD) method for estimating the transmitting ability of sires (see section 2.6.3) assumed that (a) there is a homogenous dairy cattle populations for each breed nationwide; (b) that there is no genetic trend in the population; and (c) daughters of bulls and their herd-mates are culled equally, severely.[10] As evidence on sire summaries from AI bulls in use throughout the country accumulated, it became obvious to geneticists that there were biases between different regions, between bull studs, and between bulls within a stud. It also became apparent that there were problems in the present method in using progeny data on a bull's sire and maternal grandsire. With frozen semen available and with dairymen able to choose the sires used, it became evident that cows were no longer randomly bred—dairymen were using some sires more than others—and there was not a homogenous population for each breed on a nationwide basis, as some bulls were used more heavily than others. Herd-mates of a bull's daughters thus do not represent a random sample of the breed population, but vary with the genetic merit of other sires used in the herd where a bull's daughters are being evaluated. The Modified Contemporary Comparison ('74 MCC) utilizes the PD of the sires of herd-mates to adjust for the influence of earlier used sires in evaluating the deviations to calculate the PD of the bull being summarized. Further, genetic trends in a breed population can be accounted for in MCC calculations by establishing a fixed genetic base for computing the PD. This fixed genetic base is used in pedigree evaluation. Also, young bulls sampled for AI use are not assumed to represent a random sample of the breed. They are selected on the basis of pedigree information that usually includes several generations of high PD, high repeatability bulls, and outstanding cows. The MCC computation procedure, in order to avoid errors due to the genetic diversity among bulls, assigns bulls to a genetic group based on pedigree information. The average of each bull's daughters is then regressed to the average of his assigned genetic group rather than to breed average traits.

The USDA-DHIA Modified Contemporary Comparison was adopted for USDA-DHIA Sire Summaries in 1974. The accuracy of both sire summaries and cow indexes is enhanced. All sire summaries released by the USDA are now computed, using the MCC (see Table 2.6).

2.6.3 The Predicted Difference

The Predicted Difference (PD) is the average amount of milk by which the daughters of a bull deviate from the average of the herd-mates in herds with breed average yield. PD is a genotype measure and reflects the genetic *transmitting ability* of a bull more accurately than any method developed to date. The *transmitting ability* is defined as *one-half of the breeding value*. Breeding value is measured in terms of an animal's complete genotype, but transmitting ability is a measure of the average genetic material in a random one-half the genotype present in an egg or a spermatozoon.

The PD is calculated on the basis of herd-mate records using the USDA-DHIA Modified Contemporary Comparison. In calculating PD, a regression formula is used to adjust the daughter–herd-mate differences for effects of the number of progeny, number of production records per daughter, number of herds in which daughters are located, herd averages, genetic trends in the herd, and environmental correlations among daughters within herds. Necessary adjustments are also computed for differences between the average yield of the herds in which the bull's daughters are tested and the breed average.

The equations for computing PD were developed by geneticists and biometricians at Cornell University, and the Animal Physiology and Genetics Institute, USDA, with the assistance from workers at various agricultural experiment stations in America and abroad.[11] The use of PD for USDA-DHIA Sire Summaries, now widely used by workers in the dairy-cattle-breeding field, began in 1965. Prior to this date, the New York and other AI organizations were using methods similar in principle for estimating transmitting ability. As early as 1963, the American Breeders Service (ABS) in De Forest, Wisconsin, under the leadership of Robert E. Walton, now ABS president, initiated the use of Estimated Daughter Superiority (EDS). Based on concepts similar to those in PD, the EDS provides a reliable estimate of the transmitting ability of AI sires.

The PD is one of the most significant developments to date in estimating the transmitting ability of a bull. While most emphasis is placed on the transmitting ability of bulls and cows for milk production, the PD is also beneficial in ranking sires and in calculating other important economic traits.

1. *Predicted Difference for Fat Percent.* This is the expected average deviation of a bull's progeny from herd-mates of the progeny in breed average herds.

2. *Predicted Difference for Income.* This is an index added to USDA-DHIA Sire Summaries to predict gross income in relation to breed average, based on the bull's PD for milk and fat. PD for income, based on current milk and fat (in some cases, solids-not-fat) prices, enables a direct comparison of individual bulls to breed average in the same manner of as do the PDs for milk and fat. It is an aid in making selection among bulls of the same breed and is closely correlated with the PD values for milk and fat. The ranking of sires on PD for dollar income is now in general use because it reflects the value of all milk components as a basis for payment to producers.

3. *Predicted Difference for Type (PDT).* The Predicted Difference for Type is an estimate of the genetic transmission of characteristics for conformation (type) to offspring. The PDT is expressed as an average of the type rating score of a bull's progeny based on breed averages, or breed average dams, for comparisons. This average score is usually expressed as a percentage. For example, 86.6 percent is indicative of a "Very Good" classification rating. In case of Holsteins, the PDT is based on points (plus or minus) above or below the breed average. Repeatability of PDT indicates the reliability of the type score. It explains the amount of difference between type scores of daughters and dams and in ranking sires on type transmitting ability. Repeatability of the type score is proportional to the number of type-rated progeny. Dairymen stress milk yield as the most important economic trait in breeding. However, most consider conformation—particularly as it relates to the mammary system—disposition, and general wearability important traits also to emphasize in selection.

2.6.4 Repeatability of PD for Traits

The reliability or "confidence" that can be placed in PD for various traits as an estimate of a bull's transmitting ability, as compared to other bulls within that breed, is called *repeatability*. Repeatability is expressed as a percentage and can vary from a statistical value of 0 to 1.0. The higher the repeatability, the greater the confidence that can be placed on the PD as an accurate estimate of a bull's transmitting ability. The repeatability factor is derived from computations made in applying the PD formula. The accuracy of repeatability for a sire's estimated transmitting ability depends upon (a) number of daughters in the sire summary; (b) number of herds in which tested daughters are located; and (c) number of lactation records per daughter. Repeatability increases as the number of daughters, number of herds involved, and number of lactation records per daughter increase. For example, it is more desirable in sire sampling to have twenty daughters tested in twenty herds (one daughter each) than twenty daughters in three herds than with a distribution of 17+2+1 (Table 2.5).

The values in Table 2.5 indicate that the variation from PD for a few tested daughters located in a small number of herds is large. The repeatability of the PD is, therefore, low on the average. As the number of daughters increases and as they are distributed through more herds, the repeatability factor becomes much greater and variation from estimated future production, on the average, is much reduced. Bull studs and cattle breeders now strive to obtain "multiple herd proofs" on sampled bulls. The advantages are obvious. This is one of the concepts strikingly brought out in modern breeding methods; namely, the sampling of many young bulls in many herds for progeny testing and then selecting only the best. This is usually about 15 to 20 percent of the bulls on which progeny proof is obtained.

Table 2.5. Changes in Standard Deviations of Predicted Difference for Milk at Different Repeatabilities

Number of Daughters	Number of Herds Daughters Located	Repeatability (pct.)	Expected Standard Deviation (lbs. milk)
5	1	13.6	405
10	10	33.3	356
20	20	50.0	308
70	70	79.0	205
200	200	91.0	132
1,000	1,000	98.0	61

Source: R. D. Plowman, "Notes and Concepts Used in USDA Sire Summary Procedures," *Dairy Herd Improvement Letter* (1968): 44–202, USDA.

2.6.5 USDA-DHIA Sire Summaries: Interpretation

The assembling, tabulation, evaluation, computing, and publication on a nationwide basis of the data for estimating the transmitting ability of dairy bulls for the economic traits listed above is an immense task. This work is carried on by the USDA Animal Improvement Programs Laboratory in Beltsville, Maryland. The genetic evaluations made are based on production information obtained on the daughters of dairy bulls, which are in herds participating in the Official Dairy Herd Improvement and Dairy Herd Improvement Registry (purebred) record-keeping plans. The information used is obtained by USDA's Agricultural Research Service (ARS) from Dairy Herd Improvement Associations (DHIAs) throughout the United States that are affiliated with the National Cooperative Dairy Herd Improvement Program. As of 1 January 1978, there were 2.7 million cows, 24.7 percent of the nation's total, on the official DHIA and DHIR tests. Much greater genetic progress could be made if all cows on the DHIA test were properly identified. Failure of herd operators to identify cows by sires results in about 40 to 45 percent of the DHIA lactation records being discarded for sire summaries.

Sire summaries are made in the spring and fall of each year, and a Sire Summary List is published semiannually. The sire summaries provide much of the working material for AI sire analysts, cattle breeders, and geneticists involved in applying modern concepts in cattle breeding. A cow summary is also made, as will be described later.

The complete genetic information now available on the USDA-DHIA Sire Summary List is indicated in Table 2.6.

As will be noted, this sire summary gives complete and detailed information regarding the number of tested progeny; their production as compared to the average of their contemporaries; and the PD estimates for milk and butterfat percentage; and the fat and gross income dollars. The type scores or classification values are obtained from dairy breed registry associations and have been added to this summary.

Interpretation of the sire summary, letters *a* to *y,* is described in the lower half of the table. Never before have dairy cattle breeders had such detailed and dependable information to guide the selection of sires.

In this summary, a high PD bull is listed for each breed as summarized on 1 January 1977. The reader should refer to the PD of each bull. Note also the number of daughters tested, the number of herds in which progeny are located, the daughters per herd (Eff Dau/Hd) and the effect on repeatability. The Jersey bull in this tabulation has most of the daughters tested in one herd. While the PD for milk is +1,417 pounds, the repeatability is only 31 percent, based on this

Table 2.6. USDA-DHIA Sire Summary Format: With Bull of Each Dairy Breed

			Registration						Daughters			
Breed	Number	Name of Bull		Latest Summary Date	Stud Code	Number of Herds	Eff Dau /Hd	Number	Rec/ Dau	Milk	%	Fat
	a	b	c	d	e	f	g	h	i	j	k	l
Ayrshire	128289	Mar-Ral Hikick	*	77–01	7	97	14	261	1.6	12,276	3.83	487
Brown Swiss	154912	Nakota Pavanne Evilo	*	77–01	7	38	2	62	1.9	15,156	3.81	578
Guernsey	580104	Kellogg Minnies Choice	*	77–01	3	37	6	85	1.7	11,546	4.42	510
Holstein	1526128	Whittier Farms Apollo Rocket	*	77–01	15	446	3	768	1.5	16,885	3.44	581
Jersey	620428	Generator's Jeweler		77–01	29	4	29	35	1.0	10,819	4.66	504
Milking Shorthorn	311770	Minn Duke Darius	*	77–01	17	41	5	119	2.1	11,773	3.80	447

Interpretation of Above Summary

a–Registration number
b–Registration name
c–Presence of an asterisk indicates that 51 percent or more of progeny in the summary were registered
d–Latest summary date
e–AI stud in which bull is located
f–Number of herds in which the daughters are located (No. of Hrds)

g–Effective distribution of daughters per herd (Eff Dau/Hd)
h–Number of daughters having modified contemporaries (Number)
i–Average number of records per daughter (Rec/Dau)
j–Daughters' average milk (2X, 305-day, ME), lb.
k–Daughters' percent fat (2X, 305-day, ME)
l–Daughters' average fat (2X, 305-day, ME), lb.

	Contemporary Averages						Predicted Difference				Type No.	
	No.	Milk	Fat	% 1st Cul	% 1st Rip	Rpt. %	PD74 Milk	PD74 %	PD74 Fat	PD74 $$	Rpt.	Class Score
	m	n	o	p	q	r	s	t	u	v	x	y
Ayrshire	12	11,955	467	5	39	88	+1,042	-.10	+30	+92	39	85.7
Brown Swiss	12	13,319	524	5	2	76	+2,162	-.14	+67	+196	44	82.5
Guernsey	12	10,283	458	9	11	77	+1,378	-.11	+53	+136	20	82.6
Holstein	12	15,516	562	10	62	97	+1,890	-.22	+33	+144	Type Rpt. 91	Class PDT -1.44
Jersey	15	8,111	414	0	71	31	+1,417	-.24	+47	+132	5	83.6
Milking Shorthorn	6	10,941	406	4	20	81	+1,209	+.07	+52	+125	—	—

Interpretation of Above Summary

m—Number of contemporary records per daughter
n—Herd-mate average milk, lb.
o—Herd-mate average fat, lb.
p—Percent of first lactation that were records-in-progress
q—Percent of first lactation that were records-in-progress
r—Repeatability of this summary in percent
s—PDs for milk, lb.
t—PD for percent fat
u—PDs for fat, lb.
v—PD for gross income, dollars
x—Type Number—is the number of classified daughters. Used for all breeds except Holstein.

Type Repeatability—a percent figure that shows the accuracy or reliability of the type summary. The greater the percentage figure, the more reliable the PD for type (Holsteins only).
y—Classification—daughters average classification score used for all breeds except Holsteins.
Classification: PD Type—is an estimate of the genetic abilities of sires to transmit over all type to their daughters. Expressed in terms of type points above or below (plus or minus) breed average (used only by Holsteins).

Source: Dairy Breed Registry Associations.

summary. This bull, in view of the high PD, is quite promising, but he will need tested daughters in more herds to increase his repeatability (see Table 2.5) to a level comparable with the five other bulls listed in this summary.

2.6.6 Utilizing Information for Sire Selection

The PD, using MCC, is the best estimate known for evaluating the transmitting ability of a bull for economic traits. High PD, not repeatability, determines genetic progress. A bull can have a minus PD, yet a high repeatability. In the USDA-DHIA Sire Summary of January 1977, we note a Holstein bull with 1,323 tested daughters in 562 herds with a PD of *minus* 719 pounds milk. His repeatability is 98 percent. Not a herd improver, his daughters average a little over 13,000 pounds of milk, but evidently he appealed to many dairymen.

In selecting bulls with comparable PDs, results indicate that less risk is involved in choosing the bull with the higher repeatability. Breeding authorities advise dairymen to breed only a small proportion of the herd to bulls below 70 percent repeatability. Judgment is required in selecting the bull "best for the herd," but it must be kept in mind that the genetic trend for production is upward. This increase is about one percent per year, but varies with economic conditions and within breeds.[12] The future daughters of a bull with a high PD and high repeatability will be measured against a new crop of higher producing daughters, on the average, than those making up the first proof. If a bull has only a few daughters, such as applies to a single herd proof, from high-producing cows, his PD will usually decline when he is used in AI on herds of average population. On the average, dairymen have fewer failures when "high plus PD" bulls with a repeatability of 70 percent or above are used. These are the kind of bulls that the AI industry depends upon for the majority of services. The availability of a continuing progression of bulls with high breeding values depends upon selecting and proving enough sires that are superior to those replaced. A dairyman should utilize some young sires being sampled, but as a rule for only about 15 to 20 percent of the herd.

2.7 Estimating the Breeding Value of Cows

Fifty percent of an animal's inheritance is derived through the sire and 50 percent through the dam. While major emphasis in cattle breeding is upon sire evaluation, the contribution of the dam is equally important so far as a given individual is concerned. Since a bull has many offspring, but gives no milk himself, we measire transmission of his genotypic traits largely through the performance of his female offspring. The characteristics of a bull's progeny are greatly influenced by the kind of cows he is mated with. As the saying goes, "Sometimes good cows make a poor bull's daughters look good." Geneticists have derived measures to properly evaluate the dam's contribution in such cases. Heavily producing, long-wearing cows that make a profit are the end product of all breeding efforts.

In efforts to produce bulls with high-production potential, the estimation of the dam's transmitting ability is important. Modern breeding concepts apply in evaluating the value of cows for breeding purposes. Since a cow has her own production records (a phenotypic measure) and usually only a few offspring, measures of her transmitting ability are limited. There are four sources of information available in evaluating the breeding value of a cow: (a) the animal herself (production records, type classification, and other characteristics); (b) number of progeny and their performance (two or three daughters with records of production is a better genotypic measure of the inheritance [1/2] passed on by the dam than one daughter only); (c) her ancestors (particularly the sire, maternal grandsire, and her dam); and (d) collateral relatives (her paternal and maternal half-sisters or half-brothers [half-siblings] and full brothers and sisters [full siblings]). Information helpful in estimating what a cow's breeding value will be is found in any or all of the above categories.

2.7.1 Use of Lactation Records

A cow's milk and butterfat record is not a reliable estimate of her transmitting ability for yield. The cow's own production can easily be influenced from 50 to 80 percent by herd management or environment. Several records of production, made in the same herd, are a better measure of her phenotype for yield than a single record. Repeatability of successive lactation records from the same cow is about 0.3 to 0.5 if made in the same herd. To evaluate transmitting ability, geneticists have devised means to translate measures of a cow's phenotypic performance into genotypic terms, which take into account herd-mates, breed average, collateral relatives and ancestors, and which give much more reliable estimates than the cow's own production.

2.7.2 Daughter-Dam Comparisons

The Daughter-Dam Comparison is a method of expressing the yield of a daughter above or below that of her dam. Usually expressed as "daughters production-dams production-difference," this method was used from about 1930 to 1960 as the standard method for genetic evaluation of sires. Many variations were made in its use, including the Equal Parent Index, and the Regression Index, which took into account the fact that progeny tend to regress toward the breed (or population) average. The Daughter-Dam Comparison had some value to a dairyman in comparing a daughter and her dam in the same herd. With records made in different herds, the environmental effects resulted in much bias.

2.7.3 Herd-mate Difference

A comparison of an individual cow's record with the average of her herd-mates, during the same year, breed, and season, is useful in determining whether she is above or below the herd and breed average. Herd-mate Difference minimizes environmental effects and is of value to a herd operator in appraising breeding results. The cow's herd-mate deviation is given relative emphasis in relation to the sire's PD in computing the USDA-DHIA Cow Index (see section 2.7.5) and is highly useful in this respect.

2.7.4 Estimated Producing Ability (EPA)

Estimated Producing Ability (EPA) is an estimate of a cow's ability to produce above or below the herd average. It is calculated by multiplying the average Herd-mate Difference by an appropriate weighting factor (based on the number of records a cow has). For example, a cow has two lactation records averaging 14,000 pounds of milk. Her herd-mates' average is 13,000 pounds. The Herd-mate Difference is 1,000 pounds. The weighting factor is calculated using the formula $n/n+1$, where n equals the number of lactations, and repeatability is 0.5. Thus, $n/n+1 = 2/2+1 = 0.67$. The EPA is $1,000 \times 0.67 = +670$. EPA is useful in ranking cows for production and for culling.

2.7.5 The USDA-DHIA Cow Index

Utilizing DHIA and DHIR records of production, the Animal Improvement Programs Laboratory of the USDA compiles a "Cow Index List." This list is highly useful to AI sire analysts, dairy cattle breeders, and others seeking outstanding cows based on genetic merit as it provides an estimate of a cow's transmitting ability. "A cow's transmitting ability for milk, fat yield and fat percent is the level above or below genetic breed average that she is expected to transmit to her offspring."[13]

The selection of cows for the list is carried out in two stages. The first requires that (a) the cow be registered; (b) to have an average of at least three modified contemporaries across lactations; (c) the last calving date on or after date fixed for inclusion in list; (d) to be alive; and (e) if record is in progress to be at least 150 days. The second stage of the procedure is to select for publication the information on all cows that meet the required levels for average yield and estimates of transmitting ability.

The screening levels that are used and the cows that met the criteria for selection in the 1975 USDA-DHIA Cow Index List are indicated in Table 2.7.

The MCC Method is used for estimating transmitting ability for milk and fat. It is effective in reducing environmental effects of the herd, year, and season of freshening.

The Cow Index (CI) is calculated by a formula giving proper weight to a cow's estimated transmitting ability (1/2); her production records; adjusted deviation from the Modified Contemporary Average, and the PD of the cow's sire.[14] CIs are calculated for percentage of fat on the basis of the expected average deviation of a cow's progeny from genetic breed average herd-mates. The gross income in dollars is an estimate of the gross income of a cow's daughters in relation to the breed average that is based on the CI for milk and fat.

Repeatability of CI is influenced by the amount of information available on a cow and her paternal half-sisters: the weight of the cow's deviated yield and the repeatability of the sire's PD.[15] Re-

Table 2.7. Screening Levels for Mature Equivalent (ME) Yields Daughter, Modified Contemporary Deviations (MC Dev.), and Cow Indexes (CI) Required for a Cow to Appear on USDA-DHIA Cow Index List

	Sreening Levels							
	ME Yields		MC Dev.		CI			
Breed	Milk lb.	Fat lb.	Milk lb.	Fat lb.	Milk lb.	Fat lb.	$	% Cows Qualifying
Ayrshire	12,000	450	1,000	0	300	0	55	5.4
Brown Swiss	13,000	510	1,300	0	350	0	50	5.1
Guernsey	11,000	470	1,000	0	300	0	40	4.2
Holstein	15,000	530	1,500	0	400	0	60	2.7
Jersey	10,000	470	1,000	0	300	0	50	4.2
M. Shorthorn	11,000	400	1,000	0	300	0	30	9.8

Source: USDA-DHIA Cow Index List (Spring 1975).

peatabilities for CIs are usually much lower than those obtained for the sire's PD. Less confidence should be placed in individual CIs than in PDs. This is particularly true where repeatability of the CI is very low.

Value of the Cow Index. CIs give a more accurate estimate of genetic transmitting ability than a cow's lactation record or her herd-mate deviation. The information furnished by the cow's records, her half-sisters, and the sire's PD all contribute to a more accurate estimate. For example, the estimated breeding value of a heifer calf can be calculated by adding the PD of her sire and the CI of her dam. Both values are expressed in terms of transmitting ability (one-half breeding value). Example: Sire PD = +900; Dam's CI = +703; 900 + 703 = 1,603 Estimated Breeding Value of the calf. This is the estimated amount of milk by which the heifer can be expected to exceed her herd-mates on an ME basis in a breed average herd.

2.7.6 Estimated Average Transmitting Ability (EATA)

A. E. Freeman and colleagues developed a method for estimating the transmitting ability of a cow by utilizing the same concepts as employed in the USDA Cow Index.[16] This method, the Estimated Average Transmitting Ability (EATA), is computed by some regional DHI processing laboratories and is used in cow analysis by several AI businesses. The calculations are based on five sources of pedigree record information: (a) the cow; (b) her dam; (c) her daughters; (d) her paternal half-sisters; and (e) her maternal half-sisters. The information from each source is weighted in proportion to the information contributed to the EATA. For example, a cow is twice as closely related to her dam or her daughters as she is to her maternal or paternal sisters. In calculating EATA, greater weight is given the production record of a cow's daughters than to her half-sisters. Combining the cow's EATA and the sire's PD provides an estimate of the average expected breeding value. The offspring resulting from a cow with EATA of +1,200 milk mated with a bull with +PD of 1,000 for milk would have an expected breeding value (BV 2,200 divided by 2) of 1,100 pounds. Each parent transmits only a *sample half* of its genetic material to each progeny.

In summary, sire and cow evaluation procedures used in the AI program, by breed organizations, and by many cattle breeders now utilize modern genetic concepts. The use of computerized records and the application of statistical methods, geared to the needs of the improving cattle industry, during the last two decades provide a sound basis for the selection of breeding stock. The impact of these developments is reflected in higher production per animal and in increasing income to the herd operator. In time, there will no doubt be further advances in the practical application of genetic concepts that will continue to revolutionize cattle breeding.

PART II:
ADVANCES IN MODERN BREEDING METHODS

Chapter 3: The Impact on Cattle Breeding

3.1 Introduction

The wide-scale use of AI and the adoption of modern genetic concepts beginning about 1936 have had a resounding effect on cattle improvement methods. For the first time in history, mass improvement methods, based on scientifically sound principles, capable of practical application to the cattle industry are available.

While much has been accomplished in improving production, particularly milk yields per cow throughout the world, the program is yet too new to assess fully its eventual impact. It must be kept in mind that since about 1936 the technology for semen collection, evaluation, processing, freezing, and shipment has been developed to its present level. These developments came about through research at our agricultural colleges, some trial and error, and field experience. Too, the formation and establishment of artificial breeding organizations, to provide service and semen to cattle breeders, is one of the necessary developments during this period (see Chapter 1 and Part III for further details). The potential for the wide-scale use of AI in beef cattle, swine, sheep, goats, and other farm animals is recognized but has not yet achieved the application made for dairy cattle in the United States.

Large-scale breeding projects and mass improvement methods, which AI makes possible, can succeed only when the conditions are right for their application. In the cattle-breeding industry, with its many complexities and the human equation always a factor, the time and circumstances following World War II were evidently favorable. See Figures 3.1 and 3.2, which depict the growth of the AI program in the United States. The introduction of the AI program and concepts of population genetics has caused many repercussions in the long established theories that guided cattle-breeding methods for hundreds of years. Almost every phase of cattle production, the program of breed registry organizations, the role of agricultural education, and various state and federal agencies concerned with farm livestock have been altered by these developments.

It is not easy to bring a transformation in livestock-breeding practices in a relatively short time. Yet, AI and the application of modern

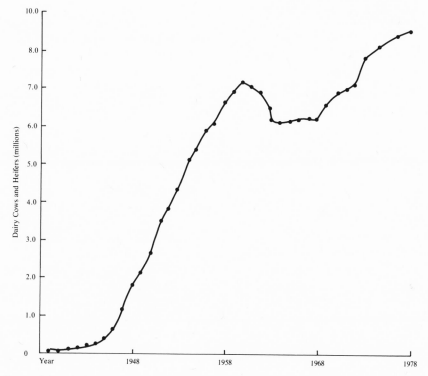

Figure 3.1. Dairy cows and heifers bred artificially to dairy bulls, 1938–1978. Source: *Dairy Herd Improvement Letter* ARS (1939–1979), USDA; and NAAB reports (1947–1979).

breeding concepts have found wide acceptance in a space of some thirty to forty years. The accomplishments and changes brought about to date can only be considered forerunners of the future. There no doubt will be improvements in AI technology and further advances in the application of genetic principles as experience dictates in the future. However, the basic philosophy as demonstrated in the improvement of milk production, poultry, and meat production will undoubtedly prevail.

The impact of AI and the adoption of modern genetic concepts as they have affected cattle and have touched the lives of many people engaged in the livestock industry to date are herewith summarized.

3.2 Increased Milk Yield Per Cow

The most dramatic demonstration of the effects of improved genetic potential, coupled with good herd management, is provided by the increase in milk yield by U.S. dairy cows since 1940. As indicated in

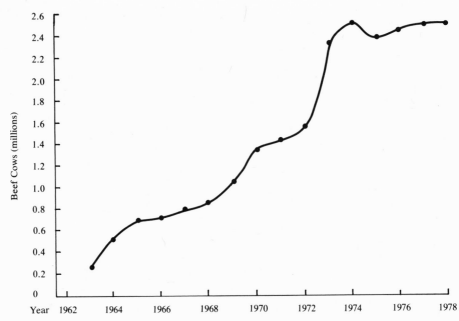

Figure 3.2.   Beef cows bred artificially to beef bulls, 1963–1978. Source: *Dairy Herd Improvement Letter* ARS (1963–1975), USDA; and NAAB reports (1947–1979).

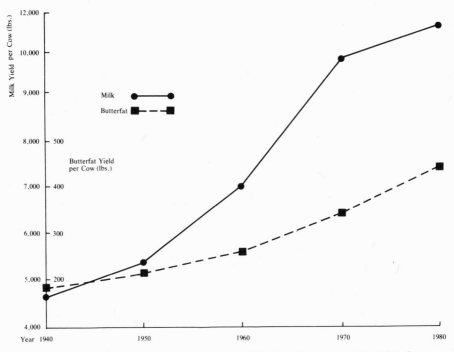

Figure 3.3.   Average production per cow of all U.S. milk cows, 1940–1978. Source: *Dairy Herd Improvement Letter* 55:2 (December 1979) Science and Education Administration, USDA, Beltsville, MD.

Table 2.1, the average annual milk production per cow increased from 4,622 pounds in 1940 to 11,240 pounds in 1978 (see Figure 3.3).

The effect of increased milk yield per cow is further demonstrated by the fact that during 1978 a total of 122 billion pounds of milk were produced by only 10.8 million cows. In 1940, it required 23.7 million cows to produce 109.4 billion pounds (Table 2.1). The fact that our national milk-production level has been maintained, and increased, with fewer cows (Figure 3.3) is a good example of the economic advantages brought about by improved breeding and efficient management.

While AI can claim much of the credit for providing the improved genetic potential for doubling the milk production per cow in a span of some thirty-five years, changes in dairy farming and management practices from 1940 to 1976 have played a major role. Herds are fewer but much larger. Dairy farming is more specialized; the farm with a few milk cows has largely disappeared, and feeding and management have undergone evolutionary changes. During 1939–1940, the average herd enrolled on the official National Cooperative Dairy Herd Improvement Program had 24.2 cows in milk.[1] During 1976–1977, the average was 77.0 cows. From 1940 to 1977, the number of dairy farms declined by about 55 percent. In 1940, there were 6.2 million farms reported in the United States.[2] Of these, 4.8 million reported milk as cream sales. For 1970, there were 2.7 million farms reported and of these only 568,000 reported keeping milk cows. By 1973, the number of dairy farms had declined to 385,000 and 24 percent reported herds of 100 or more cows.

3.2.1 Contributions from Improved Breeding

It is possible to estimate only roughly the contribution of AI and better breeding in the highly efficient dairy production program that has evolved in the United States since 1940. Geneticists estimate that only about 20 to 25 percent of the variance in production levels can be attributed to heredity.[3] The remainder is accounted for by environmental influences. I will not attempt to place a total dollar value on the advantages of producing 122 billion pounds of milk from 10.8 million cows in 1978 as compared to 109.4 billion pounds from 23.7 million cows in 1940. Prices for feed, labor, equipment, land, and other production "inputs" have increased tremendously—more than the price of milk in most instances. It is obvious that the savings in the cost of producing our national milk supply are reduced to the extent that we have roughly 13.0 million fewer milk cows to maintain. As a result, an investment to the extent of about $9 billion, if we use the

Table 3.1. The "Real" Price of Milk

Year	Average hourly wages production workers in manufacturing	Average retail price per half-gal. milk at grocery stores	Minutes of work to earn price of half-gal. milk
	Dollars	Cents	Minutes
1968	3.01	53.7	10.7
1969	3.19	55.1	10.4
1970	3.36	57.1	10.2
1971	3.57	58.9	9.9
1972	3.81	59.8	9.4
1973	4.08	65.4	9.6
1974	4.41	78.4	10.7
1975	4.81	78.5	9.8
1976	5.19	82.7	9.6
1977	5.63	83.9	8.9
1978	6.17	88.8	8.6

Source: Milk Industry Foundation, Washington, D.C., 1976–1978 records (milk facts).

conservative figure of $700 per head for milk cows, is spared the dairy industry.

Next, the yearly feed cost for cows producing at low levels (less than 6,500 pounds of milk annually) according to USDA reports in 1976 through 1977 averaged $406. Thus, on the basis of present production levels per cow, we are spared, conservatively speaking, a feed bill of some $5 billion for cows we "don't have to milk."

Housing, labor, veterinary costs, interest on investment, and miscellaneous costs are considered to be equal to feed costs for a producing cow. Thus, another $5 billion in total costs are saved. Probably some value should be allowed for the manure produced by the additional cows that would be kept if the average production were under 6,500 pounds.

In any event, it would appear that the dairy industry has some $19 to $20 billion less overhead costs per year—estimating a milk cow to stay in the herd three years—because of improved breeding and management.

The consumer gains most by the efficiency gained. The cost of a unit of milk throughout the years is shown in Table 3.1.

The "real" price of a half-gallon of milk is represented by the amount of work that a wage earner has to do to earn the money to buy it. In these terms, the "real" price of milk has declined for many years, as wage rates increased faster than milk prices. The "real" price of milk increased in 1973 and 1974, but resumed its downtrend in 1975, and continues so, as milk prices hold steady while hourly wages continue to increase.

3.3 AI Sires Improve Milk Yields

As the AI organizations gradually developed, a program emphasizing "AI proved bulls" and the application of modern breeding concepts, the milk yields of AI progeny began to excel the non-AI in most herds. This trend continues and is a reversal of the situation during the first few years of the AI program in the United States (see Chapter 2). Beginning about 1950 and steadily gaining in momentum, the AI businesses stress the heavy use of bulls that have demonstrated that their progeny will exceed herd-mates in milk yields.

Breeders of purebreds who, in some cases, looked with disdain on the bulls used in the AI program in the beginning years now use the best AI-proved bulls in increasing number. This trend added some of the higher producing, and better-managed, herds to the cow population on which AI bulls worked. Credit must be given the caliber of bulls used in AI for most of the genetic gains contributing to the higher milk production per cow in the United States the past thirty-five years. It is obvious that the only new genes entering a herd come from the sires used in that herd, unless females are purchased. This accounts for the often-made statement that "sires used in a herd account for about 90 percent of the genetic change" made in that herd. While not every dairyman utilizes AI for the entire herd, it is safe to say that at least 85 to 90 percent of the dairy herds in the United States in 1979 used AI to some extent.

Many area and state summaries of AI and non-AI daughters, which were composed over the years, indicate a superiority in milk production ranging from a few hundred pounds to over two thousand pounds in favor of the AI progeny.

3.3.1 Northeast U.S.

A summary of milk and butterfat records on over three million Holstein cows in the northeastern states, made by Cornell University,* reveals production levels as follows:

	Average Milk Production Per Cow	
Percent of AI Cows in Herd	1964	1974
Non-AI	14,290 lbs.	13,478 lbs.
81–100 AI	14,989 lbs.	15,612 lbs.

*Personal communication with R. W. Everett, Cornell University, Ithaca, New York.

Table 3.2. Holstein ME Lactation Averages of Cows in Herds of Various Levels of AI Use

% AI	1964		1969		1974	
	Milk lbs.	Fat lbs.	Milk lbs.	Fat lbs.	Milk lbs.	Fat lbs.
Non-AI	14,290	521	14,005	511	13,478	491
1–20% AI	14,557	532	14,202	512	14,106	509
21–40% AI	14,442	529	14,536	526	14,509	525
41–60% AI	14,709	538	14,776	535	15,046	546
61–80% AI	14,695	536	15,101	548	15,293	554
81–100% AI	14,989	548	15,511	563	15,612	565

Source: Cornell University data, 1975.

It will be noted that between 1964 and 1974 the non-AI cows actually lost in production, whereas the herds containing from 81 to 100 percent AI-sired cows increased 623 pounds of milk on the average during this ten-year period. The AI-sired cows, in 1974, were producing at an annual level of 2,134 pounds more milk than the non-AI. This difference is easily a $200 gross income per cow at prevailing milk prices. For a fifty-cow-milking herd, this difference in gross income per year would total about $10,000 and would clearly demonstrate the value of improved inheritance.

AI Used in Herd at Various Levels. Advantages of consistently using high PD sires for milk yields on the entire herd are pointed out in Table 3.2.

The table includes records of over three million Holstein cows. Similar comparisons were made for the other dairy breeds. The trends, as calculated for each year from 1964 to 1974, invariably showed an increasing advantage of a high percentage of AI over non-AI daughters for all breeds as time elapsed. The herds that had 100 percent AI-sired cattle continued to increase in yield and profitability as compared to lower level use of AI and non-AI. In this study, only USDA–identified–non-AI sires and their identified daughters were used. The same is true for the AI sires and their progeny.

3.3.2 Superior Sires in Active AI Use

The superiority of dairy bulls in active AI use in the United States during 1977 is set forth in Table 3.3.

Extension dairymen of Iowa State University developed this table from Bennet Cassell's USDA-DHIA Sire Summaries (Winter 1978,

Table 3.3. Average Predicted Difference (PD) Values for Bulls Summarized in Winter 1978

Breed		Number	PD for Milk	Percent Test	PD for Fat
Ayrshire	All bulls	199	+30	.00	+1
	Active AI bulls	25	+527	−.03	+18
	Non-AI bulls	147	−50	+.01	−1
	AI superiority		+577	−.04	+19
Guernsey	All bulls	533	+99	−.02	+3
	Active AI bulls	64	+592	−.06	+22
	Non-AI bulls	404	+20	−.01	0
	AI superiority		+572	−.05	+22
Holstein	All bulls	6,758	+61	−.03	−2
	Active AI bulls	906	+655	−.07	+14
	Non-AI bulls	5,098	−64	−.02	−5
	AI superiority		+719	−.05	+19
Jersey	All bulls	663	+261	−.07	+7
	Active AI bulls	107	+757	−.13	+26
	Non-AI bulls	459	+126	−.04	+3
	AI superiority		+631	−.09	+23
Brown Swiss	All bulls	267	+241	−.02	+7
	Active AI bulls	45	+791	−.03	+24
	Non-AI bulls	182	+172	−.02	+5
	AI superiority		+519	−.01	+19
Milking Shorthorn	All bulls	169	+217	+.01	+9
	Active AI bulls	10	+856	.00	+32
	Non-AI bulls	127	+207	+.01	+9
	AI superiority		+649	−.01	+23
All breeds	All bulls	8,589	+87	−.02	+5
	Active AI bulls	1,157	+661	−.05	+16
	Non-AI bulls	6,417	−32	−.01	−3
	AI superiority		+693	−.04	+19

Source: *Hoard's Dairyman* (10 October 1978): 1217, Ft. Atkinson, Wis. (used by permission).

Animal Improvement Programs Laboratory, USDA, Beltsville, Md.). The difference in PD values for active AI bulls when compared to all bulls and non-AI bulls for each breed indicates the progress being made. A plus PD of 693 pounds of milk for all active AI bulls adds further evidence of the superiority of AI over natural service for herd improvement.

3.3.3 Wisconsin AI-Sired Cows Excel

In a study of 1976 production records for 135,438 Wisconsin cows on the DHIA test, it was found that the AI-sired cows for all breeds excelled in milk and butterfat yields. The results are shown in Table 3.4.

Table 3.4. Wisconsin Records Show AI Superiority

		Milk	%	Fat
Ayrshire	AI	12,669	3.9	491
	Non-AI	12,112	3.9	460
		+557		+31
Brown Swiss	AI	13,122	4.0	519
	Non-AI	12,693	4.0	502
		+429		+17
Guernsey	AI	10,883	4.7	498
	Non-AI	10,458	4.6	475
		+425		+23
Holstein	AI	15,815	3.8	593
	Non-AI	14,623	3.7	541
		+1,192		+52
Jersey	AI	9,945	5.0	487
	Non-AI	9,393	5.0	463
		+552		+24

Source: Dairy Science Department, University of Wisconsin–Madison, 1977.

Note: For purposes of comparison, all records are set up on the 2X—305 day—ME basis.

With about 60 percent of the 1.8 million milk cows in Wisconsin serviced by AI, this gain in yield per cow amounts to millions of dollars annually for participating dairymen.

3.3.4 Effect of AI and DHIA Testing: Michigan

The effects of better breeding, through the use of well-proved AI sires, is reflected in a summary of the state DHIA average for Michigan.[4]

1. For the seventeen-year period from 1928 to 1944, before AI was in use, the average milk production per cow in DHIA was about 8,000 pounds and improved at the rate of 7 pounds of milk per year.

2. For the seventeen-year period from 1945 to 1961, when AI was in use, production per DHIA cow increased an average of 154 pounds of milk per year. The 1945 average was 8,487 pounds of milk. For 1959, the average yield had increased to 11,231 pounds of milk per cow.

3. For the twelve-year period from 1962 to 1973 with the Herd-mate Comparison in use and emphasis on high PD sires, the production per cow increased at the rate of 144 pounds of milk per year. The average per cow advanced from 11,626 pounds of milk per cow in 1962 to an average of 13,352 pounds in 1973.

4. With greater emphasis on high plus PD sires, the MCC in use, and a wider selection of superior AI sires available to dairymen, the average increase in milk yield per cow reached 385 pounds for the four-year period of 1974 to 1977.

3.4 DHIA Averages, 1940–1979

Cows enrolled in the National Dairy Herd Improvement Program in the United States show a tremendous gain in milk production due to three factors: (a) improved feeding and management and more specialized dairy farms; (b) the use of accurate production records (DHIA program) to guide management and breeding decisions; and (c) the use of AI involving high plus PD sires and the application of modern breeding concepts. The results of the combined application of these three factors are well portrayed in Table 3.5.

The production data in Table 3.5 are confined to official DHIA and DHIR testing programs. This is the most accurate measure of production per cow available and at present the only source of data used in the USDA Sire Summaries.

It is significant that, with the growth of the AI program, participation in the National Cooperative Dairy Herd Improvement Program has steadily increased. It is indicative of the realization by dairymen that herd improvement and its accompanying increased income involve the use of production information, sires that add genetic potential for yield, and good herd management. These herds provide the sires used in artificial insemination. They, likewise, lead the way in

Table 3.5. Official DHIA-DHIR Production Record of Cows, 1940–1979

Year	Cows Enrolled	Cows per Herd	Average Production			Total U.S. Cows .000	% Official Plans	%* All Plans
			Milk lbs.	Fat %	Fat lbs.			
1940	676,141	24.2	8,133	4.07	331	23,272	2.91	2.91
1945	577,200	27.2	8,592	4.03	346	25,597	2.25	2.25
1950	1,088,872	27.2	9,172	4.03	370	22,024	4.94	4.94
1955	1,333,866	32.3	9,502	3.95	375	21,581	6.18	6.18
1960	1,746,752	42.3	10,561	3.87	409	17,901	9.76	13.63
1965	2,087,581	52.1	11,976	3.82	457	15,677	13.32	18.90
1970	2,122,011	61.9	12,750	3.79	483	12,307	17.24	25.35
1975	2,432,961	75.5	13,421	3.76	505	11,220	21.69	31.00
1976	2,438,365	75.7	13,632	3.75	511	11,087	21.98	31.41
1977	2,581,164	76.5	14,435	3.73	539	11,035	23.40	33.45
1978	2,704,064	77.3	14,631	3.70	542	10,930	24.70	35.40
1979	2,792,175	77.3	14,644	3.70	542	10,853	25.70	36.80

Source: *Dairy Herd Improvement Letter* 56:5 (1980), USDA.

*All cows include Owner-Sampler, Weigh-A-Day-A-Month, AM–PM, and other unofficial dairy record-keeping plans.

sampling young sires for AI use. Thus, it is a synchronization of programs and practices that culminates in the final goal of higher production per cow and greater labor income for the dairyman.

3.5 World's Record For Milk: An AI Triumph

The Holstein cow, Beecher Arlinda Ellen, which now holds the world's record for milk production, represents four generations of AI. Ellen, bred and owned by the Harold Beecher family at Rochester, Indiana, in 1975, produced 55,611 pounds (25,270 kg.) of milk, 1,573 pounds (715 kg.) of butterfat, in 365 days, 2× milking. She began her DHIR test at five years and eight months of age. Her production for 305 days—2× milking—is 50,314 pounds (22,870 kg.) of milk, 1,406 pounds (639 kg.) of butterfat. This production level is a national and a world's record. She milked 194.5 pounds (88.4 kg.) of milk her high day and was at 125.5 pounds (56 kg.) per day her tenth month of lactation. She averaged 153 pounds (70 kg.) of milk per day for a year. The champion has had record-breaking lactations since she started out as a two year old with 20,100 pounds (9,130 kg.) M, 768 pounds (345.4 kg.) F, in 305 days, 2× milking. After finishing her world-breaking record in late 1976, she peaked again at six years, ten

Figure 3.4. Beecher Arlinda Ellen "E–91," the world-record milk producer. She represents four generations of AI breeding. (Photograph courtesy of Curtiss Breeding Service.)

months of age to produce 48,840 pounds (22,200 kg.) milk, 1,674 pounds (761 kg.) of butterfat in a year.

Ellen is a large, strongly built cow. She is officially classified "Excellent" with a score of ninety-one points.

Beecher Arlinda Ellen was sired by Pawnee Farm Arlinda Chief, an outstanding sire in AI use. He has, as of May 1977, according to the USDA Sire Summary, 13,319 daughters in 4,431 herds. His PD (May 1977) was +1,465 pounds of milk, +68 pounds of fat, and $157, with 99 percent repeatability. His 6,294 classified daughters average 81.9 age-adjusted with a PDT of +.50 with 99 percent repeatability. Her dam, Bridgecrest Skylighter Elsie—19,240 M, 614 F, 305–2×–Age 7 years, 6 months—was sired by Gray View Skylighter, a bull in AI use, as was her next three sires, Mooseheart Excellency, Pabst Sir Roburke Ragapple, and Curtiss Candy Invincible, resulting in four generations of AI breeding. Ellen has four sons, all in AI use.

3.6 AI Organizations Retrench

The cost of sampling young sires; the field introduction of frozen semen from 1953 forward; the declining dairy cattle population; and rising costs of operation all resulted in traumatic changes for AI organizations. As explained in Part III, the bull studs in America after 1954 became fewer and larger. The smaller AI cooperatives and a few proprietary-owned businesses could not muster the volume necessary to compete. There were numerous mergers and consolidations. The number of bull studs in America declined from ninety-seven in 1950 to about twenty in 1976 (Figure 3.5).

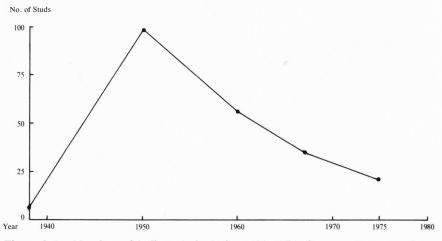

Figure 3.5.   Number of bull studs in U.S., 1938–1976. Source: *Dairy Herd Improvement Letter*, ARS, USDA; and NAAB files.
Note: Includes two all-beef studs.

No. Bulls
per Stud

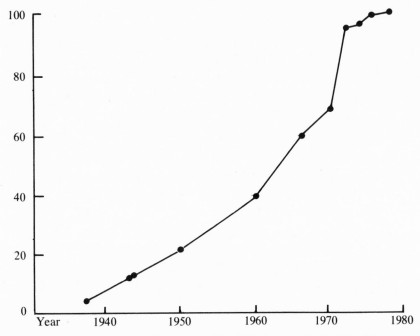

Figure 3.6. Sires per bull stud in U.S., 1938–1976. Source: *Dairy Herd Improvement Letter,* ARS, USDA; NAAB files; and reports from AI organizations throughout U.S.

With fewer bull studs, the number of bulls in use per organization increased from twenty-two in 1950 to ninety-eight in 1976 (Figure 3.6).

In addition to sires in regular service, the bull studs, engaged in sampling young bulls, each had from a dozen to several hundred young bulls in the process of progeny testing. As of 1 January 1972, a USDA report[5] indicates 2,167 dairy and 347 beef bulls in service. Of the total, 1,038 dairy bulls, or 48 percent, were coded for regular service; 1,015, or 47 percent, were coded for progeny testing only; and 114, or 5 percent, highly regarded AI-proved bulls, were coded for special or planned matings only to produce future AI sires. Most of the 347 beef bulls were in regular service.

3.6.1 Direct Service: Semen Sales

Beginning in the early 1950s and in some areas earlier, as frozen semen became available dairy herd operators were trained to inseminate cows in their respective herds. Semen was sold directly to the herd owner. This program was necessitated in part by fewer dairy herds in many localities; and the AI technician, who once serviced the area, could no longer make a living.

The unit of semen became a trade commodity and is now sold in dosages. As the trend of direct semen sales, or the "do-it-yourself" program became established, an increasing number of organizations adopted it. Inseminator-training schools were established and thousands of persons were trained. Semen is marketed by many studs to all sections of the United States and to foreign countries. The earlier program of trained inseminators carrying on the customer service for the bull stud slowly declined; at present only those studs in well-populated cow areas maintain a crew of inseminators or servicemen. Specific developments during this period are discussed in further detail in Part III.

Chapter 4: Effect of AI on the Purebred Cattle Industry

4.1 Introduction

Improvement in cattle depends upon the goals and perspectives of their owners. Where there is little incentive, there is little progress. Since the formation of cattle breeds and the establishment of herd books, or registries, breeders of purebred cattle have been strong leaders in improvement programs. What is commonly referred to as "pedigree breeding" began about 1822 with the establishment of herd books. No doubt there was some pedigree breeding long before this time, since cattle owners had always had some control of matings and had probably kept some type of records after confinement of cattle occurred.

Cattle ownership in all countries is generally regarded as a "status symbol." The early Romans used the term *pecunia,* derived from *pecus,* which means "cattle," for their money. Today, as in ancient times, a man's wealth is often estimated in terms of the cattle he owns. Cattle played important roles in the religion and mythologies of older civilizations, often being considered sacred and their slaughter forbidden. This philosophy still prevails among the Hindus in India and the eating of beef is taboo.

4.2 Role of Breed Organizations

Most of the breed registry organizations in America were formed between 1868 and 1900. However, with the influx of the "exotic" breeds beginning in 1950, many new American beef registry organizations have been formed. Some are still being formed, with about sixty beef breeds claiming recognition.

To appreciate the role played by breeders of purebreds and breed organizations, it is necessary to consider the reasons behind the formation of "recording societies" as they were first known in Europe.

4.2.1 Reasons for Breed Organizations: Breed Purity

In the formation of cattle breeds, beginning some two or three hundred years ago, each breed became known for certain uniform characteristics and the ability to transmit uniformly these characteristics. It soon became apparent that desired features and uniformity were reduced when unlike breeds were mated. People concerned with the development of a breed saw the necessity to preserve its purity and, hence, its characteristics. This led to standards for recognizing an animal as belonging to a given breed, and recording systems (herd books) were established and the use of "outside blood" was prohibited.

Early cattle breeders knew little about genetics and had few records of performance to guide them—external characteristics such as color markings, size, and conformation were the guidelines. Animals accepted for the herd register had to be from already-recorded parents; and in cases where the ancestors were not recorded, they had to be inspected and given a provisional recording. Even progeny of recorded parents had to be inspected, and if they were off-color or poor in conformation, they were refused entry. All efforts were aimed at maintaining breed purity. Also, these early breeders operated on the theory that "like begets like." However weak the system might have been, it resulted in our present-day breeds and the assurance that they breed true for certain characteristics. It was from this beginning our existing cattle breeds have been improved to our present-day levels.

4.2.2 Functions of Breed Organizations

The several functions of a breed registry organization include (a) establishing and maintaining a herd book for identifying animals, maintaining breed purity, and protecting the interests of breeders; (b) improving the breed by establishing programs to test production and to evaluate conformation (type); (c) promoting the breed and programs to improve it, promoting activities that involve "selling the breed," assisting breeders, and conducting extension and public-relations work; and (d) designing and enforcing regulations for maintaining and improving the purity of the breed.

Breeders of purebreds and their respective organizations have found it necessary to make many adjustments in their programs due to the advent of production-testing programs (such as the Advanced Register—for registered cattle only), and the DHIA for all dairy cows; due to the performance testing of beef cattle; and due to the introduction of the AI program. These organizations have succeeded in making such adjustments—too slowly in the minds of some critics—by eventually adopting practices that have economical application.

Breed organizations are controlled by an elected board of directors, which establishes policies in accord with the bylaws. Managing the breed organization and carrying out board policies are entrusted to employees. The executive secretary is usually the key employee.

Members of a purebred cattle organization, in the final analyses, determine policy and programs. Not until a sufficient number of members are in favor of a given program, bylaw, or policy change can it be enacted.

4.3 The Role of the Breeder of Purebreds

The breeder of purebred cattle is looked upon "as a seed stock" producer. His task is well described by the following lines:

> The Stockbreeder
> Yours is the task to mate and to mould
> Living things for your gain and pleasure;
> To find and to fuse the purest gold
>   Nature hoards as a hidden treasure.
>
> Yours is the heritage handed down.
> A trust without limit or measure;
> To make, not to mar, to win renown—
>   Fail not in the brave endeavor.
>
> Yours is the art and the work to blend
> Living things in beauty together;
> Yours is the power to ruin or mend—
>   The bonds that ye bind none can sever.
>
> A sacred trust are these living things
> To be carelessly dealt with NEVER!
> And faithful stewardship surely brings
>   Rich reward that shall live forever!
>
> —Dr. A. S. Alexander.[1]

The progressive breeder of purebred cattle is motivated not only by the additional income derived from the sale of breeding stock but also by pride and a deep-seated desire to create a better animal. His goal is to excel in all aspects of cattle breeding. He uses what he considers the best known methods to evaluate animals and plan matings. It has often been said that the only limiting factor in breeding better cattle is the limits of man's mind and vision.

The purebred breeder now has, as does the owner of nonregistered or grade cattle, proved methods for making maximum improvement in a given time interval. These include the application of population genetics (see chapters 2 and 3); the use of AI, which makes it possible to make almost any desired mating; production-testing programs, such as the National Cooperative DHI, and DHIR, for dairy cattle; and performance registry for beef cattle. By computer methods, sire analyses and progeny evaluation data are readily available, and conformation can be officially evaluated by the type classification or grading programs sponsored by breed organizations.

Figure 4.1. An outstanding group of two-year-old Guernsey cows. They are all daughters of one bull that is used in AI. Note their uniformity. (Photography courtesy of Curtiss Breeding Service.)

The purebred breeder produces the sires that are used for AI. The fact that, since the application of AI, milk production per cow has doubled in America and greater gains in many beef cattle are being recorded is a tribute to purebred cattle breeders. The challenge for the purebred breeder is great, but it also has its disappointments. Due to variation (see Chapter 2), sometimes the best of matings produce "duds." This is a calculated risk all cattle breeders have. The successful breeder utilizes all available information and studies cattle. He uses his judgment and experience and does not depend entirely on statistical information in guiding the breeding program.

The challenge for the purebred breeder in producing "seed stock" is well portrayed in Figure 4.2.

While purebreds are used in upgrading common stock, it is obvious that purebreds must be continually improving in merit or, after five or six generations, the grades are nearly equal. Nonregistered dairy cows, tested through the National Cooperative DHI plan, closely approximate the production levels of registered ones (see Table 4.1 that summarizes production of registered and grade Holstein cattle, both AI and non-AI in Iowa herds). This is to be expected with the successive generation use of AI bulls—and heavier culling of the grade cattle. About 15 percent of all dairy cows and heifers served by AI in the U.S. are registered. It is estimated that about 10 to 12 percent of the milk cow population is registered. The dairy breed

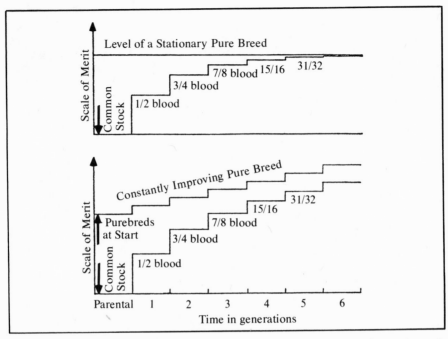

Figure 4.2.   The contrast between grading to a pure breed that is stationary in merit and grading to a pure breed that is improving its merit by a constant amount each generation. Source: Reprinted by permission from *Animal Breeding Plans*, by Jay L. Lush, 3d ed. (Ames: Iowa State University Press, 1945).

associations now have provisions for recording grade cattle with high production and good conformation.

Figure 4.2 illustrates the method whereby new beef breeds gradually attain cattle that have virtually 100 percent the characteristics (genes, "blood") of the desired breed after six to seven crosses

Table 4.1. Production of Registered and Grade, AI and Non-AI, Holsteins

Classification	No. Records	Milk lbs.	Butterfat %	Butterfat lbs.
Registered AI	7,126	14,395	3.66	527
Registered Non-AI	4,100	13,712	3.64	499
Grade AI	8,762	13,944	3.67	511
Grade Non-AI	3,456	13,302	3.63	483
All AI	15,888	14,146	3.66	518
All Non-AI	7,556	13,524	3.64	492
All Registered	11,226	14,146	3.65	517
Grade-Identified S. and D.	12,218	13,762	3.66	503
All Grade	13,451	13,714	3.66	501
Unknown Sire or Dam	1,233	13,231	3.65	483

Source: Iowa DHI Summary (1973). All records 305—2X—Me. Iowa State University–Ames. (Courtesy of Donald Voelker.)

with the semen of purebred bulls of that breed. The improvement of common stock by successive crosses to purebred bulls, "grading up," is also indicated. This sytem is utilized to improve cattle throughout the years.

4.4 Impact of AI on Purebred Dairy Cattle

Dairy farmers and purebred breeders have led the way in utilizing the AI program (see Chapter 1). At first, there were disgruntled breeders who opposed AI because it meant a loss in bull sales. There were arguments that "unless you can select the bulls used in your herd, you'll never succeed as a breeder"; "You must develop your own bloodlines"; and "You must have pride in your herd sire." These arguments had some weight before the usage of frozen semen, since herd owners had to utilize whatever semen the technican had available on a given day. With the use of frozen semen in 1953, a breeder could select the bull of his choice; furthermore, it was demonstrated that AI was producing many good cattle, and the young-sire-sampling program was thus adopted.

Today, there is a close working relationship between the AI industry and breeders of purebred dairy cattle. The sires for the AI program are produced by mating highly proved AI sires to a breeder's best cows. Breeders use the AI sires widely and have an ever-demanding market to produce more promising young bulls for sampling. A large proportion of the registered dairy cattle in America's dairy herds results from AI, as indicated in Table 4.2a.

Purebred breeders lead the way in support of production-testing

Table 4.2a. Percentage of Registered Dairy Cattle Resulting from AI

Breed	Year	Percent AI
Ayrshire	1948	5.0
	1975	61.5
Brown Swiss	1940	.0
	1974	54.7
Guernsey	1949	11.0
	1977	64.3
Holstein	1940	.0
	1976	75.4
Jersey	1940	.0
	1975	65.0
Milking Shorthorn	1940	.0
	1975	18.0

Source: Information supplied by secretaries of above-listed breed organizations (1976).
Note: The percentage of registered offspring resulting from AI continues to increase each year.

Table 4.2b. Sires of Cows on Various Testing Programs

Type of Test	DHIR	DHI	AM/PM	OS
	%	%	%	%
AI Sires	68	61	45	40
Breeders Sires	29	16	11	8
Unidentified Sires	2	23	45	52

Source: *Advanced Animal Breeder* 25:5 (1978): 5.

Note: Only DHIR (purebreds only) and DHI are official tests used for sire proving. The AM/PM and Owner-Sampler tests, as is Weigh-A-Day-A-Month, are unofficial programs.

programs so necessary for sire proving and herd improvement. A survey[2] of cows in the National Cooperative Dairy Herd Improvement Program, both official and unofficial plans, in Pennsylvania during 1977 supports this view (see Table 4.2b).

The DHIR herds (purebred) lead in the use of AI sires. The DHI (registered and nonregistered) follow closely. This survey also indicates that DHIR herds utilize sires with a higher PD and higher repeatability than those herds on other forms of production testing.

4.4.1 Effect on Dairy Breed Organizations

The dairy breed organizations supported the AI movement from its early beginning. As related in Chapter 1, the dairy breeds joined forces in 1939 by forming the Purebred Dairy Cattle Association (PDCA). Cognizant of the errors that could occur in the information for registering animals resulting from AI, the PDCA led the way, with assistance from the American Dairy Science Association (ADSA), in

Table 4.3. Units of Dairy Bull Semen Sold and Custom Frozen, 1975

Breed	U.S. Sales	International Sales	Custom Frozen	Total
Ayrshire	72,887	2,040	11,421	86,348
Brown Swiss	172,803	149,800	65,037	387,640
Guernsey	366,708	4,082	24,709	395,499
Holstein	8,616,321	793,354	444,576	9,854,251
Jersey	458,097	22,864	42,340	523,301
Milking Shorthorn	20,651	626	3,832	25,109
Red and White	51,946	7,224	520	59,690
Mixed and Unnamed	0	0	20,489	20,489
Total Dairy	9,759,413	979,990	612,924	11,352,327

Source: *Dairy Herd Improvement Letter* 52:5 (December 1976), USDA.

Note: 0—indicates no activity.

formulating regulations, developing reports, and approving AI organizations and AI technicians. The cooperation between the dairy breed organizations and the AI industry is of untold benefit to the dairy industry. The AI activity for each American dairy breed for 1975 is reflected in Table 4.3.

4.4.2 Effect on Registrations: Dairy Cattle

The chief source of income for a breed registry organization is the fee for registration of cattle and transfers of ownership. Artificial insemination, as expected, has resulted in a reduction of male registrations and transfers (Table 4.4). However, AI is credited with increasing the number of people who register an animal for the first time. The exact number is unknown, but breed secretaries consider it significant.[3]

It is obvious that the ratio of males to females registered by all breeds has been greatly reduced since 1940. Before the AI program, the ratio of males to females registered was about 1:5.0, but in 1974 the ratio was about 1:13.0. Traditionally, Brown Swiss and Milking Shorthorns have registered a higher ratio of bulls to females than the other dairy breeds; bulls of these two breeds are often used for crossbreeding in beef cattle as a means of increasing milk yields.

4.4.3 Many Registered Progeny per AI Sire

When compared to pre-AI days, when a bull with several hundred progeny was a rarity and when a sire with a hundred or more

Table 4.4. Effect of AI on Number of Males Registered

Breed	Year	Ratio of Males to Females Recorded
Ayrshire	1940	1:5
	1974	1:15
Brown Swiss	1940	1:2
	1974	1:6
Guernsey	1940	1:3
	1974	1:18
Holstein	1940	1:5
	1974	1:11
Jersey	1940	1:4
	1974	1:16
Milking Shorthorn	1940	1:2
	1974	1:4

Source: Personal communications in 1975 with secretary of each registry organization listed above.

production-tested daughters was unusual, the number of progeny registered for some AI bulls is astronomical. As a rule, with the improved evaluation methods available and sire summaries regularly released by the USDA, the bulls that rank high in transmitting milk and good conformation are used most heavily. The leading dairy sires, in terms of number of progeny registered, are shown in Table 4.5 and Figures 4.3a, 4–3b, 4–3c.

The above-listed sires have many additional nonregistered daughters. Commercial dairymen, particularly in the western states, demand high production transmitting sires but often do not keep up registration papers. The sire summaries for each of the bulls listed are based on their daughters tested in the National Cooperative Dairy Herd Improvement Program and the DHIR programs of the respective breed associations.

It will be noted that even though there are many daughters, tested

Table 4.5. Dairy Sires with Most Registered Progeny

Breed	Date of Summary	Females	Males	Total Reg. Progeny
		No.	No.	No.
Ayrshire (Selwood Betty's Commander 117936)				
	12–3–75	10,287	905	11,192
Prod.	6,240 daus. avg. 12,508M – 3.8% – 493F; 1420 Cl. Avg. 85.5			
PD (5–'76) +1284M – 0% – +51F. Rpt. 99%				
Brown Swiss (Welcome in Supreme 124652)				
	10–21–75	5,393	830	6,223
Prod.	2,284 daus. avg. +12,134 M – 4.05% – 491F			
PD(1–'78) –119M – +.03% ––1F Rpt. 99%				
Guernsey (Norgert's Royal Nance 533769)				
	9–9–75	11,536	1,436	12,972
Prod.	7,406 daus. avg. 10,410M – 4.4% – 462F; 3566 Cl. Avg. 83.6			
PD (5–'77) +416M – +0.4% – –4F Rpt. 99%				
Holstein (Paclamar Astronaut 1458744)				
	3–1–76	34,639	7,617	41,806
Prod.	30,086 daus. avg. 16,316M – 3.62% – 591F; Cl. Avg. +1.03			
PD (1–'78) +798M – –0.3% – +24 Rpt. 99%				
Jersey (Milestones Generator 602658)				
	5–3–75	10,000+	1,000+	11,000+
Prod.	8,922 daus. avg. 11,086M – 4.50% – +499F			
PD (1–'78) 1,187M – –0.43% – +17F Rpt. 99%				
Milking Shorthorn (Lilydale Lucy's Promise 10th 346424)				
	7–2–75	406	75	481
Prod.	135 daus. avg. 12,193 – 3.49% – 426F			
PD (1–'78) +1,310M – –0.06% – +42F Rpt. 83%				

Source: Data from dairy breed secretaries, sire owners, and USDA Sire Summaries. All records 305 day—2X—ME. Three of the bulls listed here are shown in Figures 4.3a, 4.3b, and 4.3c.

Figure 4.3a. Selwood Betty's Commander (Ayrshire). Over 10,000 registered daughters and 905 sons.

Figure 4.3b. Norgert's Royal Nance (Guernsey). Over 11,000 registered daughters and 1,436 sons.

Figure 4.3c. Paclamar Astronaut (Holstein). Nearly 35,000 registered daughters and over 7,000 sons. In 1977, his total offspring exceeded 200,000 head.

in different herds and under varying environments, each of the AI bulls listed in Table 4.5 ranks high in transmitting genetic production potential for milk and butterfat. Type ratings of their daughters indicate that each of these sires transmits the characters that make for good conformation and wearing qualities.

4.4.4 How Sons of Heavily Used Sires Perform

The production of bulls for AI is done to a large degree by breeders and usually AI organizations cooperating to produce future sires by selected matings. The pressure is continuous on both the breeder and AI organizations to produce higher production transmitting sires. As related in chapters 2 and 3, the best sires in sue today must produce the sons to be sampled to provide the AI sires for tomorrow. I. M. Lerner and H. P. Donald write, "Unless each generation of bulls for artificial insemination is the product of the previous generation on which money has been spent, the profit from all the work on improvement through breeding is limited to expanding the use of good bulls in commercial herds and reducing the use of poor bulls."[4]

The AI bulls with the most registered progeny (Table 4.5) have many sons sampled and used in AI. The performance of these sons is indicated in Table 4.6. As is strongly advocated, the use of high "plus

Table 4.6. AI Sires with Twenty or More Sons: USDA Summaries

		Sons					"70%" Sons	
Breed	No. of Sons	Avg. PD Milk for Sons	+ PD Milk	% Plus	No. Sons 70% Repeatability	Avg. PD Milk for "70%" Sons	+ PD Milk	% Plus
Ayrshire (Selwood Betty's Commander)								
	111	+440	104	94	18	+636	15	83
Brown Swiss (Welcome in Supreme)								
	61	−69	25	41	8	+16	4	50
Guernsey (Norgert's Royal Nance)								
	216	+98	133	62	15	+292	12	80
Holstein (Paclamar Astronaut)								
	522	+300	409	78	40	+369	31	78
Jersey (Milestones Generator)								
	71	+770	70	99	4	+846	4	100
Milking Shorthorn (Lilydale Lucy's Promise 10th)[1]								

Source: Data from USDA Sire Summaries of October 1977. Compiled by Larry W. Specht, Pennsylvania State University, November 1977. All records are 305 days—2X—ME.

[1] Insufficient sons proved to list.

PD'' bulls to produce sires for the future is well illustrated. About four of every five sons by a good ''plus proved'' sire transmits ''plus'' production. The sons of ''minus PD'' bulls, as indicated by numerous USDA Sire Summaries, on the average, are ''minus'' PD. Breed organizations and AI organizations alike stress the importance of reducing the risk in breeding disappointments by using mostly high PD sires with a repeatability of 70 percent or above (see Chapter 2).

4.4.5 Are Too Few Sires Used?

There is often concern among breeders when some sires are used very heavily in AI while other bulls, perhaps equally promising, are not utilized. For example, in the Holstein breed, which makes up about 73 percent of America's dairy cattle population, a 1974 summary[5] indicates that 125 bulls each have over 5,000 registered progeny and 1 bull sired over 6 percent of the total males registered. In the other five dairy breeds, ''popular'' bulls often sire an even higher percentage of the registered population. There are some complaints that ''the genetic base is being narrowed,'' that ''if a hereditary defect shows up the breed is ruined,'' or ''there will be deterioration due to inbreeding.''

The serious consequences mentioned above are not likely to happen in the free-enterprise system of cattle breeding and conduct of the AI program. Good judgment and competition should preclude the overuse of any single bull. The selection program must involve the sampling of many young bulls by proved sires that excel in desired traits. Any deleterious effects of too few bulls being used will first appear in those breeds with lesser numbers. There will be ample time to take corrective selection procedures should this occur.[6] The dangers of spreading undesirable, or lethal, characters through the AI program is often overemphasized. A bull in AI use is widely utilized, being mated to many females in many herds; if he carries genes for undesirable characters, it will be detected quickly and the bull will be eliminated. The damage in any single herd will be far less than if a breeder is depending on one or two sires for his entire calf crop.

Each reduction in the number of sires within a breed tends to reduce heterozygosity. This responsibility is placed upon cattle breeders and AI personnel who select sires. By careful planning and random mating, so far as bloodlines are concerned in producing future sires, the coefficient of inbreeding can be kept low. The price of progress is the use of fewer bulls, but eternal vigilance and careful appraisal of progeny on a continuing basis must prevail.

4.4.6 Effect on Milk Production

The milk production per cow has steadily increased for both registered and nonregistered cows since the AI program began. The milk yield for cows resulting from AI and identified by breed, from 1960 through 1975, are graphically presented in Figure 4.4.

The data used in the above figure consist of 3.6 million, sire-identified first lactation records over a fifteen-and-a-half-year period. There are 67,623 Ayrshire; 70,569 Brown Swiss; 321,881 Guernsey; 2,875,958 Holstein; 263,843 Jersey; and 9,987 Milking Shorthorn official DHI records summarized. During this fifteen-and-a-half-year period, the increase in milk by breeds is indicated. There are variations in the trend, particularly for all breeds except Holsteins, largely because of the influence of a few good or bad sires. This situation easily occurs when a limited number of non-AI proved bulls are used by the breeds with a lesser population. The Milking Shorthorn curve, for example, makes a rapid ascent from 1972 to 1975. This is due to the influence of several high plus PD bulls in AI use between 1966 and 1972. The USDA workers[7] who compiled the summary state that the trends for increased production in all breeds in recent years is genetic.

The tendency for bulls with a high plus PD for milk yields to be used in AI is clearly indicated in Table 4.7.

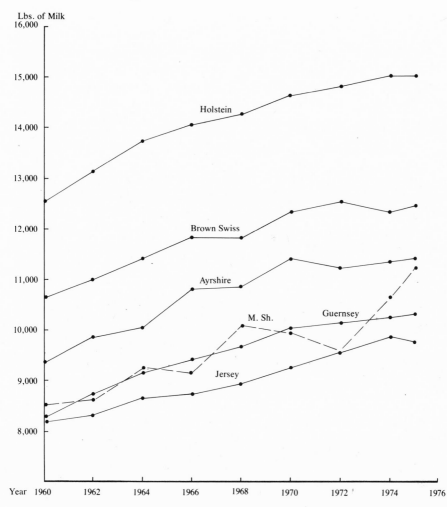

Figure 4.4. Milk production, AI first lactation daughters in six dairy breeds, 1960–1975 (305—2×—ME month of calving adjusted). Source: R. L. Powell, H. D. Norman, and Frank N. Dickinson, "Trends in Breeding Value and Production," *J. Dairy Sci.* 60:8 (1977): 1316–26.

Table 4.7. Average PD Values for Bulls Summarized in Spring 1977 by Breed and Type of Use

| | Active AI bulls | | | | Non-AI bulls | | | |
Breed	Number	PDM	Percent test	PDF	Number	PDM	Percent test	PDF
Ayrshire	25	+504	−.02	+18	375	−79	+.01	−2
Guernsey	78	+466	−.06	+16	1,269	−30	+.02	0
Holstein	908	+554	−.06	+12	6,138	−148	−.01	−7
Jersey	113	+635	−.10	+23	1,394	−47	.00	−2
Brown Swiss	39	+596	−.02	+22	485	0	+.01	+1
Milking Shorthorn	10	+743	.00	+27	166	+124	.00	+5
All Breeds	1,173	+558	−.06	+14	9,827	−104	.00	−5

Source: USDA Sire Summaries (Spring 1977).

The 1,173 bulls in active AI use have an average plus PD of 558 pounds of milk. The 9,827 bulls in natural service average a minus 104 pounds of milk and minus 5 pounds fat. The large number of non-AI bulls summarized is due to herds that use both natural and AI service; to the widespread use of young bulls to breed heifers, largely as a labor-saving measure; and the use of bulls in an attempt to get cows in calf that do not settle to AI. Most of the non-AI bulls are unproved and never will be. It is obvious they cost dairymen about 660 pounds of milk per daughter per year. At 1978 milk prices, this is over $50 less gross income per year on each non-AI daughter.

4.4.7 How Dairy Breed Officials View AI

For over forty years, the dairy breed registry organizations have made changes and adopted new programs brought on by the impact of AI and modern breeding concepts. The executive secretary of each dairy breed was asked to enumerate the desirable and undesirable effects of the AI program on the basis of his experience. The views expressed were generally in agreement:

Desirable Effects

1. Makes available many superior bulls. Provides an opportunity for breeders to develop a sound herd-breeding program as they have access to sires that under natural service conditions would not be available.
2. Improves production and type. Improves a dairyman's income. The modern progressive dairyman enjoys what a cow will produce, whether she is sired by an AI bull or not.
3. Stimulates breed-improvement programs, including young sire sampling, production testing, type classification, sire summaries, and pedigree analyses. Encourages breeder cooperation in all breed programs. . . . "People in agriculture are independent and slow to change. The purebred breeder has been a bit slow in some instances in adopting AI and breed programs. However, if all of us had used bulls on a production basis our average production and type in all breeds would be better. Basically, the AI program and its concepts are good."
4. Valuable in controlling the spread of disease, particularly venereal diseases, carried by the bull.
5. Has increased the sales price of good breeding stock.
6. Eliminates special weaknesses. When breeds adopted the USDA Sire Summary, AI businesses were forced to discontinue the use of minus PD bulls.

Undesirable Effects

1. Too much reliance on a few bulls.
2. Not enough support for young-sire-sampling program. Breeders of the

so-called colored breeds will have to develop their own young-sire programs.

3. Too many breeders stop being students of their herds and are inclined
"to believe a bull is good enough just because he was selected by a bull
stud."

4. Some breeders use a bull on the basis of a high PD regardless of the
type of his progeny.

5. AI organizations promote to sell a high volume of semen at a high price.

6. The opportunity to use "any bull in the country" has resulted in some
dairymen using so many bulls in a single herd that so much heterozygosity
results that a consistent breeding program is difficult.

7. Competition among bull studs results in some unrealistic prices being
paid for inadequately proved bulls. Should price all bulls on the basis of
semen sales after AI proof is adequate.

8. The greatest handicap AI has is people. Often the chief goal of an AI
business is to make a profit. The use of frozen semen and keen competition among AI concerns tends to minimize this situation.

It is interesting to note that no complaints were voiced against
fewer bull registrations. Breeders and dairy breed organizations pay
the cost of progress by using fewer, more carefully selected, registered bulls. The bulls selected now command good prices and the
challenge for the breeder continues at an accelerated pace.

Most AI organizations are motivated by a strong desire to improve
cattle and, consequently, the welfare of their owners. As a business
venture, the AI organization that has no reasonable profit does not
make improvements in sires and services, or stay in business. Few
make more than modest returns on capital. The numerous consolidations of bull studs (see Part III and Appendix A) in order to make ends
meet indicate that AI organizations are strongly service oriented.

4.4.8 "A Cow, a Bull, and a Man"

Cattle breeding may be looked upon as an interwoven skein of
events involving people and animals. The dramatic moments are
many, and sometimes they are lost in recorded history. How a cow, a
bull, and a man changed the destiny of three of the dairy breeds is
beautifully told by Charles Rogers.[8]

Never before in the history of cattle breeding has one cow, one bull, or a
man had such a profound influence on shaping the destiny of a breed of
dairy cattle as Lawrence Colebank has had on the Guernsey breed.
... While Jane of Vernon put type in the Brown Swiss breed and Selwood Betty's Commander put milk in the Ayrshire breed, the breeds
remained essentially the same ... not so with the Guernsey cow. Lawrence Colebank [classifier for the American Guernsey Cattle Club, of
Peterborough, New Hampshire] went up and down the United States
classifying high the cows that excelled in size, upstandingness, and pro-

duction until he wrought a new breed in just about everything but name . . . in a span of 15 years.

The above events have all transpired in the past twenty-five years or less. The influence of Jane of Vernon, through her sons and many grandsons, was spread to multitudes of cattle through AI. The same is true of Selwood Betty's Commander with over ten thousand reg-istered daughters, and many sons in AI (see Tables 4.3 and 4.4). The contributions of Lawrence Colebank in establishing an upstanding stretchy, good-sized cow, with much dairy character as the desired Guernsey conformation serves as a guide for AI organizations and breeders. Because of AI, the influence of a cow, a bull, and a man will remain contributing factors for generations of cattle yet to be born.

4.5 The Impact of AI on Beef Cattle Breeding

Artificial insemination is becoming increasingly important in beef cattle improvement programs. Two of the most important steps in making genetic improvement are the use of performance testing and maximum use of performance and progeny-tested sires. AI and modern breeding concepts provide the means to make more rapid improvement in the economic traits of beef cattle, just as they have increased the genetic production potential of dairy cattle. Both com-mercial and purebred beef cattle producers are moving, in increasing numbers, to scientific methods in their breeding programs. The extent of AI in terms of units of beef semen from 1970 to 1976 is summarized in Table 4.8.

4.5.1 Importance of Beef Cattle

There are approximately 116.3 million head of cattle, all ages, in the United States, valued at over \$27 billion.[9] About 97 million are

Table 4.8. Units of Beef Semen Sold and Custom Frozen, 1970–1976

Year	Domestic Use	International Sales	Custom Frozen	Total Units
1970	1,926,930		979,635	2,906,565
1971	2,076,788	159,121	1,055,409	3,291,318
1972	2,382,955	213,590	1,114,903	3,711,448
1973	2,719,052	348,733	1,092,334	4,160,119
1974	2,869,569	385,153	1,675,843	4,930,565
1975	1,712,264	167,798	1,425,958	3,306,020
1976	1,368,618	179,110	1,273,946	2,821,674

Source: *Dairy Herd Improvement Letter* 52:5 (December 1976), USDA; and NAAB reports.

classified as beef with about 39 million cows that have calved (see Table 2.2, Chapter 2, for all cattle numbers). There are about 19 million head of dairy cattle, all ages, in the U.S.

Beef cattle are maintained on nearly two million of the 2.8 million farms in the United States. The gross receipts for cattle sold for slaughter approximate $20 billion annually. While the United States has only about 9.7 percent of the world's cattle population, it produces 24.2 percent of the world's total beef and veal.[10] Beef is America's first choice for meat, and consumption per capital is about 120 pounds annually.

The principal function of beef cattle is to convert grass, coarse forages, and grain by-products, usually with minimum amounts of grain into nutritious meat for human use. It is estimated that about 47 percent of the land area of the continental United States is grazed all or a part of the year. The topography of much of the grazing land is too rough for growing crops. Were it not for the ability of ruminants (cattle and sheep) to convert grass and other roughages into human edible foods, our level of nutrition would suffer. While many beef cattle are finished on rations high in grain, the feeding period is relatively short. At least 60 to 75 percent of the growth and weight gains in beef cattle come from roughages that cannot be utilized by humans.

Beef production involves large ranches of thousands of acres with thousands of cattle and many smaller farm setups, usually termed *cow-calf* operators, with cow herds ranging from thirty to as many as two hundred head. The smaller herds prevail in the corn belt region and are often a part of the general farming program. The larger herds are found chiefly in the western United States where landholdings are immense, and grazing, part of it on federally owned lands, provides most of the sustenance. Beef cattle are an important part of the agricultural economy in every state in the nation. The leading states in cattle numbers as of 1 January 1978 are Texas (16.8 million), Iowa (7.6 million), Kansas (6.4 million), Missouri (6.4 million), Oklahoma (5.65 million), and California (4.75 million).

Cattle raisers and beef ranchers, as a group, are independent, hardworking folks. They have pride in their profession and want no government interference or subsidies. Through the years (see Chapter 2, Table 2.3), they have battled droughts, blizzards, disease, market fluctuations (cattle at an average value of $17.78 per head in 1934; $92.00 in 1954; $321.00 in 1974; $159.00 in 1975), and consumer preferences. Their way of life is not easily changed. However, the modern-day operator is looking for ways to make improvement. He appraises each new innovation in terms of its practicability for his operation. The somewhat slow adoption of AI in many beef operations must be considered due in part to the past experience and philosophy of beef raisers in general.

4.6 Improving the Economic Traits in Beef Cattle

The same genetic concepts, when combined with artificial insemination (large increases in the number of progeny of the "best" sires), as outlined in Chapter 2, are adaptable for beef cattle improvement. While progress has been made in improving beef cattle over the years, much greater progress no doubt could have been made if dependable yardsticks for measuring genetic variation in the most important economic characteristics had been used much earlier, and even today.

Long ago, William Thomson, Lord Kelvin (1824–1907), the famous British physicist, said, "If you can measure what you are talking about and express it with numbers you know something about it, but if you cannot measure it and describe it with numbers your knowledge is of a meager and unsatisfactory kind."

The above statement applies all too accurately to the methods generally employed in beef-breeding programs, with emphasis on aesthetic or show-ring winnings and pedigrees until a few years ago. While there is nothing wrong with aesthetic standards or pedigrees within themselves, if a beef breeder's goal is to produce good beef efficiently, as sold across the scales, he needs reliable figures to guide his breeding and management methods. The 1936 USDA Yearbook of Agriculture belabored the above points strongly and urged better criteria for measuring and evaluating animals from the standpoint of efficiency of feed utilization.[11] Investigators have not been idle in this area, and scientific methods for evaluating meat animals have long been available. In terms of numbers of beef cattle involved, however, their application was meager until about 1954, when the first state beef cattle associations began and performance testing became popular.

4.6.1 Heritability Estimates Beef Cattle

Geneticists have found that many of the variations in important economic traits of beef cattle are due to inheritance. These differences are expressed as heritability and can be measured. Heritability is defined as the proportion of phenotypic differences among animals for a given trait that is transmitted (inherited). Heritability values range from 0 (not due to gene action) to 100 percent (completely due to gene action). The larger the heritability for a given trait, the greater the opportunity to improve that trait. Most of the economic traits in beef cattle are fairly high in heritability as indicated in Table 4.9.

It should be noted that most of the traits involved in weaning

Table 4.9. Estimates of Heritability for Economic Traits

Trait	Heritability (%)[1]
Birth Weight	40
Preweaning Daily Gain	30
Weaning Weight	33
Weaning Score	30
Feedlot Gain	50
Feed Efficiency	47
Final Feedlot Weight	60
Slaughter Score	45
Pasture Gain	35
Yearling Pasture Weight	40
Mature Cow Weight	70
Maternal Ability	40
Fertility	10
Carcass Traits	
Rib-eye area	70
Fat thickness	45
Cutability	70
Tenderness	60
Carcass grade	30

Source: J. S. Brinks, "Improving Your Herd Performance through Selection," AICA Genetics for the Cattle-man, American International Charolais Association, Houston, Tex. (1971).

[1]The above heritability estimates are averages taken from several studies by Brinks.

weight, yearling weight, and feed efficiency are significantly high in heritability. Fertility, affected by many environmental factors, as well as inheritance in some cases, is low in heritability estimates. The important point is that beef producers can make intelligent selection for improvement of the traits that spell dollars, if they have the necessary performance records. The intensity of selection for a given trait determines the speed of improvement for that trait.

4.7 Performance Testing: Beef Cattle

Performance testing is measuring the growth rate of an animal in terms of weight and age. To be meaningful in selection of breeding stock, performance information must be on a comparative basis. Beef bulls on performance test are ranked just as dairy sires are, and the best can then be chosen for use.

Records of performance are useful primarily in providing basis for comparing cattle that are handled alike in a given herd. These records obtained by herd owners, with the cooperation and assistance of the personnel of state or county beef improvement associations, usually consist of weaning weight (205 days adjusted), post-weaning gain, yearling weight, carcass grade, and cutability. Ease of calving, reproductive efficiency, and other information is often calculated. The Beef Improvement Federation[12] (see 4.7.1) has developed uniform procedures for computing adjusted weights, weight ratios, and other data for performance records. For estimates of genetic value, it is necessary to adjust records to standard basis depending on age at weaning and age of dam (mature dam equivalent) and yearling weights.[13]

Table 4.10 gives an example of the form in which performance data are tabulated and presented.

Progeny Testing: Beef Bulls. Objective sire evaluation is necessary if the bulls that transmit the highest potential for genetic gains are selected. This is particularly true where a single sire in AI use may influence thousands of progeny. The performance record of the individual, his ancestors, half-brothers and sisters, and his progeny can be used to evaluate the differences in breeding value. The usual method of progeny testing a beef bull is to randomly select a group, usually ten head, of his sons or steers and to place them in a test station, where they will be handled under uniform conditions to obtain the average daily gain and other comparative information.

To progeny-test many bulls, with uniformity in procedures, and to enable comparison of bulls with progeny in several herds, the National Sire Evaluation Program has been developed.[14] Its purpose is to provide breeders with information on "expected progeny difference" between bulls. A progeny test can be conducted in a breeding herd, or a commercial herd where carcass evaluation is important, or on a large area basis with many herds involved. The present-day-progeny-test programs conducted by breed organizations or breeders, with the use of bull studs, usually involve the use of reference sires. The reference sires are designated sires, with many progeny from a random sample of cows in many herds, to which other bulls being progeny-tested can be compared. AI plays an important role in getting a large number of progeny in many herds from the sires tested. For bulls evaluated in a reference-sire system, the Expected Progeny Difference (EPD) is calculated for each important economic trait. The

Table 4.10. Example of Performance Test Report

Herd No.	205 Day Adj. Wng. Weight	Number of Contemp. at Wng.	205 Day Wng. Wt. Ratio	ADG (lbs.) on 140 Day Test	Number of Contemp. on Test	ADG on Test Ratio	365 Day Adj. Wt.	Number of Contemp. to Yrlg.	365 Day Adj. Wt. Ratio
117	627	17	102	3.86	15	118	1264	15	110

Source: *A.I. Digest* (May 1975): 8.

EPDs are expressed as ratio differences and are adjusted for progeny numbers. Thus, any EPD can be compared fairly with another. The information obtained from most progeny-testing programs includes calving ease score; birth weight; length of gestation; 200- or 205-day weaning weight; average daily gain on feed ratio; yearly weight ratio (365 day, 452 day, or 550 day); yearling weight ratio; and carcass evaluation. Table 4.11 is indicative of the type of information that can be obtained from the progeny test on a given sire.

Table 4.11. Example of Progeny Test Information, 1971–1974 (700 calves from 11 sires in 25 herds)

Bull A61	Trait	No. of Progeny	No. of Herds	Trait Ratio
Calving ease		61	25	103
205d. Wng. Wt.		59	25	98
365d. Yrlg. Wt.		53	22	97
Retail Cuts/Day		26	22	100

Source: Ron Lindquist, "Interpreting Beef Testing Terminology," *A.I. Digest* 23:5 (1975): 9.

The AI organizations are engaged in performance testing and progeny testing of beef bulls. These efforts are often in cooperation with purebred breeders, breed associations, and commercial producers. The goal is to have the majority of AI bulls' progeny proved and to have semen sold on the bull's merits in terms of transmitting economic traits at a high level.

4.7.1 Early Development of Performance Testing

Beef cattle improvement programs began on a scientific basis in the early 1930s when research workers indicated that the heritability of certain economic traits in cattle is sufficiently great to provide a sound basis for selection. Early investigations on performance testing were carried on by agricultural experiment stations and the USDA. Its merits were extolled for many years before beef cattle breeders, in general, considered it important.

No organized plan, such as the "Cow Testing Association" for dairy cows that started in 1905, was initiated for beef cattle until 1954 with the formation of the Virginia Beef Cattle Improvement Association. Since that time, largely as a result of the efforts of agricultural extension workers, at least forty-seven states have formed beef improvement organizations.[15]

4.7.2 Performance Registry International

The Texas Agricultural Experiment Station was the early pioneer in performance testing of beef cattle, beginning at the Balmorhea substation in 1941.

The first national improvement program of beef cattle was initiated in 1955 by the formation of Performance Registry International by a group of progressive Texas cattlemen. It all began when several West Texas cattlemen visited the Texas A&M Experiment Station at Balmorhea, where research on beef cattle performance was being conducted. It was agreed that such a study be conducted in the Panhandle area.[16] The first bulls were tested at Pan Tech Farm in 1950. L. A. Maddox, Jr., was employed by Texas Technological College as the farm's animal husbandman. Bulls consigned to the test by cooperating owners were evaluated on weight gain and feed efficiency during a 140-day test period. After the third test in 1953, a public auction of the performance-tested bulls was held, the first of its kind in the United States. It attracted considerable attention and in 1953, M. H. R. Ritchie, director of the International Livestock Exposition, requested an exhibit on performance testing at the "International" in Chicago. In view of the interest and questions from cattle breeders as to where a "proved high gaining bull" could be found, the Texas group decided that some type of national registry for performance-tested bulls should be established.

In 1955, at a meeting in Amarillo, Texas, the American Beef Cattle Performance Registry Association was formed.

The first presidents of the organization were Colby Conkwright, Clyde Bradford, and Max Blau. M. R. Calliham, then at West Texas Technological College, in Canyon, was the first secretary. The organization was aided by other interested cattle breeders in New Mexico, Oklahoma, and Colorado. The name of the organization, as Canadian cattle breeders became involved, was changed to its present name, Performance Registry International (PRI).

The purpose of PRI is to record, after proper certification, those sires that meet high performance test goals for gain. A Certified Meat Sire Program was initiated and has gained prominence. PRI was the first to produce and to use "production" pedigrees.

PRI has, and is, supported entirely by the income from breeders who utilize its services. For many years, the "going was rough" due to limited income. Veteran cattle breeders such as "Ferry" Carpenter of Hayden, Colorado, and Clarence Burch of Mill Creek, Oklahoma, were among the many who fought to keep PRI solvent. At the 1961 annual meeting of PRI, "Ferry" Carpenter, in a dramatic drive, raised $17,000 in pledged services for PRI. The present executive secretary is Glenn Butts, with the PRI office at Fairland, Oklahoma. Modern programming and computerized records are utilized. PRI does not have a large volume of business, but it deserves much credit for its pioneering efforts and for serving as a pacesetter for the cattle industry. It was largely through the efforts of PRI leaders that the Beef Improvement Federation was formed.

4.7.3 The Beef Improvement Federation

In January 1967, Performance Registry International initiated efforts to develop an "organization of organizations" that could bring about cooperation and standardization in performance testing throughout the United States. The first meeting, held 14 January 1967 in Denver, Colorado, drew participants from state cattle improvement associations, test stations, breed organizations, and college workers.

The next year, an overall organization, the Beef Improvement Federation (BIF), was formed. The objectives of BIF are (a) to bring about uniformity in measuring results of performance and reporting data; (b) to assist members and/or their affiliates in developing individual programs; (c) to develop cooperation among all segments of the beef industry in compiling and using performance records; (d) to encourage the development of educational programs emphasizing the use and interpretation of performance data; and (e) to develop increasing confidence in the economic potential of performance testing.

Membership in the Beef Improvement Federation consists of representatives of the beef cattle improvement association or similar organizations from respective states, beef registry organizations, Performance Registry International, National Association of Animal Breeders, and American National Cattlemen's Association.

From 1967 to 1969, Clarence Burch, of Mill Creek, Oklahoma, served as the first BIF president. The first secretary-treasurer was Frank Baker, who was the chairman of the Department of Animal Science at the University of Nebraska–Lincoln and now retired dean of agriculture at Oklahoma A&M University. Representatives of the agricultural colleges, breed organizations, and USDA provide technical advice and leadership in BIF. One of the early and important accomplishments of BIF was to develop "Guidelines for Uniform Beef Improvement Programs," which covers all phases of performance testing and is now widely used.

4.7.4 Extent of Performance Testing

Performance testing is carried out by central test stations, of which there is one or more in most states where beef cattle are important, and on the farm or ranch. The 1975 summary of activity for test stations on the farm or ranch is indicated in Table 4.12.

These data reveal that 72,355 bulls were tested in over 9,000 herds. Fifteen breed registry organizations accounted for 352,201 weaning and 121,201 yearling records on 15,305 bulls owned by 8,894 breeders.

Table 4.12. Summary: State Beef Improvement Programs in U.S., 1974

Herd Records		Central Test Stations			On Farm and Ranch				Carcass Evaluation		Sire Evaluation	
Weaning Records	Yearling Records	No. of Herds	No. of Bulls	No. of Breeders	No. of Stations	No. of Bulls	No. of Breeders	Total Breeders Testing Bulls	No. of Breeders	No. of Sires	No. of Sires	No. of Breeders
550,226	115,461	9,104	15,191	2,916	203	57,164	3,063	5,193	1,024	1,740	5,297	2,263

Source: Personal communication in July 1976 with Dixon Hubbard, extension animal scientist, USDA, Washington, D.C.

Performance testing is gaining in popularity and almost every breed registry organization has a program. The American Angus Association, at St. Joseph, Missouri, was the first to adopt an improvement program and was followed by the American Hereford, American Polled Hereford, and American Shorthorn associations. Some registry organizations, particularly those for the "new breeds," require performance as a prerequisite for registry. The American Simmental Association of Bozeman, Montana, recorded 88,988 animals in 1976.[17]

4.8 Beef Registry Organizations and AI Policy

At the start of the AI program, the breed registry organizations were almost 100 percent opposed to AI where the owner of the bull and the cow bred were not the same. Many breeders of purebreds declared that "AI should be outlawed." Claims were made that "AI will ruin the bull market," "spread inherited defects and ruin a herd," "the bulls aren't good enough," "the cows won't settle," and "it's impractical for a rancher."

As time wore on and progressive breeders and commercial cattlemen began using AI and found it could be made to work, the policy of the registry organizations softened a bit. Joint ownership of sires was permitted—usually four owners at one time—and transfer of ownership of shares in a bull became frequent. When frozen semen became available beginning about 1955, breeders widely separated by distance began using it. The breed organizations became more liberal by permitting a limited number of calves from one bull in a given year. Some began selling a limited number of breeding certificates. The policy prevented any widespread use of AI from herd to herd as prevailed for dairy cattle. In 1968, four of eleven beef registry associations required that the same ownership of the sire and dam be a prerequisite for registering AI calves. (For additional restrictions, see section 13.10, Chapter 13.) Some purebred breeders were extremely unhappy with the restrictions. The AI organizations were frustrated. The situation moderated after the U.S. Justice Department charged the American Angus Association in 1968 with restraint of interstate trade and commerce. The result was a modification of Angus regulations on AI, and other beef breed registry organizations adopted more liberal policies (see section 13.10, Chapter 13).

The AI regulations for the various beef breeds vary according to policies of the members and officers. All require that sires be registered, blood typed, and the semen properly identified. Many require AI Service Certificates, which are sold to breeders. Most have a policy regarding bulls that transmit inherited defects.[18]

Crossbreeding, using "exotic" breeds, starting with the Charolais in the early fifties, gained momentum as semen from many other non-English breeds was imported from 1960 forward. Quarantine restrictions prevented the importation of live animals from many European countries. Cattle producers and purebred breeders who were interested in starting new breeds had no recourse except to use frozen semen and AI. The recording procedures of the "new breeds" in the United States, such as Simmental, Limousin, Maine-Anjou, Gelbvieh, Murray Grey, and many others, resulted in the adoption of a fairly liberal AI policy. This policy put some pressure on the older established breed organizations to consider AI more favorably. The extent to which AI was used in beef cattle in the U.S. in 1975 is shown in Table 4.13.

It will be noted that the semen of the established breeds—Angus, Brahman, Charolais, Hereford, and Shorthorn—was used in goodly quantities. Among the new breeds used domestically, Simmental with

Table 4.13. Units of Beef and Dual-Purpose Semen Sold and Custom Frozen, 1975

Breed	U.S. Sales	International Sales	Custom Frozen	Total Units
Angus	517,135	12,368	280,270	809,773
Barzona	215	0[1]	0	215
Beefalo	1,002	0	2,785	3,787
Beef Friesian	228	10	879	1,117
Beef Master	826	0	1,895	2,721
Blonde D'Aquitaine	10,639	27	0	10,666
Brahman	54,429	49,212	61,683	165,324
Brangus	3,276	1,508	50,036	54,820
Buffalo	0	0	1,519	1,519
Charbray	0	0	548	548
Charolais	78,298	15,283	86,245	179,826
Chianina	51,683	4,854	11,997	68,534
Devon	594	2,050	475	3,119
Dexter	352	0	0	352
Galloway	273	0	351	624
Gelbvieh	38,127	439	3,202	41,768
Hays Converter	1,996	48	0	2,044
Hereford	163,220	12,966	169,329	345,515
Limousin	64,692	1,598	33,985	100,275
Lincoln Red	336	0	0	336
Longhorn	59	0	958	1,017
Maine-Anjou	78,840	1,186	32,536	112,562
Marchigiana	1,663	0	0	1,663
Meuse Rhine Issel	1,193	142	0	1,335
Murray Grey	5,485	0	17,177	22,662
Normande	1,757	13	0	1,770
Norwegian Red	1,766	0	8,089	9,855
Pinzgauer	17,080	6	706	17,792
Polled Hereford	135,504	31,199	47,280	213,983
Red Angus	28,334	1,352	2,307	31,993
Red Brangus	629	0	0	629
Red Dane	92	0	0	92
Red Poll	1,985	453	1,720	4,158
Romangola	675	80	0	755
Santa Gertrudis	5,893	17,454	33,888	57,235
Scotch Highland	686	0	0	686
Shorthorn	17,068	2,498	18,422	37,988
Simmental	388,384	12,980	155,162	556,526
South Devon	90	0	579	669
Tarentaise	32,584	0	173	32,757
Welsh Black	367	0	0	367
Mixed and Unnamed	5,171	72	401,762	407,005
Total Beef and Dual Purpose	1,712,626	167,798	1,425,958	3,306,382

Source: *Dairy Herd Improvement Letter* 52:5 (December 1976), USDA. Reported by bull studs and custom-freezing businesses in U.S.

[1]0—indicates no activity.

556,526 units of semen custom frozen and exported ranked third, being exceeded only by Angus and Hereford (horned and polled combined).

4.9 Breeding Programs and AI: New Breeds

The introduction of the new breeds was fostered by (a) purebred breeders who were disappointed in the yields they were getting from straightbred cattle; (b) by commercial producers who were already sold on crossbreeding but felt new genes were needed; and (c) by enterprising promotion-minded individuals who saw the "exotics" as a means of making money quickly. AI fitted into the scheme of developments. It was the only way except for importations of cattle at great cost. Canada, from the start, established two quarantine stations to handle bulls and breeding stock from Europe. Over one million units of semen was imported from Canada each year from 1971 to 1975. Some American cattle breeders imported European bulls into Canada, had them stationed there, and collected and shipped semen to the U.S. for use.

The semen from the so-called exotics was used for crossbreeding, and by the grading-up process (see Figure 4.2), new breeds started. Many of them are in this process now. There are two principal beef-cattle-breeding programs.

4.9.1 Straightbreeding

This program is utilized by operators who cling to one breed. That straightbreeding can produce efficient animals is attested to by the performance of our established and recognized breeds. The inheritance in straightbred cattle for profitable performance is limited to those genes within that breed. To the individual herd operator, the inheritance in his herd is limited by the selection of breeding stock followed. In straightbreeding, cattlemen try to avoid close breeding or inbreeding and introduce new genes by selecting sires from other herds and family lines. Performance testing to guide selection and the use of AI to breed heavily to the best production bulls available are important practices in this program.

Good straightbred cattle are needed in beef improvement programs. Their genetic contributions can be predicted. The producer who practices crossbreeding depends upon straightbred bulls. As a rule, straightbred cattle do not have the advantage of heterosis that is obtained in crossbreeding but there are some advantages in crossing family lines of superior merit. Successful crossbreeding depends upon superior seed stock furnished by the straightbred cattle. Purebred breeders have a great opportunity in this respect.

4.9.2 Crossbreeding

Until recent years, crossbreeding of beef cattle was frowned on. However, our present beef cattle have been shaped by this practice. In the settlement of the West, Shorthorn bulls were crossed with the native or Texas Longhorn cows. After several generations, Hereford or "whiteface" bulls were used. All of the breeds we have today resulted from selection following crossbreeding. Other crossbred breeds are the Santa Gertrudis, Brangus, Charbray, and Barzona in the United States; the Braford and the Droughtmaster in Austarlia; and the Bonsmara in Africa.

Research[19] has demonstrated that crossbreeding and its resulting heterosis can increase the output per cow as much as 25 percent as compared to straightbreeding. *Heterosis,* or hybrid vigor, is the term applied to the increased production of the offspring over that of parents when unrelated animals are mated. As a rule, the average performance of the parents can predict the performance of the offspring, especially when the trait is highly heritable. However, for many traits in beef cattle, where the parents are of different breeds, the offspring are superior in performance to that of the parents. A summary of reports universally recommends crossbreeding for commercial cattle production. The advantages are more pounds of beef marketed; greater vigor and survival of calves; and improved fertility.[20] The results of cumulative heterosis for pounds of calf weaned are indicated in Figure 4.5.

Experiment stations report 4.3 percent more calves weaned and 4.7 percent heavier weaning weights for two-breed cross calves (straightbred dams in all cases) than for straightbred contemporaries. For the three-bred cross, 4.7 percent more calves weaned and 5.0 percent heavier weaning weights. Reports show up to sixty pounds more weight at one year in crossbred steers than in their straightbred controls. Through systematic crossbreeding, the British breeds will produce 25 percent more than straightbreeding. Using dairy breeds (Brown Swiss and Holsteins) in crossbreeding adds more milk and growth. The use of Brahman-British breed crosses produces high heterosis and is widely used in the southern United States.

One of the popular crosses is the Hereford-Angus, to produce "black-baldie" calves. The male calves are used for steers. Black-baldie cows are mated to a bull, other than an Angus or a Hereford, to

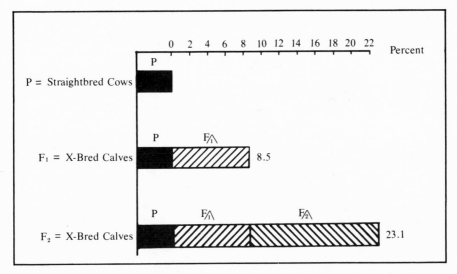

Figure 4.5. Cumulative heterosis effects for pounds of calf weaned per cow that was exposed. Source: Research statistics, USDA, Fort Robinson, Neb.; reported by Miles McCarry in *Advanced Animal Breeder* 27:7 (1979): 11.

produce the three-breed terminal cross now utilized by many commercial producers.

Heterosis Is Not Passed on Genetically. With well-planned crosses year after year, heterosis can be maintained.[21]

AI plays an important role in crossbreeding: (a) bulls maintained can largely be of one breed with a few follow-up bulls of the second or third breed to service cows not in calf by AI; (b) better sires, production wise, can be used in quantity to spare the herd operator much expense; and (c) AI provides a wide choice of breeds with performance-tested sires.

4.9.3 Breeding Programs and the New Breeds

Within the past decade, many breeds new to our continent have been imported for use in beef production (see Table 4.12 and section 4.8). There are fifty-five of the new breeds coded by NAAB for semen identification and used in performance recordings.[22] AI makes the semen of most of these breeds available. In many cases, imported bulls of some new breeds are owned by private breeders. In those instances, the semen is usually custom frozen and made available to the trade. Most bull studs now have good—usually performance tested and/or progeny tested—bulls of the more widely used new breeds.

In deciding to use one of the new breeds in a crossbreeding or an upgrading program, a cattle producer should be aware of the following points: (a) many of the new breeds are dual-purpose in their native lands, with emphasis usually on milk, but can be used to produce crossbred cows. Some of the heavier dairy breeds in America often serve the same purpose just as well; (b) some are growthy, angular breeds that might play a role in terminal crosses; (c) some of the new breeds are purely speculative. They can add little to our beef genetic pool that is not now available; and (d) time and experience will demonstrate which of the new breeds can compete under U.S. climate conditions, on the range, the feedlot, and the meat market. Table 4.13 indicates the new breeds now being used most heavily. Beef authorities and the AI industry are urging cattlemen to obtain as complete information as possible on the expected economic traits before experimenting too heavily with any new breed.

4.9.4 Buffalo (Bison): Cattle Crossbreeding

Since the turn of the century, research workers and various government agencies have attempted to crossbreed domestic beef cattle and buffalo (American bison). The objective is to combine into a crossbred range animal the hardiness and survival characteristics of the bison and the meat-producing qualities of domestic beef cattle. Animals carrying buffalo "breeding" are commonly termed *cattaloes*.

The *cattalo* is produced by crossing bison bulls on domestic cows. Males from this cross usually die at birth—partly due to difficult delivery as a result of the hump over the withers—and most that survive are sterile. The females live and are sometimes fertile. They are bred to domestic bulls. Their daughters are then bred to bison bulls to produce cattaloes.

With the upswing in crossbreeding of beef cattle, renewed efforts have been made to introduce bison crosses. Many new names, depending upon the breed crossed with the bison, have appeared. Common is "beefalo," which means any beef breed bison cross. The *Simmalo* (Simmental x Bison) is an example of specific breed cross terminology. Pictured in Figure 4.6 is the female Herefordalo #230, a Bison-Hereford first cross.

Beefalo cattle may be registered with American Beefalo Breeders, Inc., at Lapeer, Michigan. A Beefalo pureblood is generally considered to be three-eights buffalo and five-eights domestic cattle blood.[23]

One of the classic experiments in crossing bison and domestic beef

Figure 4.6. Herefordalo #230. Her sire is Geronimo, from the Walt Disney Buffalo Herd in California, and her dam is a purebred Hereford. She has several sons in AI use: Parisien Beefmaker, sired by the Simmental bull Parisien, and Extra Profit, sired by Extra #6084, also Simmental. Another younger son is sired by the Brahman bull AD Pioneer Quinca Manso #100. (Photograph courtesy of the Bull Bank, Escalon, Calif.)

cattle was conducted by the Canadian government at the Experimental Farm of Manyberries, Alberta, beginning over fifty years ago. The project was terminated in 1965 when it was felt that the potential of cattalo under range conditions had been sufficiently assessed.[24] It was found that the mortality of cattalo calves was higher than Herefords; cattalo calves did not surpass Herefords in weaning weight; cattalo were exceeded by Hereford calves by 16 percent in average daily gain in the feedlot and required 15 percent more digestible energy per pound of gain than the control Herefords. As the average percentage of bison was reduced, most of the economically important production traits improved. The investigators suggest that if cattlemen want to experiment with bison crosses, they do it on a limited scale.

4.10 The Future of AI in Beef Production

Experience gained the past twenty-five years indicates that AI can play an important role in beef cattle improvement and beef production. Progressive purebred breeders and commercial cattlemen in increasing numbers are turning to AI as a means of increasing the production per animal unit. Examples for small herds to ten thousand head or more are many and outlined in reference.[25]

Beef men have been reluctant to use AI in many cases because of the labor demands and the trouble in handling cattle. Many of these difficulties can be overcome. The introduction of the "exotics" proved to many beef men that AI can work. Experience indicates that beef AI can be successful only (a) when the herd operator seriously wants to improve herd income and is willing to take the necessary steps; (b) when he will arrange holding pastures, corrals, and chutes so the cattle can be handled effectively; (c) when cows to be bred are identified, records kept, checked for heat, and properly inseminated at the correct time; and (d) when a feeding and management regime is followed that finds cows and heifers in good flesh at breeding time.[26]

Heat synchronization, which enables the insemination of cattle on a fixed schedule, is becoming a practical reality.[27] With many cows treated and inseminated on a fixed date, much labor is saved and a calf crop of uniform age results.

In 1967, H. H. Stonaker at Colorado State University in Fort Collins, Colorado, said, "Modern beef cattle improvement, particularly at the purebred level, must benefit from the opportunity of AI. Breed competition will force this. Those breeds which use it most scientifically and rapidly have an interbreed competitive advantage. Those that fail to do this will be competitively disadvantaged. Those that restrict the use of the best known sires and restrict the availability of AI will restrict their breed's competitive advantage."[28]

Chapter 5: Technical Developments: Past and Future

5.1 Introduction

The expanded use of AI in cattle and other species, including people, was made possible through continuing research and technological developments. The contributions of the thousands of workers in this area are immense and of untold economic benefit. A worthwhile idea once born, if it is to benefit mankind, must be developed. Involved in the array of applied technology in reproduction and genetics now available, and particularly made so in the past fifty years, are not only animal-breeding specialists but also physiologists, geneticists, chemists, engineers, manufacturers, educators, and practical animal husbandmen.

5.2 The Artificial Vagina: Collection and Processing of Semen

The artificial vagina[1] as developed by early English and Russian workers, with modifications by U.S. investigators, makes possible the convenient collection of semen without undue contamination. The artificial vagina helped to make possible the great expansion of the AI program. The development of this piece of equipment is viewed by some as the most significant event in AI history.

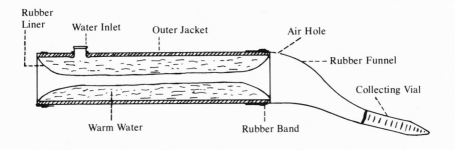

Figure 5.1. Diagrammatic view of typical artificial vagina for collection of semen. Source: H. A. Herman and F. W. Madden, *The Artificial Insemination of Diary and Beef Cattle,* 5th ed., (Columbia, Mo.: Lucas Brothers, 1974).

With the means to collect semen routinely came methods for evaluation, extension (dilution), storage, and shipping. The development of suitable semen extenders can be attributed to research workers in many countries. Only by the extension of semen, anywhere from ten to two hundred times its original volume, can many cows be inseminated by a single ejaculate. It is the means whereby one bull can sire twenty-five thousand or more calves in a single year and if sufficiently superior make noteworthy genetic gains in production.

The literature on semen extenders is voluminous.[2] Significant contributions were made by P. H. Phillips and H. A. Lardy[3] in 1940, when it was found that fresh egg yolk in the extenders protected the sperm cells during cooling and storage. As early as 1934, Milovanov and his co-workers in Russia added egg yolk to semen diluents and found it helpful. In 1941, G. W. Salisbury and his co-workers improved the extender by the addition of a sodium citrate buffer.[4] The use of heated cow's milk as a satisfactory semen diluent was reported in 1953 by D. L. Thacker and J. O. Almquist.[5] European workers were also trying various extenders including heated milk at this time.

At the present time, the two basic semen diluents are the egg yolk-citrate and heated milk. Many modifications of these diluents are in use, however.

An important contribution to semen extenders was made in 1946 and 1949 when the addition of antibiotics, penicillin, and streptomycin was recommended.[6] This procedure improved the settling rate of cows artificially bred by 10 percent and is no doubt an important factor in controlling *Vibrio fetus* and other infections that were prevalent in many herds in the early 1950s. It is now routine practice to use antibiotics in semen extenders.

The cooling of liquid semen to 40°F or slightly below in order to reduce sperm metabolism was practiced from the beginning in modern-day semen processing. It is interesting to recall, however, that horse breeders for years tried to maintain semen at body temperature when collections were made following natural service. There are many early day stories, too, of cattle semen being placed under the armpit or carried in a man's shirt pocket to keep it warm while he madly drove from the site where semen from the bull was collected to the farm where the cow was to be inseminated. The proper cooling and refrigerating of bull semen following collection, processing, and storage have become fairly well standardized the past forty years. Prior to the development of successful semen-freezing methods, ice was the principal refrigerant. Temperatures of 35°F to 40°F were maintained with ice and water and in some cases electric refrigerators. Equipment and procedures for controlling temperature at varying levels are now important facilities in the modern bull stud laboratory.

5.3 Storage and Shipping of Semen: Early Days

AI requires the storage and transportation of semen. The developments in this area, from the days of liquid semen to frozen semen, have been revolutionary.

From 1936 to 1954, liquid semen was in universal use. The usual method for shipping semen was to wrap the semen vials (usually small test tubes) in paper and fasten them to a paper-insulated can of frozen water. The can, usually wrapped in several thicknesses of paper, was then placed in an insulated "Jiffy Bag" and then into a lightweight, homemade wooden box or a corrugated paper box for shipping. This system worked fairly well for semen to be used within a couple of days following collection. Delivery of semen was by bus, parcel post, or car. Depending on distance, some technicians picked the semen up at the bull stud.

In the late 1940s, several AI organizations tried the use of airplanes for semen delivery. One, M.F.A. Artificial Breeding Association in Springfield, Missouri, perfected a shipping container with a rubber pad on the bottom to prevent breakage and utilized parachutes to drop the semen container to technicians. Noba, Inc., in Tiffin, Ohio, also used small plane delivery for a short time as did several other AI organizations. The use of light planes for delivery proved too costly for routine use and was soon abandoned. The development of the frozen-semen program beginning about 1950–1954 made it possible to maintain semen fertility over many months and the need for speedy delivery was alleviated (see section 5.4).

5.4 Frozen Semen

The use of frozen semen for artificially inseminating cattle, which began in 1952 and is still continuing, produced radical changes in the AI program. It is the most spectacular and revolutionary development to date in spreading the influence of bulls on a worldwide basis. Frozen semen, sometimes called "deep frozen semen," is used for practically 100 percent of the cattle inseminated in North America and in most other countries. Frozen semen, as the term suggests, is semen frozen and preserved at $-79°C$ ($-110°F$) in a mixture of dry ice (solid CO_2) and alcohol as a refrigerant, or semen frozen and preserved at $-196°C$ ($-320°F$) in liquid nitrogen (LN). Liquid nitrogen is the refrigerant now most used for freezing and storing semen.

The history of frozen semen dates back to the eighteenth century, when it was found that freezing stallion semen did not destroy all motility.[7] In 1897, Charles B. Davenport, an American zoologist, reported that human sperm would survive freezing. Other workers were successful in freezing frog sperm, human sperm, and that of several other species. The first successful effort in freezing bull spermatozoa came in 1950 at Cambridge University, England, when Chris Polge, L. E. A. Rowson, A. U. Smith, and A. S. Parkes found that adding glycerol to the semen diluent permitted freezing at −79°C without destroying all fertility. This discovery led to further research, in many countries, on bull-semen-freezing methods. Satisfactory methods for freezing and storing bull semen evolved and are in use today. Research in this area continues in efforts to reduce the "sperm kill" (30 to 50 percent) during freezing, and to perfect the ideal diluent and freezing procedures. Experience with present methods has established a conception rate that is approximately equal to that obtained with liquid semen.

5.4.1 Advantages and Results of Frozen Semen

The advantages of frozen semen in the AI program are:

1. Selected matings, on schedule, are enabled. Semen is available, usually at all times, and the cow owner can use the bull of his choice.

2. Frozen semen is of great benefit in the young-sire-proving program. Several hundred cows, in many scattered herds, can be inseminated in a short time and the bull retired until daughter proof is available.

3. If desired, young bulls can be sampled, a bank of fifteen to thirty thousand semen units established, and the bull slaughtered. This practice is followed in some countries. It saves maintenance costs while awaiting progeny proof. Semen storage costs are usually less than the cost of maintaining a bull. If the bull proves superior on his daughter proof, his bank of semen is used; if inferior, it is destroyed.

4. By having semen custom collected, frozen, and stored, while a valuable bull is in his prime, a herd owner is provided insurance against loss of services through accident, disease, or advancing age. Frozen semen has been stored over twenty-five years and fertility has been maintained.[8] As a rule, most frozen semen is used in one to five years.

5. By sale or exchange of frozen semen among bull studs, the superior sires—particularly of those breeds with limited numbers—can be made available to a wide array of herds.

6. Since semen in available to all shareholders, regardless of location, joint ownership of bulls—such as syndicated bulls involving several parties—is made practical by the use of frozen semen.

7. Transportation costs of semen from the bull stud to patrons are reduced, and thus delivery becomes more efficient. The technician,

distributor, or herd owner, replenishes his stock of frozen semen and liquid nitrogen every 60 to 90 days. Every-other-day delivery and semen wastage are eliminated.

8. Frozen semen permits efficient use of the semen from a given sire. While some sperm are killed by the freezing process, the loss is minimal when compared to the amount of liquid semen destroyed because of age.

9. Frozen semen is merchandised on a per unit basis. A semen unit is the amount necessary to inseminate one animal. It may be packaged in ampules, plastic straws, or frozen pellets. The dose ranges from 0.25 to 1.0 milliliter. The price per unit varies in accord with the merit and popularity of the sire. Sales transactions are simplified and bookkeeping between seller and buyer is minimal. In early days, the AI technician generally received a surplus of semen from the bull stud, and it was his responsibility to report the number of cows inseminated and to submit payment for the amount of semen utilized.

10. Frozen semen has made worldwide distribution of semen possible. International trade of semen is growing steadily. The United States is selling over one million units of dairy bull semen abroad each year (see Table 4.3, Chapter 4). The market for frozen semen in the tropics and developing countries is expanding (see Chapter 7).

5.4.2 Development of Frozen Semen Program: U.S. and Canada

Shortly after the successful freezing of bull semen by the British workers (see section 5.4) was reported, almost every AI organization in the United States and Canada, as well as many agricultural college research workers, began experimenting with frozen semen. One of the most energetic advocates of frozen semen was J. Rockefeller Prentice, the founder of American Breeders Service. He hired experts in cryobiology and put the staff at the American Foundation for the Study of Genetics in Madison, Wisconsin, to work to perfect methods for a frozen-semen program. In the early 1950s, he invited Polge from England to come to the United States to give counsel and invited AI organizations throughout the United States to attend a one-day meeting and to meet Polge.

Frozen semen was used in field trials as early as 1952 by several AI organizations. The first calf, resulting from the use of frozen semen, was born in the United States on 29 May 1953 (Figure 5.2).

The Eastern Iowa Breeding Association (EIBA) in Cedar Rapids, Iowa, was one of the first organizations to have frozen semen available. On 23 June 1953, two calves, resulting from the use of frozen semen, arrived the same day at the EIBA farm near Marion, Iowa.

It was not long before "frozen semen calves" began arriving on

Figure 5.2. "Frosty," the first calf born in the United States from the use of frozen semen. Born 29 May 1953, in the herd of John and Melford Hill in Janesville, Wisconsin, "Frosty" was sired by the American Breeders Service–owned bull Pabst Burke Tritomia Fryslan #943749. Her dam is Janey TH37867, a grade Holstein. (Photograph courtesy of American Breeders Services, De Forest, Wis.)

dairy farms throughout the United States and Canada. Once dairymen and the curious public found that such calves were as normal in all respects as those resulting from liquid semen, or sired naturally, they ceased to attract attention.

The first large AI organization that converted completely to frozen semen was the Waterloo Cattle Breeding Association in Ontario, Canada, in 1954.

5.4.3 Equipment for Handling Frozen Semen

The freezing of semen in the early 1950s involved the use of alcohol and dry ice (solid CO_2) as a refrigerant. The equipment for freezing consisted of insulated tanks, often homemade, as were the first frozen-semen-storage facilities (Figures 5.3a and 5.3b). These storage containers utilized glass wool and cork for insulation, and dry ice provided refrigeration. The proper temperature of −79°C (−110°F) could be maintained fairly well if the dry ice and alcohol were replenished every few days. Practically all freezing was done in glass ampules, though in 1955 Max Drake and B. W. Kagy placed on the market, through the Tiffinian Corporation, a plastic ampule known as "Fros-t."

Soon, the manufacturers of the dry-ice-storage chests also made available to the trade ampule holders, steel racks, and dry-ice baskets. There were several concerns in the business of low temperature storage equipment that adapted some of their products for frozen-semen storage.

It was not long before electrical refrigeration was utilized for storage. Some of the early mechanical refrigeration units held from a few thousand to sixty thousand ampules. Temperature was maintained at −84°C (−120°F) and the mechanical refrigerators were convenient. The disadvantage of electrical refrigeration was that an electrical power failure could result in the loss of an entire semen bank. In case of such an emergency, most AI organizations had motor-driven generators for standby power.

The manufacturers of the dry-ice–frozen-semen storage containers made available small containers that would fit into the trunk of the AI technician's car. These were widely used by technicians and cattle breeders. There were also small mechanical units on the market. Styrofoam storage containers and semen shipping boxes were in use

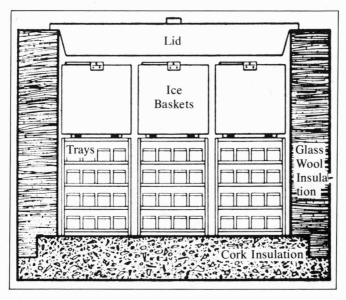

Figure 5.3a.   Cutaway side view of storage container shows proper storage arrangement. Steel racks are placed on bottom, with baskets holding dry ice placed on top. Each dry-ice basket will cover three to four racks. This gives efficient refrigeration with minimum waste. (Photograph courtesy of Meese, Inc., of Madison, Ind.; from NAAB files.)

Figure 5.3b. "The Big Bull Barn." A frozen-semen-storage box fitted with a Dewar flask. Held up to five hundred, 1 ml. ampules and used about ten pounds of dry ice per week. (Photograph courtesy of Frozen Semen Products Co., in Brienigsville, Pa.; from NAAB files.)

during 1955–1960, as were vacuum flasks, all using dry ice and alcohol for refrigeration.

Liquid Nitrogen. Beginning in the early 1950s, and in general use by 1956–1960, the storage of frozen semen was simplified and made more efficient by adapting liquid nitrogen as a refrigerant.

The utilization of liquid nitrogen for storing and shipping frozen semen must be reviewed as one of the most significant developments in the realms of artificial insemination. With temperatures maintained at −196°C (−320°F) so long as any liquid or gas is present, liquid nitrogen is a most desirable refrigerant. Semen can be shipped anywhere in the world and stored many months for future use if a supply of liquid nitrogen is available. It undoubtedly has made possible the development of international trade in frozen semen (see also the advantages outlined in section 5.4.1).

Suitable containers for storing frozen semen in liquid nitrogen appeared on the market in 1959. Of stainless-steel construction, with a vacuum between the walls and canisters for holding ampules, straws, or pellets, the liquid nitrogen containers have been constantly improved (Figures 5.4a and 5.4b).

Containers vary in size. There are small containers, usually called "semen jugs," that hold a few hundred semen units and are used by AI technicians and herd owners. Then, there are large storage units,

Figure 5.4a.   Stainless-steel container for storing frozen semen in liquid nitrogen (Photograph courtesy of Union Carbide Corp., New York, N.Y.)

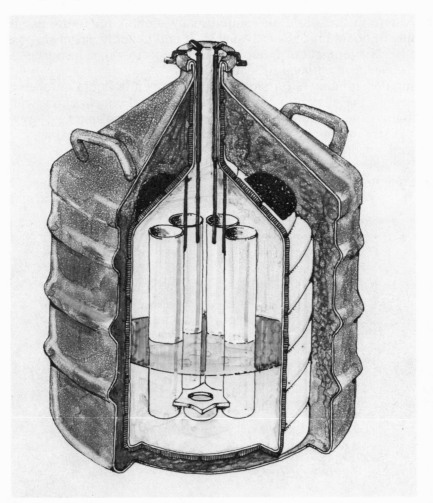

Figure 5.4b.   Inside view of LN semen-storage container. (Photograph courtesy of Union Carbide Corp., New York, N.Y.).

used by bull studs and custom-freezing businesses, that hold over fifty thousand semen doses and require refilling with liquid nitrogen about every one hundred days. Liquid nitrogen containers have replaced the dry ice-alcohol and mechanical refrigerators referred to earlier.

5.4.4 Inseminating Equipment: Semen Packaging

There have been many improvements in equipment used for insemination. The first inseminating catheters or inseminating tubes were made from small bore glass tubing. The liquid semen was drawn into the inseminating tube by means of a small glass syringe attached to the tube by a short rubber connector. With the tube in place in the reproductive tract, the semen was expelled by gentle pressure on the syringe plunger (Figure 5.5). Even though shatterproof glass tubing was used, much to the consternation of the inseminator and the cow owner, the tube would sometimes break while the cow was being inseminated. The glassware had to be washed and sterilized by heat. This was a chore few AI technicians relished.

In 1955, plastic insemination tubes and plastic polybulbs were introduced. These are disposable, inexpensive, convenient, and more sanitary than the glassware equipment. The polybulb is applied to the end of the plastic inseminating tube and draws up and expels the semen effectively (Figure 5.6).

Plastic gloves, also disposable and inexpensive, are now used by inseminators. The awkward, expensive, shoulder-length rubber

Figure 5.5. A glass inseminating tube with syringe attached. Source: "Artificial Insemination of Dairy Cows," Missouri Agricultural Experiment Station Bulletin No. 540 (July 1950).

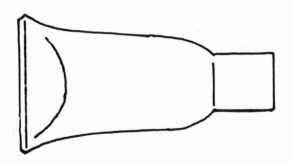

Figure 5.6.   A modern-day plastic polybulb. Used with a plastic insemination tube. Replaced the equipment shown in Figure 5.5. (Photograph courtesy of Edwards Agri-Supply, Inc., Baraboo, Wis.)

sleeve used in the early days is rarely used by AI technicians for routine work.

Semen Packaging.   Before frozen semen came into use, with each dose packaged separately, the custom was to place a few cubic milliliters in a test tube. The technician drew the necessary amount of semen, usually one cubic milliliter, each time a cow was inseminated. Glass ampules were introduced in 1955, when the practice of frozen semen gained popularity. They were provided by glassmakers who made containers for the pharmaceutical companies (Figure 5.7). The ampules are made of Pyrex glass to withstand freezing and are available in 0.5, 1, 1.2, and 1.5 ml. volume. The ampule is scored at the neck so the technician, by means of a file, removes the top portion and draws the semen from the bottom.

In recent years, the trend has been to package semen in straws made of polyvinyl chloride. The "straw system" for semen packaging

Figure 5.7.   A typical glass ampule used for frozen-semen storage. (Photograph courtesy of Wheaton Industries, Millville, N.J.)

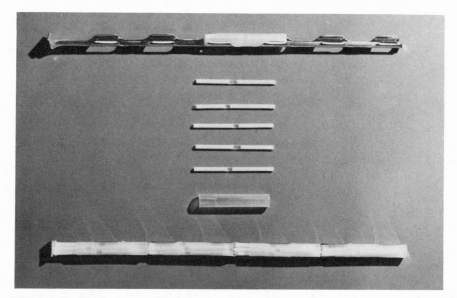

Figure 5.8. Typical straw for frozen semen. Goblet for holding straws and goblets affixed to cane for immersion in liquid nitrogen are shown at left and right. (Photograph courtesy of J. O. Almquist, Pennsylvania State University.)

was originally developed by Sorensen (1940) in Denmark, and Cassou (1950) of France made further perfections.[9] At the present time, both straws and ampules are used in America and Europe. The advantages of the straw system are (a) that many more units of semen can be stored in the same refrigerator space required for ampules; (b) there is more uniform freezing and less "sperm kill"; (c) more cows can be bred per bull; and (d) some studies show a slightly better conception rate.

Pelleting semen is another form of packaging and is used in some countries. The semen is frozen in pellets by means of dry ice. Less storage space is required for this system than for either straws or ampules. The pellet is thawed in a suitable diluent before use. The fertility of pellets seem equal to that of straws and ampules.

Laboratory Equipment. Through the years, with the ingenuity of AI laboratory managers and commercial equipment concerns, much modern efficient equipment for semen processing and handling has been developed. Included in this category are (a) automatic fillers for ampules and straws; (b) labeling machines for semen packages; (c) ampule sealers and devices for plugging straws; (d) containers for semen units in storage and in shipment; and (e) freezing baths and apparatus for handling semen units in quantity during processing and freezing. In most AI laboratories, semen processing is handled on a chainlike assembly plan. Computers are used for inventorying and record keeping.

5.5 Semen Production

Among the developments these past forty years is the knowledge gained by experience in the handling, feeding, and caring of bulls for maximum semen production with optimal fertility. Techniques in "bull psychology" and the handling of individual bulls are skills practiced to a high degree by the personnel in charge of semen production at the modern bull stud.

The development of stalls, pens, and collection room equipment to handle bulls effectively has reached a high level. Routine testing of bulls for disease and sanitary practices have become an important part of operating for bull studs.

Control of the quality of semen distributed by an AI organization begins in the bull barn. Further success depends upon the laboratory procedures, shipment, and proper utilization in the field.

5.6 Artificial Insemination in Other Animals

As previously mentioned, AI is used not only on cattle but also on horses, swine, and dairy goats. In addition, AI is used on many other species that have not been mentioned.

Sheep. AI is used for sheep breeding in many European, Asiatic, and South American countries as well as in Australia and New Zealand. The technique is similar to that used for goats and cattle.

Dogs. AI in dogs using liquid semen has been widely practiced. Now frozen semen is utilized.[10]

Turkeys and Chickens. About 85 percent of the turkey hens that produce hatching eggs are artificially inseminated at two-week intervals during the laying season to maintain high fertility. To date, frozen semen has not been very successful in turkeys. Chickens are artificially inseminated in Japan, Russia, and a few other countries.

Honeybees. The production of mated honeybee queens utilizes AI. The semen is collected from selected drones and injected into the oviducts of the queen by a special technique and apparatus.[11]

Laboratory Animals. It is now common practice to artificially inseminate many of the common laboratory animals, including rabbits, mice, rats, guinea pigs, and other rodents.

Zoo Animals. In some of the zoo animals, and also some birds, AI is used to improve conception.

The Human. Much could be written about the use of AI on people. It was in use as early as 1776 as a means of correcting sterility problems.[12] Spallanzani in 1776 reported the effects of freezing human spermatozoa. Emphasis on freezing human spermatozoa coincided with the discovery that glycerol could serve as a cryoprotective agent. It is estimated that from ten thousand to one hundred thousand births a year in the United States result from AID (artificial insemination with donor semen) as a means of overcoming male infertility. J. K. Sherman reports that about fifteen hundred births worldwide in 1976 resulted from the use of human frozen semen.[13] The longest reported period frozen human sperm have been preserved is ten years. It is proposed, and in effect in a few places, that human sperm banks utilizing frozen semen be maintained under the supervision of the medical profession.

5.7 The Future and AI

Increasing knowledge and the perfecting of new techniques in cattle reproduction forecast many possible revolutionary developments in cattle breeding of the future. Only a few, relating to cattle and involving AI application, will be mentioned.

5.7.1 Effective Control of Ovulation and Estrus

One of the drawbacks to a more successful conception rate in dairy cattle serviced by AI is failure to detect the cow in heat. In beef cattle, heat detection requires additional labor, and it is a problem to have a majority of animals bred at the desired time. Many beef operations, where AI would be practical and by the use of genetically superior sires to increase the pounds of beef produced per cow, fail to use it because of the labor and management problems involved. Likewise, many dairy heifers are bred to young unproved bulls because of the labor and management demands required for heat detection. These problems may be greatly reduced in the years ahead if the research now in progress, which looks very promising, materializes in controlled estrus and breeding schedules.

While a variety of products, fed orally, injected, or implanted for synchronization of estrus are being tested, they may be grouped as (a) *progesterones* and (b) *prostaglandins*. Results to date indicate that, in properly fed and managed cows and heifers, the time of insemination, for the synchronized animals, can be confined to a five-day period with a high level of pregnancies.[14] It has been found that treated animals need not be checked for estrus but instead inseminated at a predetermined time in accord with the method of synchronization followed. Early results are encouraging.[15] Some of the estrus synchronization compounds now available depend upon approval for use by the Food and Drug Administration (FDA). With FDA approval in November 1979 of prostaglandins for use in normally cycling beef cows and heifers, and in non-lactating dairy heifers, there no doubt will come an increased use of AI in cattle breeding. This development could progressively change beef-breeding programs in the future.

5.7.2 Embryo (Ova) Transplants

Most of the emphasis in cattle breeding has been directed toward the genetic influence of the male because of the large number of progeny involved. The development of the embryo, or fertilized ova, transplants enables a cow to produce many offspring in a single year, thus increasing her genetic impact. The progress in this field the past decade has been tremendous and further developments are in store.[16]

To obtain more than the usual one egg (ovum) shed by a cow during the estrous cycle, superovulation is brought about by hormone treatment. The administering of the necessary dosage of pregnant mare serum gonadotropin (PMSG) on the sixteenth day and human gonadotropin (HCG) on the twenty-first day of the estrous cycle allows a cow to ovulate from fifteen to twenty eggs (Figure 5.9). By means of AI, the eggs can be fertilized, flushed from the reproductive tract of the cow, while in the blastocyte stage, and implanted into recipient cows (embryo or ova transplant). The recipient should be of less value than the donor cow and should simply serve as an incubator. The estrus cycle of the recipient cow must coincide, usually by synchronization, with that of the donor cow.

The earlier technique for recovery of fertilized eggs and their implantation required surgical procedures. Many herd owners did not care to risk valuable animals as donors. Present-day techniques involve the successful flushing of the fertilized eggs from the donor cow. Implantation of the recovered fertilized egg, in the recipient animal, is done by nonsurgical means similar to AI techniques. These developments have done much to accelerate the practice of embryo transplantation.

Albert Mills, of Portable Embryonics, Inc., at Zachary, Louisiana, is one of several animal surgeons who do embryo transplants on the farm.[17] Mills has a laboratory assembled in a large trailer that enables him to travel from herd to herd. Strictly nonsurgical techniques are used for egg collection and transplants. He generally flushes four to six eggs from the donor cow, from which two or three pregnancies are estimated. The herd owner supplies the host or recipient animals.

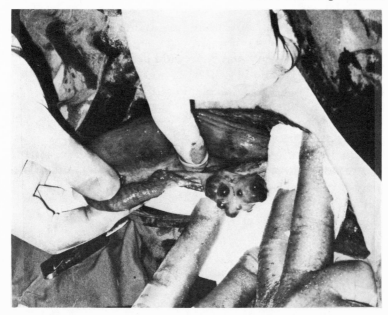

Figure 5.9. Horn of exposed uterus showing an ovary with numerous ovulation points following treatment to produce superovulation. (Photography courtesy of the Carnation-Genetics Division, Hughson, Calif.)

With frozen embryos available, it is possible they will be used for transplants in the future.

Frozen embryos from sheep, goats, cattle, horses, and several other species have been successfully preserved for months.[18] Carnation–Genetics at Hughson, California, which is an AI firm, reports confirmed pregnancies from embryos (obtained nonsurgically) that were stored for six weeks in a liquid nitrogen tank (−196°C or −320°F).[19] Modern Ova Trends, an embryo transplant business at Norval in Ontario, Canada, announced the birth of a Holstein heifer on 20 April 1977. The calf resulted from an embryo that had been frozen at −196°C, thawed, and transferred into a crossbred recipient. This calf is believed to be the first calf born in North America from a frozen-thawed embryo. With suitable culture methods, now fairly well established, frozen embryos will probably be stored and will be as readily available as is frozen semen.

The embryo transplant technique offers an added advantage: the sex of the expected calf can be predetermined before the fertilized ova is transferred. Researchers in Canada report that by biopsy methods the sex chromatin can be identified in sixteen-day-old fertilized ova.[20] It was possible to diagnose accurately the sex of twenty-two of thirty-three embryos examined. Conception rate was acceptable.

The benefits the cattle-breeding industry can derive from successful sexing techniques of bovine embryos are immense. It means that

the desire for a bull calf out of an outstanding cow may possibly be fulfilled on order.

The number of progeny one cow could produce in a year as a result of superovulation and embryo transfers is probably unknown. However, Alberta Livestock Transplants, Ltd., in Alberta, Canada, reports that the Swiss Simmental cow Big T Furka, owned by Big T Stock Farms, of Pennant, Saskatchewan, produced thirty calves in one year from two transplant operations.[21]

The development of the embryo transplant technique, plus the use of AI, provides an opportunity for selecting and mating cattle not dreamed of twenty years ago.

5.7.3 Sexed Semen

For many years, investigators throughout the world have attempted to separate sperm bearing the Y chromosome (male producing) from those bearing the X chromosome (female producing).[22] Obviously, for a cattle breeder to choose the sex of a calf in advance would be of considerable economic importance. To date, no method of separating sperm into the X and Y categories has proved reliable. Investigations are continuing with new approaches and with the advantage of sophisticated apparatus in the laboratory.

A recent development, the reported identification by microscopic methods of fluorescent F-bodies, a leg of the Y chromosome, has given encouragement to researchers.[23] This technique aids in detecting when separation of the X and Y chromosome-bearing sperm has actually occurred. Field trials are being conducted in several laboratories but to date no definite conclusions have been obtained.

Sexed semen, if available with a high degree of fertility and producing a significant change in the sex ratio, would be preferred to the method of embryo biopsy described in section 5.7.2. It would be less expensive and much more convenient. Most scientists agree that some day sexed semen will become a reality. It could be one of the most significant developments in the cattle-breeding field.

5.7.4 Improved Methods of Assessing Semen Fertility

With the development of the AI program, research in both the male and female aspects of reproduction has yielded much new knowledge. We believe it is safe to say that more accurate scientific information has been amassed on mammalian reproduction during the past forty

years than in all the years preceding. Much has been learned about the intricate nature and mechanism of spermatozoa. Most of the routine assessments of semen fertility, however, are empirical. The conception rate of the bull is still, and probably will continue to be, the final measure of semen fertility.

Research is underway on more scientifically based methods for evaluating semen quality. For example, studies by several investigators on alterations of the *acrosomal cap* of spermatozoa during processing, freezing, storage, and thawing of semen provide a quantitative measurement of fertility.[24] The acrosomal cap, or its associated membranes, contains enzymes that digest the zona pellucida (outermost covering of the bovine ovum). This phenomenon is presumed to aid penetration of an ovum by a spermatozoon and hence fertilization. Bulls that produce sperm with abnormalities of the acrosomal cap have been shown to range low in fertility. Some bull studs are now evaluating semen in the laboratory by quantitative measurements of acrosomal cap retention.

Studies are underway on the immunological properties of semen to determine if the cow establishes antibodies against sperm or components of the semen diluent. Both cattle breeders and scientists ponder the question of why a cow not settling after several inseminations to a given bull will often settle on the first service if semen from a different bull is used.

Researchers at Washington State University in Pullman, Washington, find that adding blood serum from a cow to the semen diluent causes spermatozoa to agglutinate or clump.[25] The clumped spermatozoa seem to be more fertile.

Another interesting development is the use of X-ray fluorescence spectrometry for monitoring tagged inks in semen packages.[26] This technique can help in identifying counterfeit or fraudulently labeled semen.

Biochemical evaluation of the fertilizing capacity of semen by utilizing the measurable chemical and physical properties of spermatozoa is under study at several universities. Involved are various enzyme reactions affecting the sperm and the egg. These investigations gain headway each year as more precise instruments and techiques become available.

Investigations toward perfecting the ideal semen extender (diluent) continue and scientists are providing new knowledge each year. Our present semen extenders have developed largely by trial and error, because of the complexities of spermatozoa, seminal fluids, properties of bovine reproductive tract mucus, and the role of enzymes and hormones in reproduction. In the years ahead, no doubt semen extension methods will change as scientific progress is made.

5.7.5 Pregnancy Determination

Cattlemen have long hoped for a quick and convenient method for determining early pregnancy. The palpation method, thirty-five to sixty days following service, is commonly used and is a regular practice in well-managed herds. Various biochemical and biological approaches are being investigated as other means of detecting pregnancy.

One such test is the measurement of progesterone levels in the milk as affected by the corpus luteum. Milk tested for progesterone levels from twenty to twenty-four days following insemination can be used as a diagnostic test for pregnancy. This test is also helpful in checking cows that do not show signs of cycling but ovulate—"silent heat." The progesterone tests require the use of isotopes and expensive equipment. It offers promise for use in laboratories where hundreds of milk samples could be assayed.

Another development is the detection of substances produced by the growing embryo and present in the cow's blood.[27] Antibodies against embryos can be produced and an antigen prepared that will detect pregnancy of a few days' duration. This test may eventually be developed to the point where milk or blood from the cow can be placed in a test tube with the antibody plus an indicator and a quick pregnancy diagnosis made.

5.7.6 Electronic Identification: Temperature Monitoring

One of the most interesting modern approaches in livestock technology is electronic identification of animals and temperature monitoring. The Los Alamos Scientific Laboratory (LASL) at the University of California in Los Alamos, New Mexico, has developed a small transponder that can be implanted under the skin of an animal.[28] By means of a microwave beam, transmitted by a small antenna, the animal's code number and temperature can be recorded. This system can be easily automated for dairy cows and the information placed on computers. Although it is still in a developmental stage, it is expected to have wide usage in quickly identifying animals in the dairy herd, the feedlot, sales barns, and slaughter houses.

A neck chain transponder is used in dairy herds to trip feeding devices so a specific cow receives the exact amount of the allotted concentrate ration.

Cattle can be screened for disease by the development of a new system by LASL called the Enzyme-Labeled-Antibody (ELA). Its use will permit the electronic detection of diseased animals in minimum time.

Temperature recordings by means of a fixed permanent microwave antenna can be made on animals as they pass through the milking parlor or in a cattle chute. Temperature of an animal is useful in detecting disease, mastitis, and other infections, and it can be used as a measure of stress. As a rule, cows in heat are nervous and often show temperature irregularities, thus the system may be valuable in heat detection of confined cows.

5.8 Summary

With the technical advances being made in the livestock field, only a few of which are mentioned above, it is a safe assumption that many future changes are in store for cattle breeders and the AI industry. The efforts of scientists and cattle breeders are directed toward the efficient production of food and fiber. If we are to feed a hungry world, someone must produce more food, fiber, and services if we are to maintain a high standard of living. Consumption cannot be increased indefinitely by redistributing wealth through the jockeying of monetary units. It is by increasing the output per animal, or worker, that standards of living are raised. The scientific breeding of cattle is so directed.

Chapter 6: Worldwide Use of Artificial Insemination

6.1 Introduction

The use of AI, as a means of increasing animal production through improved genetic potential, is gaining worldwide. This is particularly true in Europe, in the North and South Americas, New Zealand, Australia, and some of the Asiatic countries. AI is being utilized in many underdeveloped countries, where meat and milk are scarce, to improve the native stock. The approach is usually to grade up the native cattle by utilizing the best of the indigenous cattle for breeding purposes, and crossbreeding, utilizing semen imported from other countries. In all countries, AI is used most widely for cattle although in some it is important for swine and sheep also.

6.2 Surveys: AI Usage

One of the earliest reported surveys on the total number of cattle bred artificially was in 1956–1960 by W. Belanski, of Poland.[1] He reported 48.4 million head in sixty-three countries. Later surveys by Y. Nishikawa, of Japan;[2] Telesforo Bonadonna, of Italy;[3] and my own data from 1964 to 1976 indicate an increase of about 40 million head in the past fifteen years. The results of a survey for the years of 1970–1973 by Bonadonna are indicated in Table 6.1.

The above summary is at best a rough estimate as reporting from China, India, Africa, and some Central American countries was partial at the time of this tabulation. We suggest that the number of cattle and buffaloes bred artificially in the world is more nearly 125 to 130 million in 1976, or about 8 to 10 percent of the total.

Table 6.1. Application of Artificial Insemination in Cattle: Based on Statistical Data Obtained in Various Countries

Continent	Cows AI in 1970	Cows AI in 1971	Cows AI in 1972	Cows AI in 1973
Europe Total	58,781,762	59,871,491	62,198,754	63,952,130
North and South Americas Total	13,184,492	13,868,149	14,831,162	15,409,617
Asia Total	4,817,529	4,752,012	4,648,077	4,758,396
Africa Total	412,769	248,996	320,447	343,551
Oceania Total	1,508,739	1,478,739	1,960,000	2,034,000
Overall Total	78,700,291	80,219,387	83,958,440	86,497,694

Source: Telesforo Bonadonna, "VI International Enquiry into Artificial Insemination in the World (1971-1973)," *Zootec. Vet.* 30 (1975): 1–108.

6.3 World Total Cattle

For 1975, it was estimated that there were 1,343 billion head of cattle and buffalo in the world, about 1 head for every 3 persons.[4] The cattle population by continent and leading countries in number of head, and the estimated percentage of cattle artificially inseminated, is shown in Table 6.2.[5]

6.4 The AI Program in Selected Countries

AI is utilized in almost every country in the world that has a significant animal economy. Not only cattle, but water buffalo, swine, sheep, and goats are involved in varying degrees. In some countries, AI is important in the breeding of horses and poultry. We will briefly discuss the AI program in some selected countries to illustrate the wide versatility of the program.

Frozen semen packaged in ampules, straws, or pellets is widely used in many countries. Since substantial amounts of semen are imported into many countries, they, particularly Europe, Canada, and the United States, have developed considerable overseas export trade.

Australia. There are about 45 million cattle and 150 million sheep in Australia. Most are located in the coastal areas where grazing is available throughout the year. About 1.5 million cows, largely dairy, are artificially bred annually. AI is also used for sheep, but its use is limited by labor costs and seasonal breeding problems.

There are at least ten organizations, mostly cooperative breeding societies, that provide AI service. The King Ranch has an AI unit at Milton Park, in Bowral, New South Wales, largely for its own breeding operations. There is also a Stock Inseminators Association of Tasmania.

The AI program is rather strictly regulated by the various state governments through Stock (Artificial Insemination) Acts. Rigid requirements pertaining to sire health, semen handling, and conduct of the AI program are in effect. Dairy cattle resulting from AI are registered by the breed societies with few restrictions. The beef breed societies are more restrictive. Nearly all inseminators are laymen. Proved sires are used heavily and a young-sire-sampling program is sponsored. Semen imports are limited as a precaution against bluetongue and other diseases. Some beef breed semen is exported to the United States and to European countries.

Denmark. The Danes were the first in Europe to establish a nationally organized AI program. The first AI cooperative was organized in

Table 6.2. World Cattle and Buffalo: Extent of AI, 1973–1975

Continent and Country	Number Cattle and Buffalo	Leading Countries in Number (Total)	Cattle Breeding Age (Percent AI)
	Thousands	Thousands	
North America:	195,669		
Canada		15,018	Dairy 55.0
			Beef 10.0
Mexico		28,071	8.0
United States		131,826	Dairy 60.0
			Beef 3.0
South America:	209,858		
Argentina		58,000	9.0
Brazil		91,136	9.0
Chile		23,222	10.0
Uruguay		11,000	—
Europe:	243,666		
Denmark		3,100	90.0
France		24,700	73.0
Germany, West		14,423	70.0
Ireland		6,500	20.0
Italy		8,058	—
Netherlands		4,745	69.0
United Kingdom		14,836	62.0
Austria		2,582	45.0
Finland		1,704	99.7
Spain		4,375	25.0
Sweden		1,919	80.0
Switzerland		1,977	43.0
Germany, East		5,585	—
Hungary		2,020	98.0
Czechoslovakia		4,570	94.0
Poland		12,815	84.0
Rumania		6,000	58.0
Yugoslavia		6,100	70.0
USSR		109,100	75.0
Africa:	157,530		
Egypt		4,400	—
Ethiopia		25,315	—
Rhodesia		5,670	—
South Africa		12,600	—
Uganda		4,200	—
Asia:	491,027		
China, People's Rep.		93,150	—
India		240,555	2.0
Israel		272	—
Japan		3,525	98.0
Phillipines		7,300	32.0
Thailand		10,102	—
Turkey		13,830	—
Oceania:	44,941		
Australia		34,500	—
New Zealand		9,880	65.0
World Total	1,342,691		7.0 to 8.0

Source: Compiled by the author from various published reports and personal surveys.

1936 by Edward Sorensen, of the Royal Agricultural and Veterinary College in Copenhagen, and Jens Gylling-Holm, agricultural adviser at Tranebjaerg. The cooperative had 220 members, and 1,070 cows were inseminated the first year. The program grew rapidly, with 78 "breeding societies" and 138,000 cows bred in 1940.

In 1937, the rectovaginal technique, or cervical fixation method, for artificial insemination of cows was developed by Danish veterinarians. This method, now universally used, permits deposition of semen into the mid-cervix region and the uterine horns. Under field conditions, its use improves fertility about 10 percent above the vaginal deposition method.[6]

Denmark has a cattle population of about 3 million head with over 90 percent of the animals of breeding age serviced by AI. Since 1952, the number of AI organizations (breeding societies) has decreased from 106, the all-time high, to 46 in 1972. During 1973, these organizations serviced 1,338,437 head with a nonreturn rate of about 70 percent, and 94 percent of the cows were impregnated after one or more services. The decline in number of breeding societies is due largely to the use of frozen semen. Both straws and pellets are used.

Denmark has an extensive young-sire-proving program.[7] Matings to the best producing cows, from good cow families, to the superior AI bulls, to produce future sires are encouraged. The young sires are sampled lightly. About 60 percent of the cows are tested for milk production through the Danish Milk Recording Societies. The Danish milk-recording program started in 1895 and is one of the oldest in existence. The "cow testing" program (now DHI) started in America in 1905 and is patterned after the Danish system. Only about one-fourth of the bulls in AI are past five years of age. However, progeny-tested sires are used for at least 75 percent of the matings. Herds in Denmark average about 15 cows, and about 1,300 cows are annually serviced per AI bull. The Danes favor relatively small bull studs. Only three service over 100,000 cows annually. The philosophy is that with a fairly large number of scattered small bull studs, more young bulls are sampled and greater genetic progress results. Danish cattle breeders work very closely with the AI societies and insist that only the best bulls available are used.[8]

Great Britain–United Kingdom. Great Britain has some 14.8 million cattle of all ages and all breeds. No country in the world compares with the United Kingdom for the number of cattle breeds developed. At least fifteen breeds, most of which provided the foundation for the cattle industry in the United States, Canada, and in other countries, were developed in the United Kingdom.[9]

Approximately 65 percent of the cattle of breeding age, particularly the milk breeds, are enrolled on the AI program. There is considerable crossing of milk and beef breeds to produce animals for meat. The number of beef dairy crosses fluctuates from time to time depending upon relative milk and beef prices. When the organized AI program got underway in Great Britain about 1944, emphasis was on milk production. World War II had stimulated dairying enormously, and milk from 95 percent of the cattle of producing age was sold in England and Wales.[10] In 1947, beef bulls were introduced into the bull centers when beef imports from Argentina were uncertain. While initiated as an emergency measure, the beef dairy crosses have become a feature of considerable economic importance.

Early investigations of AI in Great Britain were carried out by workers at the Cambridge University School of Agriculture and the National Institute of Dairying, in Reading, from 1939 to 1943. Efforts were hampered by World War II. The Cambridge and District Cattle Breeders' Society was formed in 1942 to operate the first nonprofit cooperative-breeding scheme, using Shorthorn and Friesian bulls. Some critics in 1942 felt that Britain was lagging behind in adopting a wide-scale AI program.[11] There was some criticism that Britain sold its better bulls abroad and that there was too much disparity between pedigree breeders and the ordinary commercial herd.[12] In 1944–1945, an extensive and well-organized AI program was launched in England and Wales. The program was set up under the control of the Milk Marketing Board (MMB), a government agency comprehensively involved in cattle breeding, milk production, and marketing. Headquarters are in Thames Ditton, of Surrey, England. Leading the way in developing the program, in addition to cooperating farmers and breed societies, were Richard Trehane, chairman of MMB, and Joseph Edwards, scientific adviser to the board and in charge of production. The bull centers are mostly farmer-owned cooperatives.

Between 1944 and 1951, twenty-three bull centers were established, each with about thirty bulls, in areas with at least 60,000 cows. Hundreds of subcenters were organized to supply AI service in local areas. There are now twenty-three MMB centers and seven non-board centers. The non-board centers account for about 20 percent of the total cattle inseminated. During 1972–1977, cattle numbers in Britain declined, but in March 1977 there were 3.6 million cows, about 27 percent of which were beef, in the nation's breeding herd.

With the development of the AI program, and the recovery from the effects of World War II on feeding and management, milk production per cow continues to make steady gains (Figure 6.1) with an average increase of about three thousand pounds of milk per cow over the past twenty years.

Among the dairy breeds, Friesian and Ayrshire lead in numbers inseminated with services to dairy bulls accounting for 65 percent of

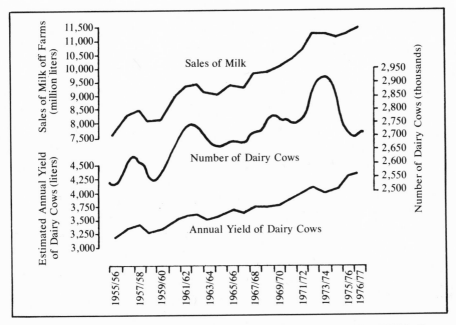

Figure 6.1. Sales of milk, number of cows, and milk yield from 1955 to 1977 in Great Britain. Source: MMB, breeding and production records for 1976–1977.

the total in 1976–1977. Dual-purpose breeds account for only 1.9 percent of the total with Welsh Black, South Devon, and Shorthorn in most demand. Services to beef bulls (24 breeds available) account for 35 percent of the total. In popularity, Hereford leads by a wide margin (386,000 head in 1976–1977) with Charolais, Aberdeen Angus, and Simmental following in that order.

During 1973, about 82,000 swine were artificially inseminated.

Great Britain has an extensive National Milk Recording (NMR) plan similar to the DHIA program that is used in the United States. During 1976–1977, a total of 952,903 cows representing 35 percent of the total dairy cows were enrolled on the NMR program. A young-sire-sampling program is conducted for the dairy breeds with about 150 to 175 young bulls being introduced annually. Most inseminations are to progeny-proved sires.

A comprehensive program of progeny-testing beef bulls is also conducted with some two hundred bulls under continuous test. About 70 percent of all beef inseminations are to progeny-tested sires.

Frozen semen is packaged in pellets and straws. An extensive research program including pregnancy testing, estrus synchronization, calving difficulty, reproductive physiology, and disease control is conducted as a part of the MMB activities. During 1972–1973, about

17,000 units of frozen semen were imported from Canada, France, and Italy. Britain has a growing semen export trade with about 450,000 units shipped, largely to Australia, in 1973.

China. Artificial insemination of horses, cattle, sheep, and swine in China began in the 1950s. The People's Republic of China has a population of about 93 million cattle and water buffalo.

Steady progress has been made in developing the AI program in the principal stock-raising areas.[13] In Nong-an County of Jilin Province, 21,000 (89 percent) of the mares were artificially inseminated in 1978 with a conception rate of 83 percent and 310 mares bred per stallion.

More than 1 million head of the native "yellow cattle" were bred by AI in 1977. In the Jilin Province, 195,000 (32.7 percent) of the breeding cows were aritificially bred with a conception rate of 76 percent and 1,595 cows bred per bull. About 90 percent of the frozen bull semen is processed in pellets and the remainder in straws.

In the southern areas, AI is used for water buffalo with about 40 thousand head, or less than 1 percent of the total, bred artificially in 1978. The conception rate was about 61 percent.

Frozen bull semen has been used for the crossbreeding of Yak. In the Sichuan Province, 3,499 Yak cows were inseminated with an average conception rate of 45 percent.

Artificial insemination of sheep is a common practice in the north and northwest with 6.65 million ewes inseminated during 1978. The first service nonreturn rate ranged from 15 to 81 percent with a final conception rate of 85 to 94 percent. AI for dairy goats has been developed in the Shansi Province with about 4,000 does inseminated, with a conception rate of 85 percent (75 percent first service), during 1978.

Artificial insemination of swine is a growing practice in China. In twenty-nine areas, during 1975, nearly 1 million sows were inseminated. The first insemination nonreturn rate ranged from 50 to 85 percent with a final conception rate of 80 to 96 percent.

According to Peilieu Cheng some success has been attained in the artificial insemination of deer, panda, and other wild animals.

Finland. Finland is a small country with a cold climate. It has a cattle population of about 1.7 million head. About 99 percent of the cattle of breeding age are artificially inseminated. Service is available in all of Finland and has been extended to Lapland, which is north of the Arctic Circle.

Artificial insemination in horses was started on a small scale in Finland before World War II, but it did not flourish. Beginning in 1943–1944, a few cows were inseminated experimentally. Following the pattern of many countries, in 1946 cattle breeders met and decided to form an AI society. The South-Western Finland AI Society,

the first in the country, was organized, and the first cow was inseminated in 1947.[14] During the first year, 1,452 cows were inseminated.
By 1968, there were eighteen AI societies, eight of which had bull and
boar stations.[15]

In June 1948, representatives of the AI societies met and formed
the Central Association of AI Societies. Its purpose is to direct and
guide all phases of the AI program, which is under the authority of the
Board of Agriculture.

The Finnish government in 1949 adopted an act that placed the
practice of AI under the supervision of the Board of Agriculture. To
supervise and direct the Board of Agriculture, there is an Artificial
Insemination Committee. AI can be practiced only by permission of
the AI Committee. For example, a private herd owner is forbidden to
collect and sell semen from bulls he owns. The respective breed
associations approve the bulls for AI use.

A well-regulated–young-sire-sampling program is carried out. Sons
of the highest indexed AI-proved bulls are mated to the best producing cows on the milk-recording scheme to produce future AI sires.
The young bulls are grown out, being checked for rate of gain, until
one year of age. Poor gaining bulls, and those with unsatisfactory
conformation, are culled. The young bulls selected for sampling are
taken to the AI centers, where semen is taken and frozen from each in
amounts of fifteen thousand to thirty thousand doses. Most young
bulls are slaughtered after a semen bank is established. Semen is
frozen in pellet form and stored in liquid nitrogen at $-196°C$. Several
hundred cows are inseminated with semen from each bull and after
progeny records are available only the semen from the best bulls is
preserved.

The leading cattle breeds in Finland, in terms of AI service, are
Ayrshire, 64 percent; Finncattle, 27 percent; Friesian, 6 percent; and
Charolais, 3.2 percent. Emphasis is on both milk and meat production. From 1948 to 1967, the average annual milk yield of all recorded
cows increased from 3,000 kg. (6,600 pounds) to 4,500 kg. (9,900
pounds).

In 1960, swine AI was initiated. By 1967, nearly 30,000 sows, 20
percent of all sows in Finland, were served by AI and continues to
gain. The best boars of the Yorkshire and Landrace breeds are used
for AI. About 400 to 500 inseminations are made per boar annually.
The most inseminations from one boar in a year has been 1,708.

Both cattle and swine inseminations are carried out by well-trained
lay technicians.

France.   France has about 25 million head of cattle of all breeds and
ages, with over 7.5 million, or about 73 percent, of the animals of
breeding age being artificially inseminated. Aside from its foreign

possessions, France is a country of about 47 million people with 29 percent engaged in agriculture.

The field use of AI began about 1943, but did not expand greatly until after World War II. The authorization and regulation of the AI program is under government jurisdiction. In 1966, the decree that regulated AI was revised to include a broadening of the program and to stress sire proving as well as authorization of bull studs (semen-producing centers) and inseminating centers. The AI organizations are largely farmer-owned cooperatives. The French AI organizations are banded together in a national body called the Union Nationale des Cooperatives D' Insemination Artificielle, which is located in Paris. This organization is concerned with promotion, regulation, research, and information as well as government and state cooperation in conduct of the AI program.

Frozen semen in French straws, or in pellet form, is universally used. R. Cassou, a prominent worker in the AI field, has developed a system of freezing semen in straws, commonly known as "French Straws," and the "Cassou" system is used worldwide.

About sixty-two cooperatives for AI, belonging to the National Union of AI Centers, inseminate nearly all of the cattle involved. Many of the AI cooperatives exceed 100,000 cows per year and services per sire average over 3,500 annually.

Figure 6.2.   A Friesian bull stud that is operated by Union Nationale des Cooperatives D'Insemination Artificielle in France. (Photograph courtesy of Vre M. Parez, Union Nationale des Cooperatives D'Insemination Artificielle, Laboratory Control.)

Emphasis is on the use of progeny-tested sires,[16] and a young-sire-sampling program is carried out. Over 50 percent of all services are to progeny-tested sires (Figure 6.2). The use of progeny-tested sires for the "meat breeds" exceeds that found in most countries. As early as 1967, over 49 percent of the meat breed bulls used in AI were progeny tested. All progeny testing is under control of the Technical Center for Progeny Testing with its central office for animal genetics and computer service affiliated with the National Center of Zoological Research.

Cattle breeders are not permitted to collect and sell semen from their bulls. Practically all inseminations are by laymen who undergo a three-month-training course. The AI program receives no government subsidy, but the states do partially subsidize the progeny-testing programs.

Swine AI is widely used in France with over fifty thousand sows bred artificially in 1973.

Federal Republic of Germany. West Germany has a cattle population of about 14.5 million. Artificial insemination is used for about 70 percent of the cows and heifers (see Table 6.2). The leading breeds are Simmental (Fleckvieh), Brown Swiss (Braunvieh), Gelbvieh (yellow cattle), German Friesians (Schwarzbunt) and Red-and-White (Rotbunt). Traditionally, milk for the market and calves for fattening are supplied by the same cow.[17]

While some AI had been practiced prior to World War II, the organized plan did not get underway until the cessation of hostilities in 1945. In the restoration years, Marshall Plan funds supplied by the United States played a part. J. Stanley Brownell, an extension dairyman at Cornell University and a pioneer in the AI organization, assisted in establishing several of the breeding centers in Germany. In some instances, military-horse-breeding stations were converted to AI centers.

The AI organizations in West Germany are mainly farmers' cooperatives, though some are owned by dairy plants. During the past ten years (1966–1976) in Bavaria, the number of herds has decreased about one-third but the cattle population has remained constant. The milk supply to plants has increased to about 140 percent, demonstrating the increased milk yield per cow. It is estimated that the genetic progress in milk yield amounts to about 1.2 percent per year.

The breeding program in Germany involves close cooperation between herd book societies, AI stations, and government officials. In each state, the minister of agriculture is looked upon as the director of the program. In many instances, the breed societies, of which there are many for each breed, operate an AI service. The breed societies are largely responsible for selection of bulls for AI use. Milk recording is carried out by an independent organization that is controlled by the government. A progeny-testing program for AI bulls began in 1952. The young bulls enter the station at about thirteen months of age. They are mated at random with four hundred cows in order to get fifty to one hundred first lactation records from fifty to one hundred different herds. The young bulls are performance tested for rate of gain, fattening ability, and carcass characteristics. In Bavaria, where there are about 4.5 million cattle, with over 2.2 million females of breeding age, a bull's progeny are evaluated for milk yield, fattening ability, conformation, and ease of calving. It has been found that high milk yield and beef production are not antagonistic; in fact, a positive correlation exists. The progeny-testing program in Bavaria is one of the most outstanding in the world and many countries are adopting some of its practices. Horst Kräußlich, head of the Institute of Animal Breeding at the University of Munich, Germany, directs the program.

The law pertaining to animal husbandry (*Tierzuchtgesetz*) of the Federal Republic of Germany provides that all sires in AI be of high quality. Every AI center must have government authorization. About 50 percent of the cattle are bred to proved bulls. Most of the inseminations are made by graduate veterinarians, but the number of lay technicians is increasing. Frozen semen in ampules, straws, and pellets is used.

India.   India, a country with 575 million people and 241 million head of cattle, has not to date developed a progressive AI program. About 2 percent of the cattle of breeding age are serviced by AI (see Table 6.2).

Artificial insemination was initiated in India in 1939 at the Palace Dairy Farm in Mysore, and healthy calves were obtained.[18] Under sponsorship of the Indian Council of Agricultural Research, efforts have been made to establish a field program. The first was at Izatnagar. In 1945–1947, the government of India opened four regional centers. It was demonstrated that AI can work under both urban and rural situations. From 1951 to 1956, a "Five-Year Plan" termed the "Key Village Scheme" was launched. The goal is to provide genetic improvement for cattle and buffalo. Artificial insemination is considered the key activity in the Scheme.[19] Government support has varied from time to time and the earlier government AI stations were closed as the "Key Village Scheme" expanded. Under the Scheme, six hundred key villages and 150 AI centers were established. The Scheme was expanded to over 400 AI centers and by 1959 over 340,000 animals were inseminated. In addition to the "Key Village Scheme," many of the states have AI units. Some private agencies, including the various foreign-aid programs by the U.S. government, have aided in developing the AI program. A semen bank has been established at the National Dairy Research Institute at Bangalore to provide semen from imported bulls for crossbreeding and from superior bulls of the Indian breeds. Semen from the bank is flown to different parts of the country.

The chief deterrent to an expanded AI program in India, which has the world's largest cattle population, is effective organization. A strong national program has not been developed, and to establish one is most difficult for several reasons: many small farms with a low level of productivity, lack of education among farmers, poor communication and inadequate transportation, poor nutrition of cattle, and the prevalence of disease. Development is further hampered by widespread religious beliefs, which forbid cattle to be used for human food or to be slaughtered for any reason.

The Ministry of Agriculture is attempting to overcome these problems. A Livestock Improvement Act has been adopted, which in many areas calls for the castration of scrub bulls and the maintenance of only properly licensed bulls of improved breeding; an expanded milk-recording plan is being developed; a calf subsidy plan providing government support for raising promising young bulls to breeding age is in effect; and an expanded disease-control program is in operation.

The government has also established numerous "Gosadans"—homes for useless cattle—where uneconomic cattle can be maintained until they die or are killed by tigers. These homes are in forested areas where the cattle subsist by grazing. Facilities have been provided for the utilization of cattle hides and carcasses.

While developing slowly, the AI program has great promise as a means of improving production of India's cattle. Use of the better native cattle and the export of frozen semen for some crossbreeding affords the best approach. Lack of facilities for the expanded use of frozen semen is a problem but this, as can other problems, may be gradually overcome.

Israel. Situated on the eastern shore of the Mediterranean, Israel is a small country with great desert and mountainous areas. Only a small part of the land is suitable for agriculture. The development of dairy cattle for high production in a hot, dry climate is not without problems, but the Israelis have succeeded remarkably well. Because of the AI program, the use of scientific-breeding methods, improvement in feeding and management, Israel has the highest average milk production per cow of any country.[20] The average for all milk-recorded herds is nearly fifteen thousand pounds per cow per year.

AI was first used in Israel during 1935 and 1936 in several small villages. In 1946, "ON" AI Cooperative, the largest in Israel, was formed in the western Yizréel Plain. "ON" (a Hebrew term meaning

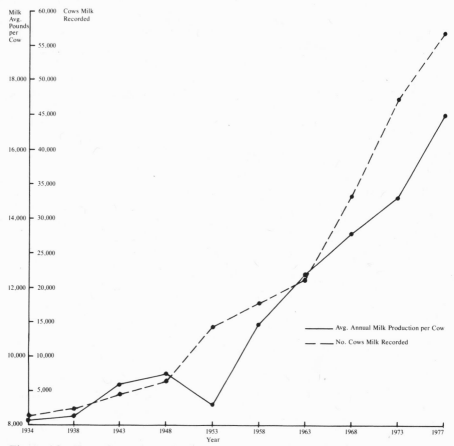

Figure 6.3 Forty-four years of milk recording in Israel. The decline in milk yield from 1948 to 1953 is due to large importations of cattle, new herds with inexperienced operators, and an outbreak of hyperkeratosis. The cows are heavily fed on grain by-products. Milk records are on an annual basis and computed to 3.3 percent of fat-corrected milk.

"male power") expanded to include about 60 percent of the nation's dairy cattle. The other 40 percent of the country's cattle are serviced by the "Hasherut" AI Cooperative in the south.

Israel has about 270,000 head of cattle of all ages. There is one dairy breed, the Israel-Friesian. Artificial insemination is used 100 percent. Over 60 percent of the dairy cows are milk recorded by the Israel Cattle Breeders Association.

The AI Cooperatives are nonprofit organizations, owned by their members, and controlled by a board of directors. Any cattle owner can become a member. The government does not subsidize or manage the AI program. The progeny-test records are analyzed on electronic computers by the Ministry of Agriculture, which is similar to the role of the USDA in the DHI and Sire Evaluation program in the United States. The government provides efficient veterinary service for the entire cattle population, handling epidemic diseases and providing protective vaccinations.

Sires for AI use are produced by mating three or four highest-ranking–AI-proved sires with cows from the official "Elite-Cow Register" that comprises some two hundred living cows. From this list, dams are chosen, which are daughters of top-proven AI sires. Semen is frozen in pellets and thawed in 3 percent sodium citrate solution.

Frozen semen from the highest-proved–Holstein-Friesian sires in the United States, Great Britain, and Sweden is imported for special matings and is used in addition to Israel-Friesian to produce future AI sires. The young calves selected are brought to the center when they are one hundred days old. They are gain tested. Semen is collected and frozen from bulls that are fourteen to fifteen months of age. About 1,350 cows are inseminated so as to provide at least 100 daughters with milk records that can be compared to contemporaries. A "kept-in-waiting" program is followed for young sires from the time enough test inseminations are made until progeny tests are completed. Progeny tests are based on 122 days and 305-day-lactation summaries. Many bulls are culled on the 122-day–daughter-lactation summary. The average bull is proved at five-and-a-half years of age, but some are eliminated when they are a little over four-and-a-half years. One out of six bulls sampled, on the average, is kept for heavy use.[21]

About 75 percent of the AI services are to highly proved bulls. Genetic progress in milk production has been about 70 kg. of milk per cow per year the past decade. The Israeli authorities say, "The cost of a proven sire, by the "kept-in-waiting" system and after a selection intensity of one out of six, is calculated to be approximately IL 40,500 (Israel pounds); whereas, the returns through his intensive use (6,000 daughters) are a lifetime genetic increased production by his daughters of 3½ to 5½ million kg. milk."[22]

Italy.   There are about eight million head of cattle in Italy. During 1973, about 3.2 million units of semen were produced by the bulls in AI use and about 1.8 million head of cattle were inseminated.

While much experimental work in AI has been done by Italian scientists, beginning with Spallanzani in 1780, the organized field application began in 1948. The AI program is controlled by the Veterinary Service of the Ministry of Agriculture. Inseminating is done by about 2,200 veterinarians, and an average of 543 cows inseminated yearly per operator.

Most of the cattle are used for both meat and milk, some for draft purposes. Chianina is one of the prominent Italian breeds known in

America for its large size and meat production. In recent years, many Holstein-Friesians have been exported to Italy by the United States and Canada for milk production.

Research and development of the AI program is aided by the Societa Italiana per Il Progresso Della Zootecnia in Milan. The director, Telesforo Bonadonna, is internationally recognized for his many studies on animal reproduction and AI.

Japan. Artificial insemination of farm animals began in Japan in 1913 when it was used for horses.[23] H. Ishikawa, professor of medicine at the Kyoto Imperial University, visited Ivanov's institute in Russia in 1912 and upon his return to Japan he initiated studies on spermatozoa and reproduction in farm animals.

The use of AI in other farm animals followed: cattle, 1928; poultry, 1936; and goats, sheep, and pigs in 1938. Studies were begun with rabbits in 1930, honeybees in 1938, and, ducks in 1949.

The leader for many years in developing the AI program in Japan is Y. Nishikawa,[24] former professor in the College of Agriculture at Kyoto University. He has been particularly active in carrying out AI in horses using frozen semen.

Japan has about 3.5 million cattle. The number is steadily increasing as both meat and milk are in short supply. AI of cattle began in a small way in 1928 as a control measure for tuberculosis and in 1933 for control of trichomoniasis. The organized program, fostered by the government, got underway in 1939. The program grew slowly and was interrupted by World War II. In the postwar reconstruction period, AI received added emphasis with the adoption of the Livestock and Reproduction Law in 1950. The government granted the establishment of 580 animal health centers in 1950 and 1951 with the necessary equipment to carry out an intensive AI program.

During 1947, about 16,000 dairy cows (29 percent of the total) and 31,000 beef cows (7.4 percent of the total) were serviced by AI. In 1971, a total of 1.3 million dairy (99.3 percent of the total) and 555,000 beef cows (93.7 percent of the total) were artificially bred. At the present time, it is estimated that 98 percent of Japan's cattle of breeding age are serviced by AI. The use of AI for beef cattle in Japan is the highest for any country reported to date.

The effect of AI, and improved feeding and management, on milk production per cow from 1935 to date is dramatic (see Figure 6.4).

In 1935, the average lactation yield was 2,700 kg. (5,940 pounds); for 1971, an average of 5,300 kg. (11,660 pounds), which is nearly double the amount in 1935. It will be noted that yield per cow dropped severely between 1940–1945, the World War II years. This situation prevailed in many war-torn countries. The AI program proved in-

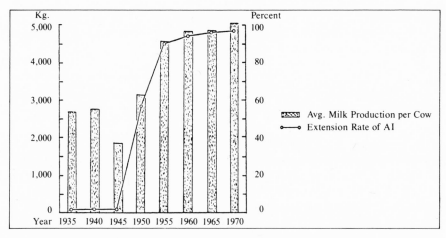

Figure 6.4. Effect of AI on milk production per cow in Japan from 1935 to 1970. Source: Y. Nishikawa, *Sixty Years of Artificial Insemination in Japan* (Kyoto, Japan: College of Agriculture, 1973), p. 20.

valuable in helping to restore and improve production in the postwar years in many countries.

Progeny-tested bulls, both dairy and beef, are emphasized in the Japanese AI program. A young-dairy-sire-sampling program is carried out as one performance and progeny testing for beef.

In 1970, there were 1,651 AI centers (keeping bulls). These are largely cooperatives, but some are owned by private operators. With the widespread use of frozen semen, the number of bull studs is becoming fewer and more cows are bred per sire. Custom collection, freezing, storage, and the sale of semen by private operators is permitted.

There are also about 100,000 pigs produced by AI in Japan yearly. Boar stations are found at some of the cooperative AI cattle stations and some are privately owned.

Goat AI began in 1948 and reached a peak of nearly thirty-five thousand head in 1960. The volume has declined due to fewer goats being raised in Japan, with only about thirty-five hundred head in 1972.

Sheep AI is not used to any extent commercially. In 1956, about five thousand ewes were inseminated, but in 1972, only a few head experimentally. The sheep population has drastically declined in the past twenty years.

The chief use of AI in poultry is to obtain highly fertile hatching eggs with the fowl kept under battery- or cage-feeding conditions. In some areas, more than 50 percent of the hens are inseminated.

Liquid semen is largely used for swine, goats, sheep and poultry.

With the development of more satisfactory methods to utilize frozen semen, Japanese authorities expect an increased use of AI in these species.

Artificial insemination of horses is used extensively in Japan. Beginning with the early work of Ishikawa in 1913, many mares were inseminated with liquid semen. Before 1930, various extenders were utilized and stallion semen was transported. In 1937, the government founded the Organization for Promotion of Horse Production. A broad educational program on horse production followed with 140 horse AI centers founded, and in 1943 over fourteen thousand mares were bred artificially. By 1961, the total was over twenty thousand annually. The early emphasis on horse production resulted from the need for work animals, military purposes, and the demand for horse meat. Since 1952, horses have been largely replaced by mechanical power, and in 1972 only a few hundred mares were bred artificially.

Japan has led the way in developing procedures for the use of frozen semen in horses beginning in 1965.[25] Horse semen frozen in straws has resulted in a conception rate of 57 percent for the six-year period of 1966–1971.[26]

Netherlands. The Netherlands has nearly five million head of cattle that serve for both milk and meat purposes. About 69 percent of the cattle of breeding age are serviced by AI.

AI was used in the Netherlands as early as 1930 as a means of combating sterility. In 1943, a few cattle owners organized AI cooperative associations, particularly in areas where low fertility was a problem. During the German occupation, both nutrition and fertility of the cattle declined. In 1946, following cessation of World War II, an expanded AI program was initiated. The goal was to improve milk production and to control sterility. About twenty-one bull studs and seven hundred local breeding societies were established. The United States helped by supplying Marshall Plan funds and technical assistance.

The AI program is regulated by the Industrial Board of Agriculture, a public corporation representing all of Dutch agriculture.[27] The inseminating is done by licensed AI technicians under the supervision of veterinarians and the Animal Health Service. Frozen semen in straws and pellets is widely used. While a private herd owner may have semen from his bulls collected and frozen, only certified technicians or veterinarians may inseminate cows.

About 65 percent of the cows are on milk-recording programs. The government subsidizes testing. A checkoff fee of a few cents per hundred pounds of milk provides the funds. All dairymen pay the checkoff fee whether or not they test.

There is close cooperation between cattle breeders and the AI program. The principal objective is to improve cattle and milk production. Young sires, many sons of the better AI-proved bulls, are sampled. Progeny evaluation is on the basis of modern concepts and includes ease of milking, protein content of the milk, and body conformation. All bulls must be registered and pass strict health requirements. The AI associations are members of the herd-book societies, and bulls going into AI use are approved, so far as pedigree and body conformation is concerned, by mutual agreement.

The AI program, milk recording, and disease control measures have largely restored the dairy industry in the Netherlands to its pre-World War II levels.

New Zealand. New Zealand is a small country made up of two main islands, isolated in the South Pacific. The climate is temperate and rainfall is adequate and fairly well distributed over the year, for exceptionally good grass production. Cattle remain outdoors all year. Milking cows obtain almost all of their feed by grazing pastures consisting of grasses and clover except for short periods of the year, when there is the addition of silage and hay made from pasture. The dairy industry, as well as beef and mutton production, is organized to produce milk and meat for manufacture and export at minimal costs.

There are about 3 million people, 60 million sheep, and 10 million head of cattle in New Zealand. About half of the cattle are dairy with roughly 2.5 million in milk.

The AI program is operated by the New Zealand Dairy Board.[28] The board has nongovernment status and represents the interest of the country's dairy farmers. It is involved in all phases of herd improvement, production-testing programs, milk prices, "bobby veal" pools, research and educational endeavors.

The New Zealand Dairy Board operates two semen-production centers—one at Newstead, near Hamilton, and the other at Awahuri, near Palmerston North. The centers are financially self-supporting.

The Newstead Center, established in 1953, produces liquid semen for inseminating most of the cows that are enrolled on the AI program. In order to have most cows freshening when grass yield is maximum, nearly all cows are inseminated during a twelve- to sixteen-week period during the spring. Most of the semen is collected in the spring and used at room temperature within three days of collection. A Caprogen semen extender containing catalase is used routinely for liquid semen.

The Awahuri Center concentrates on the production of frozen semen for nominated services, custom freezing, and export. The semen is frozen in plastic straws, containing 25 million sperm per unit and an egg-yolk-milk diluent. In 1975, about thirty-three thousand cows were bred with frozen semen.

Semen is supplied by the Dairy Board to six livestock improvement associations and veterinary clubs throughout the country. During 1975, 938,000 cows were inseminated by these organizations. Of these, 915,000 were dairy cows, representing 44 percent of all dairy cows in milk. Sixty percent of all herds with 10 or more cows used AI in 1975. Most dairy herds have over 100 cows in milk.

The majority of cows on the AI program are mated to progeny-tested bulls. The prevailing breeds are Friesian, Jersey, and Ayrshire. There are a considerable number of Friesian-Jersey crossbreds in some herds. About 53 percent of the services are to Friesian bulls; 41 percent to Jersey; and 2 percent to Ayrshire. About 4 percent of the inseminations are to beef bulls. Bulls of eight beef breeds are available and some dairy cows are inseminated with semen from beef bulls.

The Dairy Board owns about 750 bulls. However, most of the cows are bred to an annual average of 30 proved dairy bulls and 12 proved beef bulls. The board sponsors an extensive young-sire-sampling and progeny-test programs for both dairy and beef bulls. An average of 150 young dairy bulls are purchased and sampled each year. The Sire Proving Scheme seeks to identify the best 5 to 10 percent of these bulls based on the production records of 50, two-year-old daughters in about 40 herds. Bulls are evaluated by herd-mate comparisons and ranked by computing a Breeding Index (BI). If breed average is 100 and a bull's daughters produce 30 percent more milk than the average, his BI equals 130.

The heavy use of high-BI bulls is reflected by the 1975 summary of the New Zealand AI program.[29] One Ayrshire bull with a BI of 133 pounds of milk averaged 5,447 services; eight Friesian sires with an average BI of 135 averaged 51,323; and two Jersey bulls with an average BI of 134 had an average of 77,072 inseminations in a single season. The Friesian bull Welburn P. G. Butterman with a BI of 133 had 103,484 inseminations during the 1975 season.

Under a system of dairying that depends on forage feeding (pasture) and climatic conditions for pasture yields, the milk production per New Zealand cow is well below the level of cows in countries where heavy-concentrate feeding prevails. Yet the genetic merit of New Zealand bulls compares favorably with overseas bulls. Imported semen from the U.S. and Great Britain proved bulls resulted in daughters little better in milk yield and, in some cases, lower than New Zealand bulls. For example, the 1977 USDA Sire Summaries list two New Zealand Jersey bulls whose semen were imported into the United States for AI use. These two bulls had a total of 925 tested daughters averaging 3,086 liters (6,789 pounds) milk, 5.25 percent fat per lactation in New Zealand. In the United States, 79 daughters averaged 10,439 pounds of milk, amounting to 4.79 percent fat and nearly 50 percent more milk. The average milk yield per lactation for Friesian cows in New Zealand during 1975 and 1976 was 4,245 liters, about 9,340 pounds, and for Jerseys 2,640 liters, about 5,800 pounds.

Switzerland. There are approximately two million head of cattle in Switzerland. Simmental is the leading breed, followed by the Brown Highland (Brown Swiss). In recent years, Red Holstein, Holstein-Friesian, Aberdeen Angus, and several other breeds have entered the picture. Some are for crossbreeding purposes. Emphasis is on the ability of animals to produce milk efficiently and to yield a good carcass.

Switzerland did not begin a large-scale AI program until the late 1950s. In the earlier years of AI development in Europe, the Swiss government and breed societies took the position that, since many farmers depended upon bull sales as a significant source of income, the market should be preserved. Some AI was used as a means of controlling the spread of disease and reproduction troubles. Farmers liked the AI program and urged its expansion. In 1949, there were only two AI centers and about 2,000 cows were inseminated. In 1960, an additional center, the Neuenburg Center, was established; added emphasis was given the AI program and service became available in all cantons. In 1963, a total of 35,249 cows were serviced by AI. By 1973, the total was 545,000, representing 50 percent of the cows of breeding age.

The AI program in Switzerland is largely a farmers' cooperative enterprise, strongly supported and regulated by the government, with close alliance of the breed registry societies. About two-thirds of the cattle on the program are herd book recorded.

Considerable quantities of frozen semen are imported from Canada, France, and the United States to introduce new genetic potential and additional breeds. Switzerland exports considerable semen, particularly from the Simmental breed, to the United States (eighty-five thousand units in 1971) and to other countries, including India.

AI is utilized for swine and goats in addition to cattle. The AI program in Switzerland utilizes frozen semen in straws. Modern techniques and equipment are used. A well-directed sire-sampling and progeny-testing programs are carried out.

USSR. Artificial insemination methods were introduced in Russia by I. I. Ivanov, who in 1897–1907 developed the technique for mares. He regarded AI as a method that "made it possible to exploit pedigree sires more rationally."[30]

Russia is estimated to have about 109 million head of cattle, 42 million of milking age; 145 million sheep; and 72 million hogs.[31] The USSR, with its large livestock-raising enterprises on collective and state farms, provides an enormous potential for the use of AI. The goal of boosting livestock productivity as established by the party and government calls for increases in the number of livestock and continued improvement of breeds with a view toward genetic gains and production levels.

Until the revolution in Russia (from 1918 to 1920), only horses were bred by AI. In 1928, the first large-scale AI project was begun with sheep. It is reported that in 1959 some 42 million farm animals were artificially inseminated.[32] Involved were 8,934,000 cows and heifers; 32,923,000 sheep; and 15,000 hogs. During 1973, it was estimated that over 26 million cows, 44 million sheep, and 1.1 million swine were bred by AI.[33]

The USSR reportedly has fifty cattle breeds, thirty-one breeds of hogs, and sixty breeds of sheep recognized by the pedigree societies. Most of the cattle are described as dual-purpose with both milk and meat the goal. Bull calves are grown and slaughtered for meat. The average milk production per cow for the nation is estimated at 6,600 pounds (3,000 kg.). Efforts are being made to increase both milk and meat production. There is a growing trend to develop specialized milk and beef breeds on the large farms though at present most beef production is based on the use of dairy beef cattle.[34] In total beef production, Russia claims to rank second in the world with an average per capita consumption of 125 pounds (57 kg.). Efforts are being made to establish large specialized farms for breeding and fattening cattle. The chief breeds are Sichevskaya, Simmental, Swiss, and Black-and-White. The crossing of dairy breeds, particularly Red Range cows with beef bulls of the Hereford, Angus, Shorthorn, Charolais, and Kazakchsky (a Russian breed), is being carried out. Progeny tests indicate much greater gains. Over 2 million cows were inseminated to beef bulls in 1972, and it is planned to inseminate at least 10 million head annually.

Efforts to improve milk production are moving along the lines of progeny testing and sampling young bulls from selected cows and by the better sires. The Black-and-White is the main milk breed. It is being improved by the AI use of bulls from a breeding farm, "Lesnoye" livestock enterprises. Some Black-and-White improved strains have been imported from Sweden.[35] The program to date is less developed, particularly from the standpoint of utilizing the concepts of population genetics, than that used in the United States and several other countries.

The cattle-improvement program is administered by the Ministry of Agriculture through councils of "breed work" with every strain, by means of regional and district state breed amalgamations. The AI program operates largely through some four hundred AI stations—on the collective and state farms—where bulls are kept. There are hundreds of stations where AI service is provided from the bull centers.

Technicians inseminate cows in herds on the smaller farms. Liquid semen is used for the cows on some of the large collective farms that have bull studs. Semen frozen in ampules is in heavy use and straws and pellets are experimented with.

Artificial insemination appears to be the tool that, with the application of sound genetic concepts, will do much to increase milk and meat production in the USSR. L. P. Prahov states that "our country inherited backward lower productive stock raising from Tsarist (Czarist) Russia—cattle were being bred by nomadic and seminomadic methods."[36] In the program for improvement, it is probable that frozen semen from high-plus–AI-proven bulls from the United States and other countries will be imported to increase the genetic potential.

Chapter 7: Artificial Insemination in the Tropics

7.1 Introduction

The tropics, and countries substantially within the tropics, constitute a large part of the world's surface. This broad, loosely designated area contains about 475 million cattle and buffalo, 165 million sheep, and 158 million goats.[1] It also contains some of the most primitive regions of the world. The raising of cattle is a means of livelihood, but is often handicapped by severe climate, inadequate feed, insufficient water, disease, and pests.

Many of these areas termed *tropical* have a large share of the world's population and most are termed *developing countries* by persons concerned with human population and the world food supply.[2] There is a great variation in the cattle industry of the developing countries. While most of the cattle in countries such as Southeast Asia, Africa, Central and South Americas are handled in a nomadic style, depending principally upon pasture for subsistence, there are watered areas where productive and well-managed herds are found.

Improvement of cattle and other livestock in the tropics has been and continues to be a problem for all concerned. It is in these areas that greater production per animal unit, through both higher yields and better efficiency, is needed to provide food for an expanding human population.

Due to numerous factors, the AI program has not to date played a major role in improving the total cattle population in the tropics. However, with frozen semen available, air transportation, disease-control programs, improved feeding and management, protective measures to combat severe heat, and above all an expanded educational program, AI is steadily progressing in many developing areas.

7.2 How AI Can Make Genetic Improvement in Cattle in the Tropics

There are three major approaches in improving a livestock population:

1. Selection among the native cattle population for basic characters, such as milk and meat production, adaptability to the environment, growth rate, reproduction, and resistance to disease.

The best of the indigenous cattle are singled out and used for breeding purposes. Bulls used artificially can easily sire a thousand times as many calves as when used naturally. Above-average cows, mated to the best bulls that have demonstrated their superiority through many AI progeny under existing conditions, should be used to produce future sires.

2. Importation of semen from bulls known to transmit the desired characteristics for crossing with the native cattle. In many tropical areas, European breeds, *Bos taurus,* cannot survive the climate, disease, and management regime. Usually European-type cattle do not thrive when the mean annual temperature is above 65°F.[3] When crossbreeding is desirable between *Bos taurus* and *Bos indicus* types of cattle, artificial insemination provides the practical solution. Imported bulls of the European breeds usually cannot withstand tropical temperatures and remain fertile; they encounter diseases for which they have little or no built-in resistance; the cost, transportation, and maintenance of such bulls are high; and selecting one bull has the risk that he may not be a good representative of his breed in transmitting ability for the desired genetic characters. By importing frozen semen, the best of several bulls of any breed can be utilized. The only costs are for semen and for its storage and utilization. Furthermore, several exotic breeds can be tried, without undue expense, to determine which produces the best cross with the indigenous cattle under the prevailing environmental conditions.

3. Planned crossbreeding and selection. This breeding scheme combines the basic objectives of methods (1) and (2) to produce an animal that has many of the desired characteristics of both the indigenous and exotic breeds. The plan works more successfully in some tropical areas than in others depending upon the variations in the exotic stock, level of crossbreeding, climate, resistance to disease, parasite toll, and feed supply. Authorities agree that, while slow, some of the best progress in improved production can come from upgrading the indigenous cattle.[4] If the exotic breeds are too heavily used, animals lose their resistance to tropical environments.[5]

7.3 Problems in Using AI in the Tropics

1. For the most part, animals in the tropics range over large areas, and confinement, even in small pastures, is not practiced. The control of cattle is dictated more by available forage and water—and by the herdsmen—than by the fences and paddocks found in intensive farming areas. Under these conditions, it is difficult to identify animals, detect estrus, and to confine the animals for insemination. Most will be served by a bull either before or after insemination.

2. In many areas, there is a strong prejudice against artificial in-

semination, particularly among the uneducated and primitive peoples. All too often a man's status is measured by cattle numbers and not by the quality of his cattle. The custom is to permit many bulls to run with the herd.

3. Breeding diseases are a problem and are poorly controlled under a natural-service regime. All of the common diseases affecting reproduction are found in tropical areas, including brucellosis, trichomoniasis, vibriosis, and vaginitis. AI has been introduced into many privately owned herds in tropical countries to control diseases.

Cattle in the tropics are often plagued by foot-and-mouth disease, rinderpest (cattle plague), anaplasmosis, and other diseases spread by insects. Ticks are a problem, not only spreading disease but also causing anemia and hide damage. Ticks thrive in the rainy tropical climates where most livestock are found. *Bos taurus* types of cattle are particularly susceptible to tropical diseases and tick infestations, reacting more severely than *B. indicus* and indigenous stock.

4. Communications are often poor. Good roads are lacking; telephones are few; many natives cannot read or write; and cattle herds are usually many miles apart. Electricity is lacking, transportation is poor in the "back country," and liquid nitrogen for semen storage is often a problem.

5. Much of the AI practiced in the tropics involves the use of *Bos taurus* semen for *B. indicus* females. Difficulty is encountered in detecting estrus in *B. indicus* females as the normal behavior found in *B. taurus* females is suppressed. As a rule, the estrus period is of short duration in *B. indicus* and the ovulation time varies greatly. Ovulation frequently occurs without signs of "heat."[6]

B. indicus bulls are sluggish and lack libido. Most semen is obtained from Zebu bulls by electro-ejaculation and is used widely in the tropics.

6. Effective herd management and feed production are problems in efficient cattle production, hence the wide-scale use of AI in many tropical countries. Most of these countries have a rainy season and a dry season. During the rainy season, forage is lush and cattle thrive. During the dry season, the forage becomes woody and low in nutrition. Cattle lose weight. Body gains and milk production reach a low ebb. In many areas, animals are three to four years old before calving the first time. Fertility is impaired during times of low-level nutrition. The use of silage, or green forage, would do much to provide a more uniform feed supply in many countries.

Sanitation and disease control, particularly immunization, is lacking in many areas as is effective insect control. Better animal husbandry is a prerequisite in most tropical countries before improvement through breeding, including the use of AI, can be a strong factor in increasing production per animal.

7.4 Extent of AI Use in the Tropics

During the past forty years, many attempts have been made to establish AI programs in tropical countries. In terms of a nationally organized program, because of the disadvantages listed above, few have succeeded. Organized programs have been generally successful only in Jamaica, Kenya, Brazil, Colombia, and to a lesser extent in India and the Philippines. As a rule, from less than 1 percent to not over 10 percent of the females of breeding age are involved.

For the most part, the AI program in tropical areas is confined to those herds with better-than-average management and facilities, and where the herd owner wants to make improvement. In every country where AI succeeds, the demand for service comes from cattle breeders. In tropical areas, the more advanced native cattle owners and persons engaged in agriculture as a business are demanding higher-producing cattle to keep pace with the expanding economy. In most instances, government assistance is given to advance the AI program.

Africa. In Kenya, where there is a thriving agriculture carried on largely by Europeans, artificial insemination was practiced as early as 1935. Since that time, the AI volume has expanded steadily.[7] The program has now been extended for the benefit of African peasant farmers. Semen from native cattle, as well as imported frozen semen, is used. In 1946, a Central Artificial Insemination Station was established in Kabete, near Nairobi, and numerous subcenters were established. Most of the European breeds in addition to Sahiwal, Africander, and others are available. During 1973, about 50,000 head of cattle were serviced and in South Africa the total reached about 245,000.

Frozen semen, often from British, American, and Canadian bull studs, is imported for use on the European breeds or crossing with the native cattle. Some of the successes and failures with frozen semen are outlined in an account written to the author by J. D. Struthers of Southern Rhodesia:

> During the month of December, 1961, the first shipment of deep-frozen semen to be imported from the United States arrived in Central Africa. Frozen in dry ice the shipment consisted of 400 ampules of semen from Santa Gertrudis, Charolais, Black and Red Angus, Brahman and Polled Hereford bulls. The semen was shipped from Eastern Iowa Breeders, Cedar Rapids, Iowa, and handled by the Agricultural Company of Pan America. . . . Ten days after arrival, the entire shipment was lost without a single ampule having been used due to failure of a shipment of dry ice to arrive. The dry ice was being sent from Kitwe, some 400 miles away.
>
> After this failure, we purchased liquid nitrogen equipment and reordered another shipment of semen. It arrived January 11, 1962. Inseminations were made beginning in January, 1962, and continuing until August

of the same year. From 200 inseminations, including second and third services, 112 cows or 56 percent were diagnosed pregnant.

In succeeding trials, we found that we got only a 35 percent first conception rate when cows inseminated were nursing calves. After weaning, cows inseminated resulted in a 65 percent diagnosed pregnancy rate. We attribute the poor conception rate while the calf is nursing due to the strain and the cow in producing milk and a falling plane of nutrition due to matured grasses.

Our experience indicates that a satisfactory conception rate can be obtained when AI is practiced during the normal breeding season which is the rainy season. Thereafter, supplementary feeding of both protein and energy would be required.[8]

Brazil.   Artificial insemination of cattle, sheep, swine, and to some extent horses has been practiced in Brazil since 1938, when an experimental AI station was set up near Sao Paulo. The AI program is under government supervision. There are both government-sponsored and privately owned bull studs in operation. In addition, there are many firms engaged in the importation and sale of frozen semen.

Brazil has about 96 million head of cattle, the most of any country in South America, with about twenty-six million of breeding age.[9] Zebu (*B. indicus*) make up the majority with many crossbred Zebu-European animals. There are the usual rainy and dry seasons with animals gaining in condition a part of the year and losing weight when pastures become woody. Most of the cattle are in large herds. Many haciendas, handed down from Portuguese land grants since the fifteenth century, range from a few hundred to fifty thousand or more acres. The beef herds (Zebu) range from a few hundred to ten thousand or more animals. Around the principal cities, Sao Paulo, Rio de Janeiro, Minas Gerais, and others, many good dairy herds of straight European breeding and *B. indicus* and *B. taurus* crosses are located.

There are twenty-nine national AI centers in Brazil (1976) under government sponsorship, and about thirty-five commercial semen-producing businesses were granted federal approval.[10] Many of the commercial concerns are corporations owned by one or more veterinarians. There are nearly one hundred licensed semen dealers. During 1976, about 1.5 million head, or about 6 percent of the cattle of breeding age, were served by AI. About 80 percent of the semen used is produced in Brazil, and 20 percent is imported. The United States supplies about 58 percent of the semen imports with Holandesa (Friesians—Black and White, and Red and White), Polled Hereford, Santa Gertrudis, and Hereford breeds the leaders. There are about fifteen licensed semen import firms.

Brazil probably has the best Zebu cattle in the world and their owners are proud of them. The Nelore breed leads in number with Gir, Guzerat, Indubrasil, and Sindi in lesser numbers. Some breeds are hornless and denoted as "Mocho," while some excel in milk and are known as "Leiterio," such as the Gir Leiterio.

Brazil has a growing dairy industry and Friesians, Brown Swiss, and Guernseys are in demand. The country is outstanding in its dairy manufacturing industry. Considerable amounts of cheese are exported, and milk is an important food item.

As do most tropical countries, Brazil needs more testing for production in both dairy and beef cattle. It is an aggressive country, however, and with a stronger educational program and a stable government, great strides in cattle improvement, aided by AI, can be expected.

In South and Central Americas, as well as some of the larger tropical islands in the Pacific that maintain a sizable cattle population, conditions are tropical or semitropical. The foregoing discussions on Africa and Brazil with respect to utilization of AI will apply in a general way to all, as it does to India (see Chapter 6). Some of these countries have made a greater progress in the use of AI than others (see Table 6.2).

7.5 AI of Water Buffalo

Artificial insemination is used for the domestic buffalo, *Bos bubalis,* generally known as water buffalo. There are about 106 million water buffalo in the world. They are of great economic importance in the Far East, the Middle East, and in tropical or semitropical areas in southern Europe, and in the Central and South Americas. Water buffalo are used for draft, milk, and meat. There are wild and domesticated water buffalo. The Cape buffalo in Africa is an example of the wild species (*B. caffer*). The domestic buffalo is classified in two main categories: the swamp buffalo and the river buffalo.[11]

There are about 51 million domestic buffalo of the river type in India. India has the largest population of buffalo in the world. China and Egypt also have many. They are important for milk and draft purposes. Buffalo produce about 50 percent of the milk in India and are superior in yield to the average Indian milk cow (about four hundred to five hundred pounds of milk) per lactation. Buffalo average about eleven hundred pounds of milk per lactation, and some breeds (the Murrah) produce from three thousand to five thousand pounds, with some individuals over ten thousand pounds.[12]

The swamp buffalo are found in the Philippines, Singapore, and other countries in the Far East. They are low in milk production but useful for draft and meat. Their feet are large, and they are slow moving with great draft power, thus ideally suited for cultivation of the swampy and water-logged rice paddies.

Buffalo are maintained in these warm regions because they thrive and produce much better than cattle and are more economical to maintain. They vary widely in genetic characteristics and the opportunity for selection of desired traits is great. Artificial insemination is used, particularly in India, to improve milk-producing qualities.

Artificial insemination in water buffalo is carried out much the same way as in cattle. Semen is collected by the artificial vagina, which the buffalo bull readily serves, and processed as for cattle semen. Buffalo semen can be frozen and stored in liquid nitrogen.

The buffalo female reaches sexual maturity at two to three years of age, depending on the level of nutrition. The average duration of estrus is usually about twenty-four hours or more. Insemination is performed using the rectovaginal technique or the speculum.

AI conception rates in India buffalo as reported vary from 50 to 66 percent the first insemination.[13] In the Egyptian buffalo, conception rates varied from 35 to 55 percent. The fertility varies with the level of nutrition and season of the year. Most buffalo develop estrus and are inseminated during the cooler months of the year.

With milk-recording programs available and the ability to select the superior-progeny-tested sires, AI can be used with buffalo to make systematic improvement. Progress is under way at government experiment stations in India and Egypt. This approach can produce animals that are adapted to the environment—a necessary requirement for all tropical and developing countries.

7.6 Best Approach for AI Use in Tropics

Experience[14] to date indicates that the best approach in using AI for cattle in the tropics is in terms of individual herds. For reasons outlined above (7.3), AI programs involving large areas and "across the board" herds seldom succeed. To be successful, the individual herd approach requires that a number of conditions be satisfied.

The first condition is for the herd owner who desires to improve his cattle to be willing to make the herd-management adjustments necessary for effective AI. This often involves corrals and equipment for handling cattle as well as management changes.

The second requirement is a source of semen from bulls that can be expected to transmit the genetic characteristics desired at high levels. Frozen semen can be obtained from within a country's bull studs or imported. In any event, sires with progeny proof or good-performance records should be used. Little is gained by using mediocre bulls. This situation often is true with native bulls where no assembled production records are available.

Third, trained persons, or veterinarians, are needed to inseminate the cattle. Many tropical or semitropical countries conduct inseminator-training schools, usually directed by veterinarians and often under government-supported programs. In the larger herds, the inseminator, often working under the supervision of a veterinarian, is a full-time employee devoting all his time to herd reproduction, including the necessary management for good results.

The necessary supplies for insemination and preservation of semen are the fourth requirement. Usually liquid nitrogen for semen-storage units can be obtained, even if imported. The modern-semen-storage unit requires refilling with liquid nitrogen only about five or six times per year.

Identification or numbering systems for all cattle and a breeding- and calving-record-keeping program are a fifth major requirement.

Proper nutrition levels to enable normal reproduction on a yearly basis are also an important condition. Good growth of the young stock is necessary to save time in reaching the desired breeding size.

Another important requirement is adequate disease and insect control with sanitation practices to control sickness and impaired fertility. Where herds are not too large, a trained inseminator could work additional herds. As success is attained and educational developments in the area progress, such programs can gradually become regional.

7.7 Summary

In tropical areas, AI can do much to increase the genetic potential for better animal production. Provisions for an adequate feed supply, disease control, and improved animal husbandry must accompany the use of better bulls through AI if the desired results are to be achieved.

The use of frozen semen from well-proved bulls, imported where necessary, can save years of time and effort in establishing a sound genetic base. However, tropical countries must establish their own progeny-testing programs to utilize fully the best of the indigenous breeds and to measure the results of crossbreeding. The continued growth and advancement of the AI program would seem to be an essential part of agricultural development in tropical areas and in all "developing countries."

PART III:
THE NATIONAL ASSOCIATION OF ANIMAL BREEDERS, INC., AND THE AI PROGRAM IN THE UNITED STATES

Chapter 8: Early History and Beginning

8.1 Introduction

The National Association of Animal Breeders (NAAB) is an organization composed of AI businesses throughout the United States and Puerto Rico with associate members in other countries. Members consist of farmer-owned cooperatives and privately owned corporations or proprietary organizations. Voting membership is restricted to organizations in the United States. The growth and development of the NAAB coincides with the growth of the AI program in the United States and with the worldwide expansion of AI as a means of improving cattle.

8.2 Beginning of NAAB

The NAAB began in 1946 when managers and directors of the early AI organizations in the United States felt the need for an interchange of ideas on this new but rapidly expanding program.

As a new industry, the plan of organization for AI businesses, techniques employed by the program, and the general scheme of operations had to be learned by experience. Technological information regarding semen evaluation, dilution (extension) holding time, transportation, and insemination techniques was amassed by research workers at many of the agricultural colleges beginning in 1936. Much credit is due these pioneer researchers for paving the way for a practical AI program. However, field experience and field-derived data were badly needed. Managers of AI organizations soon found that the management of college dairy herds, under controlled conditions, often provided more encouraging results than were obtained in farmers' herds. In the beginning years of the AI program, few men were trained for positions as organization managers, laboratory managers, field supervisors, salesmen, and inseminators. Therefore, the exchange of ideas and information by the people engaged in carrying out the program in those early days played an important part in developing the industry to its present level.

8.3 Cooperatives Lead the Way

The free exchange of ideas and knowledge gained in the field, unless competition and secret formulae regarding products and techniques exist, is characteristic of farm-oriented groups. All but two or three of the AI associations operating in 1946 were farmer-owned cooperatives. Competition was not considered important. At that time, AI operators did not use frozen semen, unit-dose packaging, long-time storage, and long-distance shipping. With liquid semen in use, every organization was limited in the distance it could successfully operate in a routine manner. As a rule, semen was used within forty-eight hours after collection and few AI organizations operated over large areas. Service areas usually included a county, several counties, an entire state, or parts of two states if cow population and convenience so dictated. There was little conflict regarding the territories that would be serviced. By mutual agreement, the cooperatives would not encroach on each other's territories without an understanding that such moves were in the interest of better service and economy for the patron.

The AI program in the United States began as one of "free enterprise." This system prevails today, and the organization and development of the NAAB is no doubt responsible in a large measure for maintaining this freedom.

8.4 The First Managers' Meeting

On 2, 3, and 4 December 1946, representatives of various AI organizations met at Tiffin, Ohio, to discuss AI methods and practices in use.[1] The conference, entitled "A Managers' Meeting of Artificial Breeding Association in the United States," was sponsored by Noba, Inc., of Ohio, with Max Drake, long-time manager of the company, as the acting chairman.

Twenty-five managers representing seventeen states attended the meeting. Topics of discussion ranged from a brief history of the scope of operations and techniques used by individual organizations to financing, service charges, methods used for paying inseminators, semen diluters, and so on.

A temporary committee of the managers was appointed on the second day of the meeting with the following appointments: Kenneth Wallin, Wisconsin, acting chairman; John Fawcett, Missouri; D. C.

Master, West Virginia; D. C. Reid, New Jersey; Maurice Johnson, New York. Drake was also elected to become a member of the committee.

A set of resolutions, drafted by the committee members, was presented to the managers as follows:

> The Temporary Committee of the Managers submits the following resolution for the group's consideration:
>
> **Whereas, it is felt to be highly desirable to—**
>
> 1. Promote cooperative relationships among artificial insemination businesses.
> 2. Standardize methods of procedure.
> 3. Develop and carry out a Code of Ethics.
> 4. Hold further national conferences of managers for mutual benefit.
>
> Be it resolved that an Executive Committee be elected to carry out plans of managers of artificial insemination businesses.

It was also voted, during this meeting, that the members of the temporary committee of the managers serve as the executive committee until the next annual meeting, at which time permanent officers would be chosen. J. W. Pirie, of Cedar Rapids, Iowa, was added to the executive committee, and Drake was appointed executive secretary. The responsibility of the executive committee was to draw up a plan of organization and to formulate a constitution and bylaws for the new organization. After a motion was made and carried that the resolutions be adopted, the NAAB was on its way.

8.5 The Guelph, Ontario, Meeting

On 23 June 1947, in connection with the annual meeting of the American Dairy Science Association at Ontario Agricultural College, in Guelph, Ontario, about seventy-five managers, college workers, and others attended what was called "a Managers' Meeting." Most members of the executive committee were present. At this meeting, the objectives for an organization, as outlined at the managers' first session in December 1946, were enlarged upon.

It was suggested that the name of the organization be "The National Association of Artificial Breeders." Provisions for membership, associate membership, voting privileges, quorum, officers, and finances were discussed and placed on the agenda for a meeting of the managers to be held 29 September 1947 at Cedar Rapids, Iowa.

The proposed outline for an organization, which was to be presented at the Cedar Rapids, Iowa, meeting was entitled "Tentative Outline of Proposed Organization of Artificial Breeding Work" and set forth the purpose of the proposed organization.

8.6 The Second Meeting of Managers and Directors[2]

The second meeting of managers and AI organization directors was held at Cedar Rapids, Iowa, on 29–30 September 1947.[3] J. W. Pirie, president of the Eastern Iowa Breeding Association, was in charge of arrangements. This might well be called the NAAB organizational meeting.

Considerable discussion developed as to whether the organization should include managers only or both managers and directors of member organizations. Another topic of concern was that managers and directors should hold separate meetings to discuss their specific problems and then meet together for a discussion of mutual problems. It was also suggested that a "manager's organization" be formed first and later an overall organization be developed if it was found to be needed. After a lengthy pro-and-con discussion as to whether the organization should be for "managers only," the original premise, or whether it should involve owners and directors of AI businesses, Maurice Johnson, of New York, moved that the organization should include managers and directors. The motion was seconded by William F. Schaefer, Jr., manager of the Northeastern Pennsylvania Artificial Breeding Coop. (NEPA), in Tunkhannock, Pennsylvania. The motion carried, and, as a result, NAAB was formed covering all facets of the then existing AI program.

Following the vote as to the type of organization, Sam B. Rose, of Ohio, moved that the membership be on a voluntary basis, with a simple set of bylaws. This motion was seconded by Kenneth Wallin, manager of the Badger Breeders Cooperative in Shawano, Wisconsin. The motion was carried.

The bylaws were adopted at this meeting by a majority vote:

National Association of Artificial Breeders—Bylaws

The fundamental purpose of artificial breeding of dairy cattle is to locate and spread the influence of sires known to have desirable transmitting ability.

Purpose:

1.a. To promote cooperative relationships between artificial breeding organizations through procurement and exchange of sires.
   b. Semen exchange in emergencies.
   c. Assist in procurement of supplies.
2. To facilitate an exchange of information through organization of publications and reports, and possible publication of a National Journal.
3. To study and recommend uniform procedures.
4. Develop and promote a "Code of Ethics."
5. To promote a closer working arrangement with the purebred associations through their purebred dairy cattle association.[4]

The name "National Association of Artificial Breeding" was also officially accepted and prevailed until 1964, when the word *Animal* was substituted for *Artificial*. It was also voted that "conception reports" (nonreturn rates) be on a 60–90-day basis based on first services. In the earlier years of the AI program, considerable attention was focused on the nonreturn rates reported by inseminators as well as that compiled for individual bulls in AI use. No doubt, at this time, interest in high nonreturn rates from AI was accentuated by the fact that many herd owners questioned whether or not it could equal the results generally obtained from natural service. As time passed, it was found that nonreturn rates, while a useful relative guide to cows in calf, were some 8 to 12 percent higher than actual pregnancies.

8.7 First Board of Directors and Officers

The duly elected directors and officers of the board as selected at the organization consisted of: Max Drake, of Ohio, president; John Fawcett, of Missouri, vice-president; J. W. Pirie, of Iowa, secretary-treasurer; Maurice Johnson, of New York; D. C. Master, West Virginia; Douglas Reid, New Jersey; and Kenneth Wallin, Wisconsin.

8.8 A Free Enterprise Program

In the beginning of the NAAB, it is noteworthy that the leaders were free to choose what seemed to be the most logical course for expansion in the use of AI. Programs for the use of AI had already been set up in many European countries before NAAB was founded. All of these were fairly well regulated by governmental edicts and agencies. In most countries, it is the policy that laymen should not be engaged in conduct of the AI program so far as the technical aspects, sire selections, and field services are involved.

The approach for AI in the United States received strong support and assistance from extension dairymen at the various agricultural colleges; college researchers who provided much of the technical information; and progressive dairymen who were experienced in herd improvement through the DHIA and Purebred Bull Association programs. There are no federal or state regulations on the AI program, except those that pertain to the health of sires that are transported into another state. Some four or five states have instituted a technician (inseminator) licensing act.

In general, the AI program has been self-regulating and the NAAB can claim much of the credit. The "Foreword," by Max Drake, of the first managers' meeting no doubt helped pave the way for a free-enterprise program. It reads as follows: "As we counsel together, let us keep in mind we are in a young and growing business. We can do what we will to move forward in our chosen field."

In these early endeavors, the contributions of many people, particularly dairy extension workers and county agents, should be recognized. As we trace the development of NAAB and the growth of the AI program in terms of states and organizations (Appendix A), many of the major leaders will be discussed.

Chapter 9: Growth and Progress of the Program and the NAAB

9.1 Introduction

The thirty-year period from 1944 to 1974 has been referred to by some as "the agricultural revolution." In no period of our nation's history has agriculture undergone so many technological economic and social changes. These developments have had a dramatic effect on the AI program. Likewise, the growth and development of the NAAB was influenced in no small part by the economic picture in the cattle industry—both beef and dairy—during these years.

The trend, as is well recorded, has been one of fewer and larger farms and fewer dairy herds but each with a greater number of cows.[1] In 1934, there were nearly 7 million farms averaging 155 acres in size. In 1950, there were about 5.4 million farms, averaging 215 acres in size in America. By 1974, the number had decreased to 2.6 million farms averaging 428 acres. At the time the AI program was initiated, there were many family-sized herds of five to ten cows. By 1970, these had nearly all disappeared. It is estimated that in 1950, there were 3.6 million farms with some milk cows.[2] By 1973, this number was estimated at 385,000 and steadily declining.

The change in dairy herd size is reflected in the number of cows in herds on the National Cooperative Dairy Herd Improvement Program. In 1940, there were twenty-four cows in a herd; in 1950, twenty-seven cows; in 1960, an average of forty-two cows; in 1970, sixty-two cows; and in 1978, seventy-eight head.

The change in dairy cattle numbers is shown in Table 2.3 (Chapter 2). The buildup in total cattle, largely beef, from 1930 forward is indicated in Table 2.2 (Chapter 2).

As is reflected in the table, the peak number of cows and heifers of breeding age (average seventeen months) was reached in 1945. Much of the increase in dairy cattle numbers occurred during the World War II years, when dairy products were in heavy demand. At the same time that dairy cattle numbers were declining, the number of cows artificially inseminated was increasing rapidly (Table 9.1). Following World War II, the increase in the number of cows on the AI program in the United States increased from 360,732 in 1945 to 2.6 million in 1950.

Many of the smaller dairy herds, when their owners were faced with the necessity of installing bulk tanks, building new milking facilities, and with more costly labor, simply sold out. A large number of such dairymen bred their milk cows to beef bulls and thus gradually switched to a beef operation. As indicated in Table 9.1, in 1961, it was estimated that 435,592 milk cows were inseminated with beef semen and in 1964, nearly 1.2 million. From 1950 to 1970, total cows in milk declined nearly 50 percent. Many inseminators were forced to seek other employment as the herds they had been servicing simply "dried up." It was in these years, too, that competition among AI organizations for business expanded.

Advancing technology, particularly the use of frozen semen, had a pronounced effect on the AI program. With frozen semen stored in single-dose units, shipping was facilitated, and since 1955 most AI businesses have become nationwide operators. The practice of selling semen by the unit or "dose" developed.

The decline in the number of dairy cows, the cost of converting to a frozen semen program, and reduced income forced many smaller organizations, particularly farmer-owned cooperatives, to merge or consolidate as a means of survival.

9.2 NAAB's First Annual Meeting

Approximately twenty-seven organizations were represented at the first annual meeting of the NAAB, which was held on 13, 14 September 1948 in Ithaca, New York. Officially, this meeting was the first to be called the "annual meeting" under the adopted name of the National Association of Artificial Breeders.

About one hundred persons registered, including twenty-seven managers. Also attending were representatives from the Ontario Association of Artificial Breeders, which marked the beginning of a long and pleasant working relationship between the AI organizations in Canada and those in the United States.

Plans were made at this meeting to incorporate the organization. The incorporating committee had previously met in Shawano in the spring of 1949, and, with the aid of the attorney for the Badger Breeders Cooperative, prepared the articles and bylaws for incorporation in Wisconsin.

J. W. Pirie, the secretary-treasurer, reported twenty-two paid-up members in the association. Income for the year amounted to $545.00, and the treasury showed a balance of $618.51. After some discussion, it was agreed that some means for paying operational expenses must be developed. The idea of assessments on the basis of first-service cows was discussed. The matter was left to the board of directors for further recommendations.

Table 9.1. Status of Artificial Breeding Program in the United States, 1939–1972

Year	Studs	Sires in Service — Dairy	Sires in Service — Beef	Sires in Service — Total	Sires in Service — Average per Stud	Herds[1]	Dairy Cows Bred to — Dairy Bulls	Dairy Cows Bred to — Beef Bulls	Beef Cows Bred to Beef Bulls	Total Cattle Bred	Cows Bred per Sire
1939	7	—	—	33	4.7	646	—	—	—	7,359	228
1940	25	—	—	138	5.5	2,971	—	—	—	33,977	246
1941	35	—	—	237	6.8	5,997	—	—	—	70,751	299
1942	46	—	—	412	9.0	12,118	—	—	—	112,788	274
1943	59	—	—	574	9.7	23,448	—	—	—	182,524	318
1944	56	—	—	657	11.7	28,627	—	—	—	218,070	332
1945	67	—	—	729	10.9	43,998	—	—	—	360,732	495
1946	78	—	—	900	11.5	73,293	—	—	—	537,376	597
1947	84	—	—	1,453	17.3	140,571	—	—	—	1,184,168	815
1948	91	—	—	1,745	19.2	224,493	—	—	—	1,713,581	982
1949	90	—	—	1,940	21.6	316,177	—	—	—	2,091,175	1,078
1950	97	—	—	2,104	21.7	409,300	—	—	—	2,619,555	1,245
1951	94	—	—	2,187	23.3	548,300	—	—	—	3,509,573	1,605
1952	94	—	—	2,324	24.7	671,100	—	—	—	4,295,243	1,848
1953	96	—	—	2,598	27.1	755,000	—	—	—	4,845,222	1,865
1954	93	—	—	2,661	28.6	805,000	—	—	—	5,155,240	1,937
1955	79	—	—	2,450	31.0	845,900	—	—	—	5,413,874	2,210
1956	79	—	—	2,553	32.3	900,400	—	—	—	5,762,656	2,257
1957	75	—	—	2,651	35.3	946,000	—	—	—	6,055,982	2,284
1958	71	—	—	2,676	37.7	975,372	—	—	—	6,645,568	2,483
1959	64	—	—	2,460	38.4	930,059	—	—	—	6,932,294	2,816
1960	62	—	—	2,544	41.0	910,000	—	—	—	7,144,679	2,808
1961	56	—	—	2,486	44.4	863,781	7,047,148	[2]435,592	—	7,482,740	3,010
1962	56	2,036	420	2,456	43.9	862,150	6,837,681	[2]911,006	—	7,748,687	3,155
1963	51	2,158	401	2,559	50.2	621,141	6,468,545	969,748	235,289	7,673,582	2,999
1964	50	2,140	398	2,538	50.8	654,311	6,165,599	1,117,395	464,959	7,747,953	3,053
1965	[3]46	1,867	449	2,316	50.3	591,859	6,301,178	963,657	615,147	7,879,982	3,402
1966	[4]44	1,949	439	2,388	54.3	540,265	6,413,453	873,127	695,181	7,981,761	3,342
1967	35	2,012	364	2,376	67.9	458,782	6,259,425	788,933	672,819	[7]7,871,265	3,313
1968	[3]33	2,028	352	2,380	72.1	407,375	6,423,786	714,850	795,242	7,933,878	3,334
1969	[3]30	1,955	390	2,345	78.2	387,979	6,590,147	694,916	924,381	8,209,444	3,501
1970	31	1,911	364	2,275	73.4	369,197	[5]6,693,216	615,322	[5]1,258,446	[8]8,566,984	[5,6]3,641
1971	[6]24	1,958	349	2,307	96.1	350,611	6,759,215	525,956	1,357,918	8,643,089	[6]3,620
1972	[6]26	2,167	347	2,514	96.7	—	—	—	—	—	—

Source: *Dairy Herd Improvement Letter* 48:4 (June–July 1972), ARS, USDA.

[1] Prior to 1963, number of herds largely reflected membership rather than those actually serviced.

[2] Probably includes some beef-to-beef inseminations.

[3] Includes one all-beef stud.

[4] Total cattle bred in 1967 includes 150,088 first services, by state only, where breakdowns were not reported.

[5] Revised data after publication of 1971 artificial-breeding report.

[6] Based on data where bulls were reported.

9.3 Study Committees Appointed

On 8 November 1948, the NAAB board of directors met in Chicago. Since concern was expressed regarding the lagging DHIA testing program, it was agreed that each AI organization should be encouraged to do everything possible to expand DHIA testing at the local level. At this meeting, standing committees were established in order to resolve some of the problems facing the association. The following committees were created: semen shipping equipment committee; semen preparation and preservation committee; inseminator's kits committee; bull management committee; research committee; inseminating techniques committee; nominating committee; and the incorporating committee.

9.4 Exchange of Information Emphasized

At the first annual meeting, the board urged each member association to cooperate with the standing committees by furnishing information and material for their studies. It was agreed that the committee reports would be published and sent to every member, after they were presented to the association at the annual meetings.

In these early years, there was a dearth of information concerning almost all phases of the applied AI program. For the most part, equipment such as shipping containers, inseminating tubes, artificial vaginas, and inseminator's kits were homemade. On another front, the organization and conduct of a successful AI program were being developed by experience. The information collected by these early committees and the free exchange of ideas among workers in the field enabled the NAAB and the AI program to progress.

Throughout the years, and still true today, the NAAB has had a strong array of committees and special representatives (see Appendix B). In addition to the information summarized and the programs fostered, the committees and special representatives have done much to bring about strong cooperation between the NAAB and every facet of the livestock industry, colleges of agriculture, state and federal authorities, and the farm press.

Beginning in 1949, the proceedings of the annual meeting, also designated as the annual convention, has been published each year. This practice has made a wealth of information available to the livestock industry in addition to providing a record of action taken at such meetings.

9.5 Second Annual NAAB Meeting

Dyersburg, Tennessee, was the site of the NAAB's second annual meeting on 17, 19 August 1949.[3] The total attendance as recorded amounted to 135 persons: forty-four member associations from twenty-six states had registered.

It was reported that an average of 22.6 bulls were maintained per organization. Only 9.4 bulls on the average were proved sires. The average 60–90 day nonreturn rate was reported to average 60.7 percent. Much attention was given to feeds for bulls, bedding, bull exercisers, frequency of semen collection, and similar items. Only five of the seventy organizations responding douched the sheath of the bull before collecting semen and only half washed the sheath. However, nearly all of the organizations reported that they regularly tested the bulls for brucellosis, trichomoniasis, and other infections.

Much discussion centered on the topic of semen preservation, including diluters (semen extenders), temperature of the artificial vagina at collection time, sterilization of equipment, evaluation of semen quality, and the best temperature for semen storage and shipping. The committee on insemination techniques emphasized in minute detail the sanitation and sterilization procedures that should be used for equipment.

The research committee presented a comprehensive report and listed twenty-two areas that needed investigation. These included semen diluters, "repeat" breeder cows, semen fertility evaluation, cardinal principles for success in operating an artificial-breeding business, and the need for shortcuts in determining the transmitting ability of a bull on a sound basis before he goes into AI use. The committee felt that the NAAB, at that time, was not in a financial position to do this research but should implement research by appealing to state and federal agencies.

9.6 First Articles of Incorporation and Bylaws Adopted

At the business meeting of the second annual meeting, Wallin, chairman of the incorporating committee, presented revisions to the proposed articles of incorporation and bylaws.

After minor alterations, the articles of incorporation and bylaws were adopted on motion by Kermit Carey, of Michigan, and seconded by W. R. Amidon, of New Hampshire.

On 17 September 1949, J. Rockefeller Prentice, president of the Wisconsin Scientific Breeding Institute, in Madison; Dr. Edwin Carlson, of the East Central Breeders Cooperative, in Waupun, Wisconsin; and Wallin met at Shawano to complete the incorporation papers by having them signed by three residents of Wisconsin. Wallin acted as recording secretary. At the 1950 annual meeting of the association, the articles and bylaws as amended were formally adopted. While revisions have been made in the articles of incorporation and bylaws from time to time, the guiding principles set forth in 1949–1950 have adequately served the association.

9.7 First Complaint on Unethical Practices

At the board meeting on 12 December 1949, a member of the NAAB appeared and charged that another NAAB member had moved into his organization's area, placed a technician in the territory, and was advertising for business.

Following discussion, the board decided to develop and present at the next annual meeting a "Code of Ethics" for the industry. While a "Code" had been discussed as a means of stating principles for fairness and honesty, and working relations among NAAB members, a formal statement had not been drawn up.

A committee consisting of Max Drake, Kenneth Wallin, and Fred J. Hatler was charged with the task of drafting a proposed "Code of Ethics."

At this meeting, the secretary was instructed to draft a suitable "certificate of membership" to be issued to each of the forty-seven-member organizations enrolled.

9.8 Cooperation With the Purebred Dairy Cattle Association Proposed

At a board meeting on 16, 17 February 1950, several letters and reports from the Purebred Dairy Cattle Association (PDCA) were reviewed. PDCA reported that it had been necessary for breed associations to return a high percentage of registry applications for "test tube" calves to the applicants for corrections. It was agreed to notify all NAAB members of this situation. Some discussion evolved as to PDCA and NAAB jointly employing a secretary to handle the affairs of each association. It was voted that the executive committee for each association (PDCA and NAAB) be invited to meet together to discuss the matter further.

On 21 June 1950, at a board meeting, at the home of Maurice Johnson, it was decided to draft two budgets: one relative to hiring a secretary jointly with PDCA; another relative to NAAB hiring a secretary to serve its needs and not in partnership with PDCA. It also developed that PDCA had raised some questions regarding the charging of special fees for AI service to certain bulls. It was decided to appoint a committee of three to work with Floyd Johnston, chairman of the PDCA committee on rules and regulations pertaining to the artificial breeding of registered dairy cattle. The following were appointed: William F. Schaefer, Jr., chairman; Richard Kellogg; and Marshall Carpenter. This was the beginning of a long working relationship between the PDCA and the NAAB.

9.9 Third Annual Meeting and Convention: Wisconsin

Fifty-six member organizations were represented at NAAB's third annual meeting of 21–22 September 1950, at Land O' Lakes, Wisconsin.[4] Three hundred people attended.

President Drake, in his opening report, called attention to the fact that the organizations present represented about 90 percent of all the artificial breeding being done in the United States:

> As you know, we are greatly varied in our organizational makeup. We are big and little, we are privately owned, cooperatively owned, with partnerships and corporations—not for profit—in between. Now it isn't always easy to fit an overall organization around such groups. In the past few years, however, we have come to know each other better. It is my belief that today the time is here for us to move forward together. As a national organization, we must consider ourselves mature and responsible for our own conduct. Any organization must sooner or later adopt its own rules of conduct and a sound financing plan—or perish. We are at that point.

The first topic of discussion was the matter of setting up provisions to finance NAAB. The treasurer's report showed that NAAB had a deficit of $263.98. There were no provisions, other than membership fees, to finance the organization. As a solution, it was decided that an assessment of member organizations, based on the number of first service cows, should be adopted. The board set assessments at one-half cent per first service cow for the fiscal year 1950–1951. It was agreed by the members to pay the secretary-treasurer of the organization $50 per month and allow one dollar per hour for secretarial help. All members also agreed that the expenses of the directors attending board meetings be paid.

The feasibility of an executive secretary jointly employed by PDCA and NAAB was also discussed. Johnson explained that the NAAB board had decided it would be impractical for one man to serve two organizations and felt that PDCA and NAAB should keep their lines drawn on business matters. He emphasized, however, that it was necessary for both organizations to work closely together in solving mutual problems. Howard Greene, of Wisconsin, moved that considerations for hiring a secretary jointly be dropped. The motion was seconded and carried.

Although it did not cover many situations that developed later, particularly as competition between cooperatively owned and proprietary, or privately owned, artificial-breeding organizations grew, the following "code of ethics" was presented at the meeting:

> **1.** All members shall conduct their affairs to carry out the aims and objectives of the National Association of Artificial Breeders.

2. All members will use only truthful statements concerning their own association and other associations.
3. Misleading as well as false statements will be considered as a breach of the "Code of Ethics".

Greene recommended and moved that the proposed "code of ethics" be adopted. It was seconded by Stanley Wiggin, of New Jersey, and carried. The first statement of the code proved to be only the beginning of an NAAB "code of ethics." As will be pointed out, frequent additions and deletions were made as technological advances and merchandising practices developed.

The various standing committees reported on semen preparation and preservation, inseminating kits, shipping containers, insemination techniques, and research. A special report by the research committee covered a survey of investigations underway on all phases of reproduction and AI at the various agricultural experiment stations.

It was agreed at this meeting that "an additional charge would not be added to the AI service fee as a provision for registration of offspring from special bulls." This practice violated the PDCA rules and was strenuously objected to by the dairy breed associations.

One of the speakers at the meeting, Prof. Raymond Albrectsen, of Cornell University, New York, presented a report on "Production Results from the Artificial Breeding Program." He compared results in New York where proved bulls and "analyzed young sires" were used. The proved sires produced daughters averaging 148 pounds of milk and 12 pounds of butterfat more than their dams. He emphasized the importance of evaluating bulls on a wide cross section of the dairy cow population and pointed out some of the pitfalls when a bull was proved in only one herd. The daughter averages of the "analyzed young sires" were nearly equal to those of the proved sires. This report, the first of its kind at an NAAB meeting, was the forerunner of many discussions on sire evaluation and sire proving as better selection tools became available.

9.10 Financing Plans Develop

Several meetings of the board of directors were held following the third annual meeting to deal with problems of finance and plans for the rapidly growing association. At a January 1951 meeting held in Chicago, cash assets of the NAAB were $106.20 and bills owed (largely as a result of the 1950 annual meeting) totaled $718.36. The secretary was advised to send every member a statement for assessments covering the fiscal year of 30 June 1950 to 1 July 1951.

At another board meeting held in Cleveland, Ohio, on 27, 28 May 1951, the treasury had a balance of $2,618.58 as a result of assessments, at the rate of one-half cent per first service cow, collected since the last meeting. It was voted that members who did not pay assessments within ninety days after the end of their fiscal year would be listed as delinquent.

On 6–7 August 1951, the board met in Montreal, Canada.

The Canadian representatives expressed an interest in developing a strong association of the AI interests in Canada. The fee and membership arrangement of the NAAB was discussed, as was the interest of the American and Canadian groups working together.

The finances had improved, with $5,568.27 free balance in the treasury.

It was also agreed that members were free to erect exhibits, regarding their AI program, at the NAAB annual meeting.

9.11 Fourth Annual Meeting: Massachusetts

The fourth annual meeting was held 23–26 September 1951, at Swampscott, Massachusetts.[5] Hosting the meeting was the Artificial Breeding Cooperatives of New England, perhaps more commonly known as the Artificial Breeding Council of New England and was made up of the artificial-breeding cooperatives from Massachusetts, Connecticut, Maine, New Hampshire, Rhode Island, and Vermont. The meeting was attended by about 250 persons. Fifty-seven AI organizations from thirty-one states were represented. There were sixteen guests from Canada.

At the business meeting, a budget was adopted and plans made for hiring a full-time executive secretary for NAAB. The budget called for an annual expenditure of $18,421.88, with $15,000 allocated for an executive secretary, including salary, travel, and office expense. The treasury showed a balance of $6,944.16 on 1 September 1951. The revised "code of ethics" was reviewed, item by item, and adopted on motion by A. C. Baltzer, of Michigan, and seconded by Lester Reynolds, of Connecticut.

The terms of directors for Johnson and Hatler had expired. Elected to the board were R. W. Seath, of Owatonna, Minnesota, and J. W. Nordyke, of Missouri. Officers elected were Wallin, president; Schaefer, vice-president; and W. R. Amidon, of New Hampshire, as secretary-treasurer.

9.12 Interim Board Action, 1951–1952

The board of directors met several times during the year. Items considered included changes in the PDCA uniform breeding receipt that would eliminate the necessity of the signature of the owner of a cow at the time of insemination; the appointment of committees; questions from the field regarding the code of ethics adopted in September 1951 (especially item 8, regarding the hiring of technicians employed by another organization); plans for hiring an executive secretary; and developing a program for the 1952 annual meeting.

It was agreed that the association should have a representative at the International Congress on Animal Reproduction and Artificial Insemination to be held at Copenhagen, Denmark, the summer of 1952. Drake was chosen as the representative to attend the Congress.

At a board meeting on 13–14 May 1952, held in Chicago, the treasurer reported a balance of $11,413.48. At that time, there were only four delinquent assessment accounts. It was agreed at this meeting that the association strongly support the National Brucellosis Program (a contribution of $1,000.00 was recommended to the NAAB delegates by the board in 1950).

9.13 1952 Annual Meeting: Missouri

The fifth annual convention[6] was held at Springfield, Missouri, on 21–24 September 1952, with the MFA (Missouri Farmers Association) Artificial Breeding Association as host. There were about three hundred people, from thirty-two states and Canada, in attendance. While the volume of cows artificially inseminated continued to grow, consolidations among some of the smaller AI organizations had begun; there were only fifty-four member organizations represented by delegates. J. Warren Nordyke, manager of the MFA Artificial Breeding Association, served as program chairman.

Considerable action focused on association affairs, presentation of committee reports, several speakers, and panel discussions on sire evaluation methods, but also a bit of "Ozark humor," and philosophy at the informal gatherings. President Wallin summarized the growth and development of the AI program from about 7,000 cows in 1938 to over 3 million head in 1951.

9.14 Business Meeting: One of Action

There was probably no annual meeting of the association equal to that held in 1952 with respect to basic decisions that would affect the future.

The code of ethics as prepared in 1951 was reviewed point by point. There was a great deal of discussion over the manner in which sire proofs would be published. The committee recommended that "in published daughter-dam comparisons or daughter averages, no daughters with completed records of 305 days or less will be omitted.'' (*Records* is interpreted to mean lactation). In the arguments that ensued, an amendment to the above statement was offered as follows:

> Since the use of the sire's index tends to mislead dairy farmers, indices shall not be used as published information except as recognized by the purebred breed registry associations, in published daughter-dam comparisons of daughter averages. All records will be used and factored to a 305–day–2X milking–M.E. basis.

J. Rockefeller Prentice, among others, strongly objected to the terminology of the amendment and raised the question: "Is it unethical to use any index or use any index not recognized by a breed registry association?" He favored the Equal-Parent Index and stated that his organization would continue to use it.

After lengthy discussion and debate, a motion was made that the amendment to the Code of Ethics be tabled and the matter be referred to the committee on the code of ethics and discipline. The motion passed. Item 5 then was adopted as originally presented. On motion, the revisions of the 1950–1951 code and additions were adopted.

> **1.** All members shall conduct their affairs to carry out the aims and objectives of the National Association of Artificial Breeders.
> **2.** No sire will be purchased or leased under any agreement providing extra payment if his daughters in the owner's herd achieve a stipulated production and/or type level.
> **3.** No sire will be put into service for payment of an additional service fee as a condition of registration of his offspring.
> **4.** The source and date of daughter-dam comparisons, or daughter averages will be indicated when published. No daughters with completed lactation records of 305 days or less will be omitted.
> **5.** Publication of non-return information will be on a 60–90 day basis, and will be labelled as such.
> **6.** No good purpose is accomplished by extravagant claims of the merits of sires, or of the efficiency of organizations. All members will use truthful statements concerning their own association and support such statements with adequate proof. The purchase price of sires used by N.A.A.B. members shall not be quoted in their printed statements.
> **7.** It shall be unethical for any member of this organization to interfere by unfounded criticism with techniques and practices developed by another member organization.
> **8.** No organization shall employ a person who has been dishonest or willfully negligent with the rules of The Purebred Dairy Cattle Association.
> **9.** Any complaints or questions on the interpretation of this Code shall be submitted to N.A.A.B. or authorized committee thereof, upon request of N.A.A.B.

10. If any of the provisions of this Code shall violate any State or Federal law, neither N.A.A.B. nor any of its members shall be found thereby or shall be deemed to have agreed thereto.[7]

This code remained in effect until September 1959. During these years, however, with increasing competition among organizations, Item 6 was frequently violated and was looked upon by some as illegal from the standpoint of limiting trade and denying individual freedom.

Generally speaking, the NAAB code of ethics has served as a guide for conduct among the NAAB members. It is a voluntarily enforced code. Experience over the years indicates that no NAAB member took any pride in the fact his organization was guilty of violations and much of the success of the free-enterprise system of AI in America is closely associated with the desire of the industry to "police itself."

The budget and assessments, as presented by the board at the 1952 meeting, came under great discussion and debate. The budget proposed increasing assessments to one cent per first service cow in order to hire an executive secretary, establish an office, and carry on the work of the association that was proving to be a heavy burden for the elected officers.

However, an amendment to the bylaws was proposed that would, in part, read: "And the dues shall be based upon the number of (first-service) cows inseminated by each member organization, and in no event shall exceed one cent per cow, or a total of $2,000." It was favored by J. Rockefeller Prentice, who, in part, said:

> I do not feel we can afford a 1¢ assessment, or a $2,000 maximum. If that assessment is raised to 1¢, we are going to have to spend $6,000. I do not approve of the way the money has been spent. I am speaking for the American Breeders Service. If the six studs we operate were members—if the higher budget is adopted, at 1¢ per cow, that is going to mean $6,000 which we cannot afford. Therefore, we will not, if that kind of budget goes through, with a $2,000 maximum be able, as an entire group, to support this work. I would like to see the maximum at $1,000.

A quick count showed that eleven other member organizations, either on a half cent or one cent assessment would be affected by a $2,000 maximum. There was prolonged discussion, with some of the cooperative members voicing opinion that those organizations that inseminated the most cows should be able to pay the most.

After further debate, it was voted to leave the bylaws as stated "and the dues shall be based upon the number of cows inseminated by each member organization and shall in no event exceed 1¢ per cow." Thus, the amendment was defeated and no ceiling placed on assessments.

Secretary W. R. Amidon read the proposed manual of functions for an executive secretary and the proposed budget for 1952–1953, which totaled expenditures of $25,000. The balance in the treasury at the time of the annual meeting in 1952 was $14,127.43. There was discussion on the pros and cons of hiring an executive secretary with suggestions that "he could be a part-time man"; "he can use his home for an office this first year"; and so forth. A motion was made by D. G. Wood, of West Virginia, that the board of directors be authorized to hire an executive secretary. George Ranney, of Vermont, seconded the motion. A standing vote indicated thirty-four delegates for and ten against. The motion carried.

A budget of $25,000 was adopted for the 1952–1953 fiscal year. It was also voted to appropriate $1,000 for the National Brucellosis Program.

At the reorganization of the board on 23 September 1952, W. R. Amidon was elected president, W. R. Nordyke, vice-president, and R. W. Seath, of Minnesota, elected as secretary-treasurer until such time as an executive secretary was hired. At this same meeting, the board voted to set assessments at three-fourths of a cent per first service cow.

At the board meetings following the 1952 meeting, it was decided on 6 November 1952 to hold the 1953 annual meeting in East Lansing, Michigan.

At a meeting of the board in February 1953, Prentice stated, in a letter dated 12 December 1952, that the business of the Northwestern Artificial Breeding Association, in Duluth, Minnesota; the Indiana Artificial Breeding Association, in Carmel; and the Southeastern Artificial Breeding Association, in Asheville, North Carolina, had been taken over by the American Breeders Service of Chicago. Because of this action, he requested that the membership of each one of these organizations be withdrawn from NAAB, to be effective 1 March 1953. Prentice's request was granted leaving only one of his organizations—Wisconsin Scientific Breeding Institute, in Madison—a member.

Chapter 10: The Developing Years, 1953–1958

10.1 First Executive Secretary Hired: Headquarters Established

At a meeting of the board of directors held on 9–10 April 1953 in Pittsburgh, Pennsylvania, several of the persons under consideration for the position of executive secretary were interviewed.

After some discussion, it was decided that an unofficial ballot be taken on which each director would list the candidates in order of his preference. As a result of this ballot, Max Drake moved that a unanimous ballot be cast for Harry A. Herman, professor of dairy husbandry at the University of Missouri–Columbia, for the position of executive secretary.

"Dr. Herman," as explained by Sam B. Rose, of Findlay, Ohio, and first historian for NAAB, "came to the organization with a background of experience involving over twenty-four years teaching and research in dairy production and field and laboratory experience in artificial breeding beginning in 1937. . . . He served for years as Secretary, or as Chairman, of the Purebred Dairy Cattle Association Research Committee. As early as 1938, he helped draw up the first form (the Between Herds Certificate) for use in registering cattle resulting from artificial insemination."

In July 1953, Herman took over the duties as executive secretary and proceeded to establish the association's headquarters in Columbia, Missouri. The first office was located in the Exchange National Bank Building, until 31 December 1956.

10.2 The NAAB News

The NAAB published its first newsletter, *The NAAB News,* in September 1953. It appeared just in time for the 1953 annual meeting. *The NAAB News* was the official publication of the organization for over three years. It was published quarterly the first year, then bimonthly until 31 December 1956.

On 1 January 1957, the name *NAAB News* was dropped and *A.I. Digest* was adopted. The *A.I. Digest* became a monthy publication and continued so until May 1977. Its circulation increased to about fifteen thousand copies per month. It was circulated throughout the United States and in nearly one hundred foreign countries.

The *A.I. Digest* carried industry news, reports of meetings, and activities throughout the industry; research findings and new methods; and breed association highlights. The publication also carried advertisements for supplies, cattle, and equipment.

Early in 1977, the name of the publication was changed to *Advanced Animal Breeder*. It is published nine times yearly.

10.3 The Sixth Annual Meeting: Michigan

The sixth annual meeting of the association was held on 13–16 September 1953 at East Lansing, Michigan. About 330 people registered, including 25 Canadian representatives, and one delegate from Puerto Rico, Felix O. Espada.

At this meeting, the research committee gave its first formal report on frozen semen.[1] There was great interest in frozen semen and many organizations were experimenting with it. The American Breeders Service in Chicago, Illinois, was the major organization in this field for the United States. Few people at this meeting realized the great impact frozen semen was to make on the AI program and on the operations of the NAAB in the years to come.

10.3.1 Research Support Voted

At this meeting, Dr. Herman made his first report as executive secretary and submitted two budgets. One budget called for assessments at three-fourths of a cent per cow as had been adopted by the board during the 1952 annual meeting in Springfield, Missouri. The second budget called for assessments of one cent per first service cow with one-fourth of a cent earmarked for support of research. The research program was to be administered by the NAAB research committee. The idea was to provide grants to aid research workers who were pursuing projects of special interest to the AI industry but needed additional financial aid. The second budget, favoring research support, was passed.

At the election of new directors and officers, the following members were chosen: E. D. Humphrey, of Ohio, and Kenneth Wallin, of Wisconsin, as directors; and J. Stanley Earl, of Unadilla, New York, as vice-president.

10.4 Research Program Policy Adopted

The board in 1953 established the following policy on the research program:

1. The chief function of this organization, from a research standpoint, will be to coordinate and correlate research in artificial insemination and ani-

mal reproduction so as to avoid needless duplication. The plan will be to work harmoniously with member associations and agricultural experiment stations in furtherance of experimental work of mutual interest.

2. To assemble and disseminate the results of research to member associations.

3. To supply limited funds for aid in carrying on research and specific projects needing activation.

4. To initiate research endeavors along lines most needed, but not being vigorously investigated, so far as the industry is concerned.

5. The research program will be administered by the Research Committee of the NAAB. It will be the function of this committee to plan and promote the program as outlined above.

10.5 Pertinent Board Action, 1953–1954

During 1953 and 1954, the NAAB initiated and organized many programs that provided service to member organizations and industry. These included (a) a revision of the bylaws to establish an associate membership including organizations outside the boundaries of the United States; (b) establishment of permanent committees; (c) authorized the supplying of *The NAAB News* (later the *A.I. Digest*) to members on a ratio proportional to assessments paid; (d) agreed to include paid advertisements (dealing with supplies, equipment, and cattle) in *The NAAB News*; (e) proposed awards for inseminators on the basis of cows serviced and years of tenure; (f) established a member newsletter contest and decided to encourage commercial exhibits at the annual meeting; (g) financially supported six of the nineteen research projects that were submitted before the board; and (h) approved twenty-one applications for membership in 1953–1954. A reserve fund, involving money not required for current operations of the association, was also established. In addition, the board approved the idea of adopting a suitable slogan and insignia for the association; the U.S. Department of Internal Revenue also declared NAAB as a tax-exempt organization on a "not for profit basis."

10.6 Seventh Annual Convention

The seventh annual meeting, or convention, of the NAAB was held 26–30 September 1954 in Harrisburg, Pennsylvania. More than 535 persons registered, making the 1954 convention the largest to date. Representatives were present from forty states, Canada, and Puerto Rico.

C. R. Henderson, of the Department of Animal Industry at Cornell University in Ithaca, New York, presented a paper entitled "Selecting and Sampling Young Sires." This paper is of historical importance in the annals of the AI industry. It marked the first time the importance of sampling young sires, and the dependability of AB[2] proof was presented in a comprehensive manner to the NAAB members. The young sire selection and sampling program suggested paved the way for the more sophisticated siring-evaluation methods now in use throughout the AI industry.

Another matter of discussion focused on frozen semen. The results of a survey indicated that twenty-one members of the NAAB were experimenting with frozen semen in the field. Wisconsin Scientific Breeding Institute, in Madison, led all organizations in cows inseminated with frozen semen, reporting twenty thousand cows with an average 60–90 day nonreturn rate of 50 to 68 percent for first services. Herman summarized the current research on frozen semen underway in the United States and Europe.

At this meeting, the association adopted a resolution urging that the term *diluter* be replaced by the term *extender* when referring to processed semen. Comments indicated that some dairy farmers were inclined to view "diluted semen" with suspicion, and the term itself suggested that some persons might construe it to mean an inferior product.

E. R. Carlson, of Waupun, Wisconsin, the NAAB representative to the American Veterinary Medical Association, reported on health regulations for bull studs as recommended by the AVMA committee. The recommendations covered general sanitary practices, bovine tuberculosis, bovine venereal trichomoniasis, and bovine vibriosis. The NAAB adopted the sire health recommendations as presented. These recommendations have been updated and revised over the years but have served as an invaluable guide, as well as lending much credence, to the AI program.

The first formal complaint filed by one member against another member for violation of the NAAB code of ethics was aired, in private session, by the board. Representatives of both organizations attended this session and presented their arguments. The point in question was the placing of inseminators by member A in the generally recognized territory of member B. Rather than take disciplinary action, under somewhat conflicting circumstances, the board recommended that the two parties, the NAAB executive secretary, and the dairy extension representatives in the state meet and arbitrate the situation. As a result, the matter eventually was settled amiably and fairly.

This incident proved to be the forerunner of many similar situations as competition between organizations developed and dairy cow numbers lessened. In practically every instance, difficulties were settled by arbitration. In a few cases, however, legal steps were taken. As time wore on, it became increasingly evident that in a free enterprise system no one had a claim on any patron's business except as he might maintain it through high-quality service, loyalty, and competitive pricing. This was a rather bitter lesson for some of the cooperative members to acknowledge. Gradually, the acceptance of the free enterprise system, which characterizes the AI program in America, came about. Complaints regarding violations of the NAAB code of ethics as related to the competition for business became fewer.

T. E. Patrick, of Louisiana, and F. G. Stevenson were elected as the new directors succeeding R. W. Seath and Paul Heller, of Connecticut. J. Stanley Earl, of New York, was elected president and Ben Simonson, of Oregon, was elected vice-president.

10.7 Activities and Conventions, 1954–1956

At a meeting of the NAAB board in Chicago on 9–10 November 1954, the following action was taken:

1. Final authorization to establish a reserve fund. Approval was given to start this fund by placing $10,000 with the Boone County Building and Loan Association, in Columbia, Missouri, at 3 percent interest.
2. Appointed a special committee, headed by William F. Schaefer, Jr., of Pennsylvania, to study "Means and Methods for Recognizing Meritorious Sires Used for Artificial Insemination."
3. Approved the encouraging of working relationships with AI organizations and programs in countries outside the North American continent.

10.8 Insignia and Slogan Approved

Earlier, it had been suggested that the NAAB have an insignia and a slogan. The executive secretary and the publications committee canvassed the membership for ideas. At the 1954 annual meeting, the suggested insignias and slogans submitted to that date, in what had developed as a "contest" among member organization employees, were displayed. There were eighteen ideas submitted for an insignia and seven ideas for a slogan. The delegates at the convention voted on their first, second, and third choices for an insignia and for a slogan. These results were reviewed by the NAAB board at its November 1954 meeting.

The judges for the publication contest had selected what they considered the first three insignias and the first three slogans among those submitted. The judges and the popular vote gave first place to the insignia submitted by Henry T. Soborg, of Southern Illinois Breeding Association, in Breese.[3]

Figure 10.1. Insignia and slogan adopted by the NAAB board of directors in November 1954. (Photograph courtesy of NAAB).

The insignia adopted, when used with color, shows the bright rays of the deep, golden-colored sun penetrating a dark blue sky. Soborg, in submitting this slogan said, "The design expresses the dawn of a new area in cattle improvement. The rising sun depicts the flooding of new light and increased intensity in breeding better cattle."

While used in various ways and sizes, this insignia symbolizes NAAB and is recognized worldwide (Figure 10.1).

The slogan adopted was submitted by Bion Carpenter, publicity director for the New York Artificial Breeding Cooperative in Ithaca, New York. The slogan submitted read "Dedicated to better cattle for better farm living." However, the board shortened it to read "Better

Cattle for Better Living'' and approved its adoption as the official slogan for NAAB.

10.9 Significant Developments and Board Action

At a meeting of the board held in Kansas City, Missouri, on 7–8 April 1955, the following action was taken. Regarding the matter of patrons of one organization desiring to use semen from bulls owned by another member organization, the board recommended that:

1. A member of the NAAB should not sell semen directly to a technician who is employed by, or is a part of the organization, of another NAAB member without mutual understanding being established between the two organizations.

2. Where patrons of a member organization desire special matings with semen furnished by another bull stud, it is suggested that arrangements for obtaining the semen and provisions for its use be worked out between the bull studs involved. This marked the first action to provide for selected matings without violating the then existing code of ethics of the NAAB.

3. The problems of NAAB member organizations in financial stress were discussed. The cost of adding equipment for frozen semen, maintaining a battery of bulls for at least five dairy breeds and some beef bulls—though the beef bulls were largely to breed milk cows—in a market becoming more competitive each day as well as a dwindling milk cow population severely taxed the resources and management ability of smaller organizations. It was decided that seasoned managers and directors be invited to attend meetings of members suffering management and organizational problems. No committee was appointed, but this move was the forerunner of the managers' committee and the directors' committee established in following years.

4. The PDCA regulations governing artificial insemination of purebred dairy cattle, involving the use of frozen semen, were reviewed and approved. These regulations, designated to become effective 1 January 1956, were the first frozen semen regulations to be adopted by dairy breed registry organizations in America.

10.10 Eighth Annual Convention[4]

The eighth annual convention was held 15–17 August 1955 at Seattle, Washington. Over four hundred persons attended the convention. There were delegates from thirty-nine states and Canada as well as visitors from China, India, Cuba, Puerto Rico, Philippine Islands, and British West Indies.

Roy Snyder, then manager of the Waterloo Cattle Breeders' Association in Ontario announced that in late 1954 and early 1955, his organization had changed to a 100 percent frozen semen program, the first organization to make the complete changeover on the North American continent.

In his annual report, the executive secretary noted that the association's roster included seventy-three members and ten associates. During the year, 5,155,240 cows in 593,190 dairy herds in the United States were serviced by the organized AI program. The *NAAB News* had reached a circulation of eight thousand copies per issue. The slogan and insignia adopted in 1954 were put to use, with the necessary engravings and mats furnished to each member.

At the business session of the convention, a resolution was adopted recommending to the Dairy Husbandry Research Branch, of the USDA, "that DHIA production averages, both AI and naturally sired animals combined, be published annually by state, by breed, by year." Another resolution that was adopted stated that "Therefore be it resolved that the National Association of Artificial Breeders revise its bylaws so that in the future this organization may be known as the International Association of Artificial Breeders." The purpose was to enable AI organizations in foreign countries to be members of the fast-growing NAAB. It was agreed that the motion should be studied by the board of directors and the executive secretary before any further action was taken. The final decision of the officers was not to change the bylaws but to permit foreign-based organizations to become associate members.

In the reorganization of the board the following officers were elected: J. Stanley Earl, reelected president; F. G. Stevenson, of MFA Dairy Breeders in Springfield, Missouri, vice-president; and H. A. Herman, as secretary-treasurer. At a later meeting of the board, these three officers were elected to constitute an executive committee. This was the first such committee for the association.

10.11 Developments of 1955–1956

The AI program in the United States continued to gain popularity. On 1 January 1956, there were 661,497 herds enrolled in 1,502 local AI associations. At this time, many AI organizations were small, maintaining only a few bulls. It was common practice, too, that local breeding units be designated as "artificial breeding associations." There were seventy-nine bull studs, as we designate them today, in actual operation. Of this number, seventy-four were members of NAAB.

Thirty-six percent of the sires in AI use were listed as "proved" on the basis of Daughter-Dam Comparisons. The dams averaged 10,575 pounds of milk containing 436 pounds of butterfat. The daughters averaged 11,265 pounds of milk and 476 pounds of butterfat for a superiority of 690 pounds milk and 40 pounds of butterfat. For the most part, sire proof was based on naturally sired daughters.

Following the 1955 annual convention, the board of directors and the NAAB staff were concerned with the following developments:

1. The format of the 1955 annual meeting met with considerable favor, in that it encouraged member participation in panel discussions, had recognized speakers outside the AI field to discuss current topics, and allowed people in attendance a reasonable amount of time to visit with each other. It was agreed that the general format of future meetings would be along these lines.

2. In September 1955, the International Charolais Association in Houston, Texas, approached the NAAB office seeking help in establishing regulations for the use of AI for registered cattle. This approach was the first from a beef registry organization to utilize AI and to seek assistance from NAAB. The executive secretary of NAAB met with the Charolais Association officers several times and rules governing AI were drafted.

3. Ten research projects were approved with a total of $12,200 expended for grants. It was agreed to support the research on *Vibrio foetus* at Cornell University.

4. The executive secretary was empowered to expand the *NAAB News* to "meet the demands of the industry."

5. The matter of canceling PDCA Inseminators Approval was becoming a problem at this time. The man disapproved by one organization often went to work for a competing business. The rules adopted by PDCA in 1954 called for reasons for dismissal of an approved technician to be stated and if charges were not upheld, the technician could be reinstated. In some cases, technicians took legal steps and threatened to sue organizations that dismissed them. It had been voted at the 1954 annual meeting that the NAAB office circulate a list of dismissed technicians to the NAAB membership. The board of directors, realizing the danger of legal proceedings, decided that "as a service to NAAB members, technicians dismissed *for cause* by a member organization may be reported to the executive secretary's office, and these names will be circulated to members."

This plan was followed and was assisted by Karl B. Musser, secretary of the Purebred Dairy Cattle Association, who sent the NAAB office a list of names of all canceled technicians that were added to those submitted by members. As time went on, this plan "backfired" because competition among organizations became greater, and often a technician was canceled because he considered changing employers.

6. A committee consisting of Wallin and Stevenson was appointed to study a retirement program for NAAB employees of long tenure. This committee reported in April 1956, suggesting that longtime employees be eligible for retirement at sixty-five years of age at one-half their annual salary including social security. This plan as such was not adopted but led to the adoption of a retirement plan in 1959.

10.12 1956 Annual Convention

The 1956 annual convention was held 19–22 September at St. Paul, Minnesota. Over five hundred representatives from thirty-four states and seven foreign countries attended the meeting. Delegates from the British West Indies, Costa Rica, Puerto Rico, Israel, and South Africa were also present, in addition to twenty-four delegates from associates and members in Canada. The program emphasized the promotion of the AI program, the reorganization of the association to make it function more effectively, the tax situation for cooperatives, sire health, sire recognition, and testing for production.

L. O. Gilmore, of the Ohio Argicultural Experiment Station at Wooster urged the AI organizations to set up programs to screen calves and to trace each abnormal calf to its sire; to study family backgrounds before purchasing a bull; and to eliminate a bull promptly once he was found to carry undesirable recessives.

William F. Schaefer, Jr., on behalf of the sire recognition committee, presented recommendations for recognizing outstanding AI sires and sought the cooperation of the USDA, dairy breed registry organizations, and the Purebred Dairy Cattle Association. After lengthy discussion, it was decided to delay action of the USDA until after the sire summary program was developed, as well as the action of dairy breed associations before putting the proposed NAAB plan into effect.

Discussions of semen preservation indicated expansion in the use of frozen semen, and at least thrity-five bull studs and universities were conducting extensive investigations on frozen semen.

Resolutions were adopted to (a) encourage cooperation among bull studs and breed associations in detecting and reporting inherited defects; (b) rescind the resolution adopted at the 1955 annual meeting stating that, when a patron of one bull stud wants semen from a bull in another organization, provisions be worked out for its procurement by the bull studs involved (declared to be impractical); (c) establish a recognition program for AI technicians; (d) express commendation for the *NAAB News* and other communications emanating from the NAAB office.

At this meeting, an informal agreement was reached between the NAAB and the USDA Dairy Husbandry Research Branch, which enabled the USDA to receive semen from NAAB member-owned bulls for research purposes. This action paved the way for utilization of sires from NAAB member bull studs in the USDA herd at Beltsville, Maryland, and other USDA herds.

Another important action was adoption of the 1954 and 1956 recommendations on sire health and management of the American Veterinary Medical Association. These recommendations, while updated from time to time, became the "Code of Minimum Standards—For Health of Bulls and Hygiene of Bull Studs Producing Semen for Artificial Insemination" that has now served the industry for over twenty years. Leaders in developing these recommendations were E. R. Carlson, Waupun, Wisconsin; David E. Bartlett, De Forest, Wisconsin; E. J. Weatherby, College Park, Maryland; C. R. Reeds, Tor-District, Ontario; Greg Raps, Des Moines, Iowa; W. D. Goeke, Hampshire, Illinois; H. J. Hill, Fort Collins, Colorado; and B. W. Kagy, Tiffin, Ohio.

The report from the international relations committee was made by Cesar Estrada Gonzalez, of Puerto Rico. This newly appointed committee sought to encourage working relationships, exchange and sale of semen, and exchange of information relative to the AI program on an international scale. In 1956, the secretary-treasurer attended two meetings in Washington, D.C., at the request of Ezra T. Benson, secretary of agriculture, to consider rules and regulations for the export of cattle and both liquid and frozen semen. The committee also suggested that a part of the *NAAB News* be printed in Spanish and French.

Elected directors at the 1956 annual meeting were Frank Heim, Jackson, Michigan, and Richard Kellogg, Columbus, Ohio.

10.13 Expanding Program: Long-Range Plans, 1956–1958

Following the 1956 annual meeting at St. Paul, Minnesota, steps were taken by the NAAB board and officers to establish additional standing committees for the organization. With the expanding activities of the association, it was agreed that committees should be operating in "all major fields of interest." The necessary committees were established and the pattern of permanent and temporary committees has been followed since (see Appendix B).

The pressure of an expanding program against a fixed income was becoming more evident each day as the NAAB broadened its scope of activities. During this time, too, the continued decline in numbers of dairy cows and the competition among organizations were causing some "belt tightening" for most members. Many were still questioning the positive effects of using an all frozen semen program. At a board of directors meeting held on 28–29 October 1956 in Detroit, the executive secretary presented a summary of the status of the AI program and of the NAAB and urged adoption of long-range goals for the AI program and the association.

The report was approved by the directors, and a motion was made that it be implemented as rapidly as possible, insofar as finances and manpower permitted. Other pertinent decisions of the board, during the year, were:

1. Requested the sire recognition committee to exhaust all possibilities to draft a workable program, in cooperation with the dairy breed registry organization, that would be set up on a breed basis and to take breed averages into account.

2. Endorsed a tentative plan, under discussion in many areas of the AI industry at the time, for the NAAB and the breed registry organizations to investigate possibilities of a joint "Breed-NAAB" program for recording and reporting sires known to transmit inherited defects. (This program failed to get started because of the legal involvements possible.)

3. Established a committee made up of board members to "study the practicability of awards at the national level for outstanding AI technicians (inseminators)." Paul Heller, manager of the Connecticut Artificial Breeding Association in Woodbridge, was named chairman.

4. Took action that, beginning 1 January 1957, the official publication of NAAB be designated as *The A.I. Digest* instead of the *NAAB News*.

5. Agreed that an assistant secretary, particularly to assist with the *A.I. Digest* and the growing workload of the organization, be hired as soon as the budget permitted and as soon as a suitable man could be located. Also, empowered the executive secretary to procure additional office space and to hire additional secretarial assistance as needed.

6. It was agreed by the board, in August 1957, that a resolution be drawn up to prohibit the quoting of the price paid for sires in *A.I. Digest* advertisements. This was done and became a part of the code of ethics.

10.14 The 1957 Convention

On 11–14 August 1957, the tenth annual convention was held in Toronto. This was the first meeting of the NAAB to be conducted outside the United States. The program at this meeting focused on better breeding. In his opening remarks, President Earl stressed the fact that only about 26 percent of the dairy cow population in North America was enrolled in the AI program. He said "our greatest challenge is to increase the percentage of cows on this program and reduce the number of scrub bulls in use."

Another point of discussion pertained to researching the AI program now in progress. Dr. N. L. VanDemark of the University of Illinois reported on research work that indicated the possibility of developing a method of semen preservation at room temperature.

Charles Krumm, manager of the Southern Wisconsin Breeders' Cooperative in Madison, Wisconsin, and Lantz Womack, of Winnsboro, Louisiana, also president of the Louisiana Artificial Breeding Cooperative in Baton Rouge, were elected to the board of directors. In the reorganization of the board, Richard Kellogg, manager of the Central Ohio Breeding Association in Columbus, was elected vice-president. The makeup of the directorship involved four managers and three directors of AI cooperatives.

A record number of 760 people from six Canadian provinces, thirty-one states, and also India, Pakistan, Turkey, and Puerto Rico registered for the convention.

Chapter 11: Two Decades of AI Progress

11.1 Introduction

History portrays the progress of programs and peoples. All too often history deals with exploits and the acts of individuals and ignores the mechanics and technology that contribute to the attainments of man. After nearly twenty years of existence, it was becoming evident that the use of AI as a means of cattle improvement was here to stay. An increasing number of cattle breeders, educators, breed officials, and editors referred to AI as "man's most valuable tool" in increasing the genetic potential for greater income.

Beef breeders, too, were beginning to use AI, particularly to produce sons from outstanding "popular" sires, but the beef registry organizations as a group continued to object to the use of AI. An increasing number of beef bulls were added to bull studs during this period, however, as dairymen were converting to beef by beef dairy matings. A total of 256 beef sires, not including Milking Shorthorns, was reported by NAAB members as utilized in 1957–1958. Accurate estimates are not available, but probably well over 500,000 milk cows were inseminated annually with beef semen. Most of the beef bulls in use by AI organizations at this time were mediocre in the eyes of beef producers. Dairymen wanted a "white-faced calf" or a "black calf" and merit of the sire used was often not seriously considered.

11.2 AI Growth Continues

During 1957, there was a gain of about 6 percent in the total number of cows that were inseminated in the United States (Table 9.1). Frozen semen played a role in the increase as it now was becoming possible for a dairyman to "use his choice of a bull" almost any time he wished.

One NAAB member, the American Breeders Service, Inc., in Chicago, reported over 1 million cows serviced. New York Artificial Breeders Cooperative, in Ithaca reported 438,000 first services, to lead all cooperatives in volume. Ten states reported over 40 percent of their dairy cows and heifers of breeding age serviced by AI. Maine led with 48.6 percent of its dairy cows enrolled; and Wisconsin, with 1,095,312 cows—42.9 percent of the total—led with the most cows serviced per state.

Membership in the NAAB had increased to 105 organizations or individuals. There were seventy-one organized bull studs and sixteen privately owned cattle breeders or rancher operations selling semen. There were eighteen associate members in six foreign countries.

11.3 New Headquarters and Larger Staff

With the expansion of the AI program and increased activities of the NAAB, the space rented from the Exchange National Bank in July 1953 became inadequate. On 2 January 1957, the association headquarters was moved to 8 North Ninth St. The building occupied provided about sixteen hundred square feet of space, all located on one floor, and easily accessible to the post office, the bank, and the printer.

Two additional full-time stenographer-secretaries were employed, along with a part-time associate editor. Dr. Paul Fisher, professor of Journalism at the University of Missouri–Columbia, acted on a part-time basis as associate editor and consultant for the *A.I. Digest,* up until 1962.

11.4 *A.I. Digest* Advertising Policy

Inquiries were frequently received from member organizations requesting advertising space in the *A.I. Digest* to offer services and semen for sale. A survey was made of the membership. About half of the members were opposed to carrying advertisements for semen or service in the *Digest*. The feeling expressed was that it should not be a vehicle for competition among members. The board of directors considered the situation from all angles. Recognizing that some revenue needed to further the *Digest* would be eliminated, the board took this action on 23 October 1957:

> It is agreed that no advertising offering semen for sale by private breeders, or artificial insemination businesses, or semen producing businesses, should be accepted for publication in the *A.I. Digest*.

It was also agreed, however, that the NAAB might run advertisements in the *Digest,* under its own name, encouraging semen sales to other countries. This advertising policy, with little modification, has been in effect ever since.

11.5 NAAB Technician Awards Established

A committee consisting of directors Paul Heller, Frank Heim, and Richard Kellogg presented recommendations for the recognition of outstanding technicians to the board, which were adopted as follows:

1. A technician must be nominated for recognition by the organization that he works for.

2. The organization must be a member of NAAB.

3. Requirements for qualifying for an award:

a. Insemination of at least thirty-five thousand first service cows. The total may be accumulated from employment in more than one organization. (A record of first service by years must accompany the nomination.)

b. Twenty years of service. This may involve service with more than one organization.

c. Must be actively employed as a technician (inseminator) at the time of nomination.

4. Nominations must be submitted to the NAAB executive secretary.

5. The award shall consist of a suitably inscribed certificate and an attractive lapel pin.

6. The award cannot be duplicated in each category.

7. A picture and a short resume for each technician recognized under this program will be carried in the *A.I. Digest.*

This plan was approved by the voting delegates at the 1958 convention. The awards program was expanded from time to time as will be related later. It has proved to be one of the most rewarding and stimulating programs adopted by the NAAB (see Appendix B).

11.6 The 1958 Annual Convention

The 1958 annual meeting was held 17–20 August at New Orleans.[1] Over six hundred people attended, from almost all states in the United States, Canada, Puerto Rico, South America, and Iran.

The following members were elected as new officers of the association: Richard Kellogg, as president; and Charles Krumm, vice-president.

R. E. Hodgson, chief of the USDA Animal Research Branch, reviewed the many projects underway in researching cattle improvement. At the time, twenty-eight bull studs were involved in a cooperative agreement that would enable semen to be furnished from their best proved sires for breeding cows in nineteen branch experimental herds that were operated by the USDA.

Herman reported over $28,000 expended for research on association projects by his organization in 1957. These projects included vibrio control, sex determination, semen preservation, evaluation of sire-transmitting potential, sire mortality, and evaluation of environmental factors affecting production.

Resolutions were also adopted (a) encouraging more adequate support for DHIA testing and sire proving; (b) approving a national ear-tagging program so that one ear tag would suffice for DHIA, brucellosis, and AI identification and eliminating the "Christmas tree" program by which cows carry three or more tags; (c) encouraging breed association programs to report inherited defects of a deleterious nature; and (d) to study and revise NAAB bylaws affecting membership and voting power.

11.7 Bylaw Changes: Convention at Atlantic City, 1959

During the fiscal year of 1958–1959, the NAAB board of directors considered the role of the association in the growing AI program.[2] Several members had expressed the feeling that there was disparity in the dues paid by bull stud members and other members. Over the years, and particularly with the development of the frozen semen program, private breeders were in several instances offering frozen semen for sale and doing their own insemination work rather than collaborating with one of the established bull studs. The directors felt that accepting members whose interests were not in common with the objectives and purposes of organized bull studs posed problems of service and control. They also stated that "membership should be extended only to an organization, maintaining a bull stud for the primary purpose of semen production for artificial insemination of dairy cattle, under contract with the Purebred Dairy Cattle Association and inseminating not less than 5,000 first-service cows per year, and approved by the NAAB board of directors."

It was further agreed that membership may be extended to other organizations on an associate basis, "providing they are in the insemination business, semen production business, or allied services, and providing such associate membership appears conducive to the best interests of the Association, and the annual minimum fee for such membership shall be $50.00 per year."

This discussion was in accord with a resolution adopted at the 1958 annual meeting which requested "that the president be instructed to appoint a study committee of not less than five members to study qualifications for membership."

11.8 Study Committee Evaluates Status of NAAB

President Kellogg appointed the entire board as a study committee, to be assisted by F. G. Stevenson, of the MFA Dairy Breeders in Springfield, Missouri, and K. E. Wallin, of the Badger Breeders Coop., in Shawano, Wisconsin. The study committee recommended that:

1. All types of private enterprises should be given fair consideration in election of the NAAB board of directors. The viewpoint was expressed that since both privately owned and cooperatively owned artificial breeding businesses make up the NAAB membership, all are part and parcel of the free enterprise system. Further, the philosphy should be that it matters not whether a man represents a cooperative or a privately owned (proprietary) business—or whether the business he represents breeds 5,000 or 500,000 cows annually; if he is recognized as one who is competent and can make a contribution to the NAAB and the AI industry, he deserves consideration for the board of directors.

It was also suggested that a system of caucus of delegates by regions for nomination of directors at the NAAB convention should be considered.

2. It may be desirable to consider delegate representation on the basis of the volume of first-service cows.

3. The Committee recommended the objectives of the *A.I. Digest* as a house organ and trade magazine, primarily to furnish industry information, on a self-paying basis. However, such a publication should not attempt to move toward the area of a general farm publication.

11.9 Bylaws Revised

At the 1959 annual meeting held 30 August–2 September at Atlantic City, New Jersey, the board of directors presented the following amendments and changes in the bylaws and code of ethics that were adopted by the voting delegates:

Section 2. There shall be two membership classifications:

a. *Members*—Eligibility shall be based upon an organization or an individual producing semen used for insemination of not less than 5,000 first-service cows, or other farm animals, annually.

b. *Associate Members* (*Non-Voting*)—Membership may be extended to organizations or individuals on an associate basis, providing they are in the semen producing business, inseminating business or allied services. However, no organization or individual shall be accepted as an associate member if qualifications for membership are met, unless the board of directors finds such associate membership to the best interests of the Association.

Discussion evolved as to the status of several businesses, now in the membership category, but producing semen for less than five thousand first service cows annually. It was generally agreed that "voting membership" should encompass only bull studs and such was the intent of the original bylaws. Those members below the five thousand cow limit would drop voting rights and be classified as associate members if they desired. At the time these changes were adopted, there were eighteen semen suppliers, beef breeders, and custom freezing businesses that could not qualify as voting members.

There were sixty-seven bull studs in the United States and Canada, all of which qualified, and twenty associate members in Canada and seven foreign countries.

Membership Meeting—Article V, Section 1, was changed to read "The Annual Meeting of the members and associate members of the Association shall be held during the late summer or early fall of the year."

Section 7. Voting
Revised to provide voting power in accord with the volume of cows serviced as follows:

"Each member shall be entitled to one voting delegate for the first 250,000 first services or any part thereof in excess of 5,000 first services; two voting delegates for the first 500,000 first services; and, three voting delegates for over 500,000 first services, with semen produced by the organization, on each question submitted for vote or candidate to be elected.

In the election of directors, each member shall have the right to vote for as many persons as there are directors to be elected.

In order to vote, a member must have at least one representative at the meeting; and such representatives must be selected either from the board of directors or from the management personnel of the member.

"Associate members shall not be entitled to vote."

Considerable debate developed over the changes of Section 7.[3] Some felt that, since NAAB was incorporated under the cooperative laws of Wisconsin, the "one member one vote" interpretation should be followed. This point was cleared with the NAAB attorneys and the proposed changes were found to be legal. There were objections that "the big organizations would control the Association" and some delegates were opposed to voting rights that were confined to "directors or management personnel" of the member. Most delegates, however, felt that as bull studs became fewer, because of mergers and consolidations, that the voting delegate body should be fairly large. Section 7 was adopted as revised by a vote of thirty-one in favor and thirteen in opposition.

Article VI—Directors, Section 2, was revised by the addition of *Duties.*

"The Directors shall carry out the desires of members and associate members. The Directors are empowered to establish and maintain a national headquarters, employ an Executive Secretary and other personnel necessary to conduct the affairs of the Association."

The second clause was added to stipulate more fully the authority vested in the NAAB board.

Under Section 4, *Qualifications,* the word *manager* was changed to *management personnel* so the statement read "only management personnel and directors of member organizations shall be eligible for nomination to the board of directors."

Item (e) Section 7, Article VI, *Notice of All Regular Meetings,* was revised to read "shall be given each director by the secretary at least thirty

(30) days previous to the time fixed for the meeting." The original bylaws stated five (5) days notice.

Also, it was voted Section 4, Article VI, on *Qualifications* of Directors, to insert "the Directors shall be elected by secret ballot, from the persons so nominated." While voting by secret ballot had been the practice, it had not been stipulated in the bylaws until this action was taken.

Article VIII—Dues and Finance, Section 2, was revised to read in part "The dues shall be based upon the number of first service inseminations involving semen produced by each member organization, and shall in no event exceed one cent per first service except the minimum shall be $25.00 per year."

Article XI—Indemnification of Officers and Directors was revised to be more specific in protecting each director or officer, or former director of the association, against liabilities arising from his being a member of the board or an officer of the association.

In connection with the bylaw amendments, considerable discussion ensued relative to making it mandatory that all segments of the industry be represented on the board. Sam B. Rose moved that the bylaws read that the board be composed of at least one member from management, one member from the farmer-director group, and one from privately owned operations. A. H. Forsythe moved that in order to insure representation from each segment of the AI industry, the board be increased to nine members and three persons—one from each group—to be elected annually.

It was the consensus that legislating directors on the board was not the way to proceed and that the objectives outlined could be accomplished with a seven-man board as easily as with nine directors. Both motions were defeated.

11.10 Code of Ethics Amended

As experience was gained in the AI program, as competition increased, and as more people became involved, the interpretation of the code of ethics produced a variety of opinions among the NAAB members. Certainly this was not an unexpected situation in the growing free-enterprise program that was beginning to assume some maturity and stature. While sometimes abused, the code of ethics had become a respected set of rules for guiding this self-regulated new industry.

The following code changes were adopted at the 1959 annual meeting:

1. Item 2, of the code, was amended to read as follows:
   No sire will be purchased or leased under any agreement providing

extra payment if his daughters in the owner's herd achieve a stipulated production and/or type level.

2. Items 4 and 5 of the code were combined to read as follows:

The source and date of daughter-dam comparisons, or daughter averages will be indicated when published. No daughters with completed lactation records of 305 days or less will be omitted.

3. Item 6 was amended to read:

No good purpose is accomplished by extravagant claims of the merits of sires, or of the efficiency of organizations. All members will use truthful statements concerning their own association and support such statements with adequate proof. The purchase price of sires used by NAAB members shall not be quoted in their printed statements.

The change was adopted without a dissenting vote.

4. Item 7, formerly Item 9, of the code was amended to read as follows:

It shall be unethical for any member of this organization to interfere by unfounded criticism with techniques and practices developed by another member organization.

11.11 Competition for Business

The adoption of Item 7 of the code represented a considerable change in philosophy on the part of the NAAB members. Item 9 included the statement "interference with or by proselyting of technicians employed by other organizations, or in any manner unfairly disrupting another organization." The "proselyting of technicians" statement was questioned for some time. J. Rockefeller Prentice vowed it to be illegal and said that he would ignore it. This he did. It was not unusual for the executive secretary to receive a phone call, early in the morning, from a disturbed member organization manager stating "ABS has hired my technician in x county and is supplying him semen" and "What's NAAB going to do about it?" In other cases, competing organizations placed a technician in the same area serviced by another member. Legal advice procured by the NAAB made it clear that there are no laws under a free-enterprise system that prohibit another party offering a man a job.

These events led NAAB member organizations, in some cases, to place technicians under contract. Such contracts could be enforced—and were—in several instances. As competition increased, the number of dairy cows were declining in many areas and the use of frozen semen spread. Binding organization-technician contracts became fewer. Times were getting difficult from a financial standpoint for many small AI cooperatives. Frozen semen permitted great expansion in the area served, and the better-managed, better-financed, and more progressive organizations invaded new territories. In most cases, AI cooperatives respected each other's trade territories. Privately owned businesses were considered "fair game" by the cooperatives. In turn, the privately owned businesses had to solicit their customers in areas already occupied, in most cases, by cooperatives. While not fully realized at the time, the revision of the NAAB code of ethics represented a maturing process for the organization. As would be inevitable in a free-enterprise system, the doors were opened for the wide-scale competition among AI businesses that prevails today.

11.12 The 1959 Annual Meeting: Atlantic City, New Jersey

The twelfth annual convention was held in Atlantic City, New Jersey, on 30 August to 2 September 1959. Nearly six hundred people from almost all states in the United States, Canada, Greece, Guatemala, Japan, and Jamaica attended.

John W. Bartlett, chairman of the Dairy Department at Rutgers University, gave a report entitled "Twenty-One Years of Artificial Breeding in America" (Table 11.1). According to Bartlett, "On a national basis, daughters of proved sires used in artificial insemination exceed the dams by 678 lbs. of milk and 37 lbs. of fat. They stand about 50 percent above all cows."

One of the highlights of the convention was a presentation given by Enos J. Perry,[4] a retired extension dairyman of Rutgers University, who is fondly referred to as the father of the cooperative AI movement in the United States. He gave a report called "Artificial Insemination Around the World." Results were presented from a summary of questionnaires returned by fifty-five countries indicating that over twenty-one million cows had been bred artificially in these countries.

The resolutions presented by Phil Higley, chairman of the resolutions committee, were adopted as follows: (a) urge the USDA to provide stronger support for the DHIA program and sire evaluation; (b) appeal to each member organization to set aside one-fourth cent to one-half cent for each first service cow to aid the NAAB coordinated research program; (c) commend the *A.I. Digest* and the role it was playing in the AI program; and (d) commend the job the executive secretary was doing in administering affairs of the association.

Table 11.1. Growth of the AI Program, 1939–1959

Year	Cows	Herds	Sires	Cow per Sire
1939	7,359	646	33	228
1949	2,091,175	316,177	1,940	1,078
1959	6,932,294	930,000	2,544	2,816

Source: *Dairy Herd Improvement Letter* 38:3, USDA.

Chapter 12: Years of Adjustment, 1960–1966

12.1 Introduction

During 1959 and 1960, the organized AI program in the United States was in its twenty-second year. There were 111 NAAB member organizations. These included 62 voting members and 36 associate members in the United States, 11 in Canada, 3 in South America, 2 each in Germany and South Africa, and 1 each in Israel, Netherlands, British West Indies, Mexico, and Australia. It was estimated that these organizations serviced about 15 million cows annually.

In the United States, 6,932,294 cows and heifers were inseminated during 1959, representing about 32 percent of the nation's dairy cattle of breeding age (see Table 9.1). There were 2,460 bulls kept in 67 U.S. bull studs. Of these, more than 300 were beef bulls. An increasing number of milk cows were inseminated with beef semen as dairymen quit milking.

The price of AI service had remained nearly stationary since the beginning of the NAAB. The usual service rate was $5.00 to $7.00, which included semen and technician services. It was customary to provide up to three services if necessary to get a cow in calf for this fee. The service fee was gradually increased as AI organizations made adjustments to the economy.

12.2 Impact of Economic Change

Beginning shortly after the end of World War II, changes in the agricultural economy were a vital force in shaping the progress of the AI program in the United States. The rate of growth in terms of new cows that were enrolled in the AI program became more gradual. Two factors were responsible for this gradual growth: (a) dairy herds were going out of business, and (b) new herds enrolled on the AI program and the remaining herds were becoming larger. From 1946 to 1955, milk cow numbers dwindled from 27 million to 21 million head (see Figure 12.1). Dairy farms declined by 22 percent during this period and were becoming fewer each year. At the same time, all costs of operations for the AI organizations were increasing. The financial squeeze on many smaller AI cooperatives was severe and the way was being paved for many mergers.

Members of the NAAB combated rising costs by applying im-

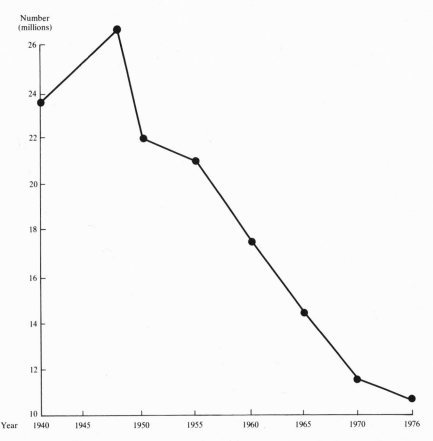

Figure 12.1.   Milk cow population during 1940–1976 in the U.S. Source: Data from agricultural statistics (1979), USDA, Washington, D.C.

proved technology, enabling more cows to be bred per sire, greater extension (dilution) of semen, enlarged service areas, and more cows inseminated per worker. At the same time, the use of frozen semen was expanding, selected matings were becoming the accustomed practice, and more AI-proved sires were in use. Dairymen were getting improved service and better sires at about the same cost as twenty years earlier. Meanwhile, almost everything else a dairyman had to purchase was nearly double in price. The AI industry was due for some drastic changes in pricing of service as well as in the mode of operation. From 1958 to 1960, consolidation, or sale, of bull studs reduced the number from seventy-two to sixty-seven. The direct sale of frozen semen to herd owners increased and more technicians were employed on a straight commission basis than had been the practice to date.

12.3 Assessments on Direct Semen Sales

With the increasing use of frozen semen, some NAAB members had started selling on an ampule or "semen dose" basis to technicians and distributors. Some semen was being sold directly to herd operators for insemination of herds under their control. Training courses for "do-it-yourself" insemination were sponsored by several NAAB members. The ampule of semen as a trade unit solved many problems: (a) it was easy to deliver and could be stored for weeks in liquid nitrogen; (b) inseminators usually inseminated only one cow per ampule (reducing the practice of reporting only a part of the cows inseminated with liquid semen as was sometimes done by dishonest operators); (c) accounting was simplified as an operator bought a given amount of frozen semen and the sales transaction was completed.

However, many of the semen distributors and technicians, some acting as independent operators, did not send individual cow-breeding receipts to the parent organization. This presented a problem in payment of the one-cent-first-service-cow assessments to the NAAB. President Richard Kellogg, in October 1959, appointed a special committee to set up a means for estimating such assessments. The committee consisted of P. I. Higley, American Breeders Service, Inc., chairman; Alvin Reuwee, Noba, Inc.; and Harry A. Herman, NAAB. This committee presented the following resolution at the 1960 annual meeting.

Section 2, Article VIII of the Bylaws of the National Association of Artificial Breeders, Inc., entitled "Dues and Finances" states in part that

Now, Therefore, Be It Resolved That, starting with dues payable for the year 1960, where semen is sold on a per ampule basis and where breeding receipts are not written or accurate data are not available for determining the number of first service inseminations, dues be paid on semen involved in such transactions at the rate of two ampules sold being equal to one first service.

The resolution was adopted.

The expanding use of frozen semen had resulted in a greater exchange, or sale, of semen between NAAB member bull studs. The question of dues payment to the NAAB on such semen had arisen. It was generally conceded that dues should not be collected twice on the same unit of semen. Therefore, the special committee introduced the following resolution that was adopted.

Now, Therefore, Be It Resolved That, where one bull stud owning member sells semen to another bull stud owning member for use where breeding receipts are written, they be allowed to determine which one of the two, either the seller or the buyer, shall pay the dues on the semen involved in the transaction and that the executive secretary of NAAB be informed in writing by the semen producing organization that the buyer is paying the dues on the first service inseminations involving such purchased semen, should this be the case.

The above resolution served as a guide for many years with the semen-buying bull stud subtracting the units of semen purchased from another NAAB member when paying dues.

12.4 Sire Evaluation Program Developments

Beginning as early as 1956, the NAAB sire recognition committee, headed by William Schaefer, Jr., had been striving to develop a program that would especially recognize the outstanding sires used in AI. Difficulties in assembling data on sires, keeping summaries up to date, and accounting for area environmental differences posed problems that the committee felt could not be properly handled by NAAB facilities. Furthermore, the workers in the USDA Dairy Herd Improvement Association had begun a sire evaluation summary. Some of the dairy breed registry associations had also established a sire recognition program.

Therefore, the committee recommended to the NAAB board that it consider disbanding the committee and instead consider establishing a new committee composed of sire analysts. The committee would be known as the sire evaluation committee and its function would be:

1. to cooperate with breed associations in connection with their AI sire evaluation programs and sire recognition programs, present and future;
2. to cooperate with the USDA in studying the evaluation of the new DHIA sire-proving program, and possible future modifications;
3. to continue the study of AI sire recognition and the possibility of a workable program as new techniques and data are available;
4. to act as a liaison for the NAAB board of directors and members in studying and disseminating the latest techniques of sire evaluation and sire-proving programs.

The recommendations of the sire recognition committee were adopted at the 1960 annual meeting. Thus, the sire evaluation committee was established.

12.5 NAAB Retirement Program Adopted

The executive committee, with Charles Krumm serving as chairman, was given the responsibility of developing a retirement plan for NAAB employees. After evaluating numerous plans and with due

consideration for NAAB finances and its half-dozen employees, the
board took the following action:[1]

> It was agreed to establish a pension trust fund, self administered, for
> NAAB employee retirement. This fund to be known as The NAAB Pen-
> sion Trust Fund. The plans are for about 10 percent of the annual payroll
> to be placed in the fund each year. Investments to be made in reliable
> interest bearing securities. Details of employee participation and retire-
> ment payments to be worked out by the NAAB attorney, auditor and the
> secretary-treasurer.

It was voted to appoint Harry A. Herman as trustee of the pension
fund and to name the NAAB executive committee as the pension fund
committee. The retirement plan was cleared through legal channels,
including the Internal Revenue Service, and became effective 30
November 1959.

The NAAB Retirement Pension Purchase Plan, as it became officially and legally known, served satisfactorily as a hedge against inflation for many years. About 1968, with a decline in the market value of
stocks and some bonds, the net value of the pension trust fund suf-
fered, although interest and dividends stayed at approximately the
same level. In 1975, the NAAB board of directors converted the
self-administered plan to an annuity type of retirement program. This
program, handled through one of the large life insurance companies,
is now in effect.

12.6 The 1960 Annual Convention

The 1960 annual convention[2] was held 21–24 August at Louisville,
Kentucky. Over six hundred people attended, representing thirty-five
states, Canada, and several foreign countries. M. N. (Matt) Dietrick
of American Breeders Service in Chicago, Illinois, was elected director for a three-year term. He succeeded Charles Krumm, who retired
as director. Lantz Womack, of Winnsboro, Louisiana, was reelected
for a three-year term. Richard Kellogg of Columbus, Ohio, was reelected president, and Womack, vice-president.

12.7 AI Program Developments, 1960–1961

As of 1 January 1961, it was reported by the USDA Statistical
Reporting Service that 7,144,679 dairy and beef cows were artificially
inseminated through the organized AI program in 1960. A rough estimate would indicate nearly one-half million milk cows and beef cows
inseminated with beef semen. With a population of 22,488,000 dairy

cows and heifers of breeding age (seventeen months old) reported, 31.8 percent were bred artificially.

12.7.1 AI Proved Sires Segregated in Reporting

During 1960, about 40 percent of the bulls used in the AI program in the United States were "proved" sires. However, the production proof included both AI and non-AI sired daughters.[3] Evaluation was based on a comparison of milk and butterfat records on a daughter-dam basis, as had been the custom for many years. In 1959, there were 963 proved sires in use, representing 38.4 percent of the total owned by bull studs. This was nearly four times as many proved sires, percentage wise, as in pre-AI days when only 8 to 10 percent of the dairy sires in use were proved. The segregation of AI progeny-proved bulls indicated that the average production level of their daughters was lower than had heretofore been believed.

Sire Comparisons: AI and Non-AI. Before 1962, sires for which dam and daughter records had been compiled were designated by the USDA as proved sires. Both AI and non-AI progeny were grouped together for the compilations prior to 1960. That year the USDA separated AI and non-AI progeny in assembling sire summary data. Also, the term *summarized sire* was adopted rather than *proved sire* that was subject to varied interpretations. Table 12.1 indicates the production levels when progeny proof was based only on AI daughters with completed DHIA records.

When the production of the non-AI progeny was omitted from the sire summary, the "proved" bulls, based on AI progeny only, dropped both in number and milk production. This difference was to be expected because most of the breeder-owned proved bulls had been used in only one herd and usually had been mated with much better-than-average cows. The report stimulated action on the part of NAAB member organizations to facilitate "young sire sampling" and

Table 12.1. Proved Sires in Service

Year	Number Proved	Percent of Total Proved	Dams		Daughters	
			Milk	Fat	Milk	Fat
			(lbs.)	(lbs.)	(lbs.)	(lbs.)
1959[1]	963	38.4	10,918	451	11,493	482
1960[2]	477	18.8	10,790	437	10,792	443

Source: *Dairy Herd Improvement Letter* 37:3 (March 1961), USDA.
[1]Proved sires in AI service including *both* non-AI and AI progeny.
[2]Proved sires in AI service including *only* AI progeny.

the development of "AI proved sires." Adding impetus to the need in AI use for sires transmitting higher-production potential was the steadily improving production per cow in American dairy herds. A summary of the 1960–1961 DHIA records, USDA, indicated that the average for DHIA-tested cows had reached an all-time high of 11,045 pounds of milk and 428 pounds of butterfat. This production per cow represented an increase of 718 pounds of milk and 27 pounds of fat over 1959 when the average was 10,327 pounds of milk, 401 pounds of fat. Milk production per cow was increasing about two percent a year. The situation called for AI bulls proved for high levels of production. NAAB members accepted the challenge, and the emphasis on AI-proved bulls with a high "plus proof" has become greater each year.

12.7.2 Increase in Mergers

The expanding use of frozen semen placed most of the bulls used in AI in a national role. In the days when the distribution of semen was limited to several counties and state areas, the use of local popular bulls and of bulls proved in a single herd by a good breeder was sufficient. However, this picture was progressively changing. Semen was being shipped from coast to coast, even exported to foreign countries, by NAAB members. The small bull stud was soon competing with the larger bull studs that owned more higher ranking proved bulls. The small stud also was unable to finance an extensive young-sire-sampling program where several breeds were involved, where cows on test were limited, and where operating costs were increased. This situation resulted in numerous mergers of AI organizations. These mergers started as early as 1950, as the number of studs decreased from ninety-seven to sixty-two by 1960.

Pooling Bull Power. Several NAAB members met the demand for higher-production bulls and for the reduction of operating costs by exchanging or, in some cases, purchasing semen from other members. One such group was Federated Breeders Foundation, which was formed in January 1961, by six cooperatives: Minnesota Valley Breeders (New Prague); Cache Valley Breeders Association (Logan, Utah); Evergreen Northwest Breeders (Burlington, Washington); Oregon Dairy Breeders (Corvallis, Oregon); and Nebraska Dairy Breeders (Fremont, Nebraska). Each organization kept its identity and continued its same service area but had more outstanding sires available.

An earlier organization set up for the pooling of frozen semen from the best bulls was United Breeders, Inc. Members consisted of Central Ohio Breeding Association (Columbus), a cooperative; Northern

Ohio Breeding Association (Tiffin), also a cooperative; and Eastern Iowa Breeders (Cedar Rapids), a privately owned AI business.

The practice of selling, or exchanging, semen, particularly among cooperative NAAB members, grew steadily and is in wide practice now.

On 1 January 1961, fifty-six bull studs were in operation because of the mergers. Several states, as will be discussed in Appendix A, were closing out their bull studs and were buying frozen semen from studs in other states.

12.7.3 Beef AI Grows

There were 435,592 dairy and beef cows inseminated with beef bull semen by AI organizations in 1961, according to the USDA-NAAB reports.[4] The majority of the cows and heifers inseminated were no doubt beef dairy matings, as many small dairymen were "converting to beef" by breeding their milk cows to beef bulls. A substantial number of beef producers, however, particularly purebred breeders, were using AI. It is interesting to note that the Triple E Ranch in Holopaw, Florida, started AI in its 3,700 cow herd in 1959–1960.[5] In February 1961, the ranch management announced it would sell all its bulls and use AI entirely. The reason was given as "the benefit of more rate-of-gain-tested-bulls and more pasture for the cow herd."

As other beef producers were showing interest, more emphasis was being placed on the use of "performance tested" beef bulls. The NAAB members purchased many of the highest indexed bulls that were enrolled in the performance or rate-gain testing programs. The sale of semen from such bulls slowly began to find a market among progressive beef producers (see Chapter 4).

12.7.4 George Nichols: Assistant Secretary

In July 1961, George Nichols, field editor for Crow Publications, in Modesto, California, was employed as the assistant secretary for NAAB.[6] Nichols, a graduate of Colorado State University, Fort Collins, served fourteen years as the field secretary of the Ohio Holstein-Friesian Association. His duties with NAAB included acting as editor and advertising director for the *A.I. Digest*.

12.8 The 1961 Convention: Syracuse, New York

Over 450 persons, from thirty-two states, Canada, Japan, Puerto Rico, and Jamaica, attended the 1961 annual convention at Syracuse, New York, on 22–23 August.

Melvin L. Kenley, of Cary, Illinois, chairman of the resolutions committee, presented five resolutions all of which were approved. They dealt with (a) conflict of annual meeting dates for NAAB, the AVMA, and the American Institute of Cooperation; (b) further consideration of proposed plans by the dairy breed organizations to increase blood-typing fees; (c) increased support for DHIA testing and sire proving; and (d) Code of Minimum Standards for the Health of Bulls and Hygiene of Bull Studs.

Having served on the board for two full terms, Jay C. Harris, of Burlington, Washington, and Paul Heller, of Montpelier, Vermont, were ineligible for reelection as directors. Earl Groff, of Strasburg, Pennsylvania, was reelected. Elected for terms of three years each were Laurel Behnke (Clintonville, Wisconsin) and Norman Allen (Schaghticoke, New York).

12.9 An Expanding Program

The use of artificial insemination continued its steady gain. Reports from the USDA and NAAB indicate that during the 1961 calendar year 7,482,740 cows and heifers over two years of age, in 863,871 herds, were bred artificially by the AI organizations in the United States.[7] This involved about 37 percent of the dairy cows in the nation.

12.9.1 Beef Bull Use

For the first time, the number of cows inseminated with beef bull semen was tabulated separately from dairy. During 1961, there were 435,592 inseminations with beef semen. Most AI organizations labeled a "beef insemination" if beef bull semen was used and "dairy insemination" if dairy bull semen was utilized. The greater portion of "beef" inseminations were to milk cows during these years.

Of the 2,456 bulls maintained by U.S. AI organizations, 420 were beef and 2,036 were dairy. An average of 3,010 cows were inseminated per bull, the highest number to date.

12.9.2 Sire Health Code Revised

During 1962, the use of healthy sires and of disease-free semen was emphasized. At the 1961 annual meeting, a resolution was adopted that required all members, both voting and nonvoting, who were engaged in the operation of a bull stud and/or custom collection of bull semen to comply with the AVMA-NAAB "Code of Minimum Standards for Health of Bulls and Hygiene of Bull Studs Producing Semen for Artificial Insemination" by 1 June 1962.

12.9.3 Swine AI Inaugurated

In May 1962, Roy Snyder, manager of the Waterloo Cattle Breeding Association in Ontario, inaugurated a swine-AI-breeding service. This project came about as the result of successful field trials, utilizing Ontario Veterinary College boars, and by June 1962, over two hundred sows had been artificially inseminated with an 80 percent conception rate. Litters averaged eleven to twelve pigs. The project was supervised by C. A. V. Barker, of the Ontario Veterinary College in Guelph.

While the Ontario workers were quite optimistic over the results, they explained that swine AI had its limitations because no satisfactory method for storing boar semen for long periods had yet been perfected. The Ontario investigators were at this time attempting to develop a satisfactory method for freezing boar semen. Efforts to freeze swine semen were also underway at various agricultural experiment stations in the United States, but no AI organization had yet offered routine service for sows.

12.9.4 Goat AI Service Offered

During the summer of 1962, the NAAB cooperated with the American Milk Goat Record Association in setting up an AI program, utilizing frozen buck semen, for dairy goats. The freezing of buck semen was demonstrated by Barker, who aided in the project. Ten registered dairy goat bucks, supplied by breeders in the United States, were shipped to Barker's laboratory in Ontario for the collection and processing of frozen semen.

The Central Ohio Breeding Association and Jessup Breeders, located in Artesia, California, arranged to store and distribute the frozen buck semen to other NAAB member organizations. The goal was to make frozen buck semen available to thousands of milk goat owners throughout the country. There was, and continues to be, a strong demand for such service.[8]

The project, however, was not too successful. Some of the bucks sent to Ontario were low in fertility—only about half of those involved produced semen satisfactory for freezing; goat owners, particularly the back-lot goat keeper with only one or two does, were not experienced in detecting heat and in getting insemination made at the opportune time. The small volume of business, and considering the problems encountered in educating goat owners, made it unprofitable. Although some fifteen hundred to two thousand does were satisfactorily inseminated during 1962 and 1963, the project was discontinued.

12.9.5 Midwest Genes: Merger

One of the largest mergers of NAAB member bull studs to date occurred 1 May 1962, when four Midwest cooperatives joined to form the Midwest Genes, Inc., in Anoka, Minnesota. The merging organizations were Consolidated Breeders Cooperative (Anoka, Minnesota); Iowa Dairy Breeders (Des Moines); MFA Dairy Breeders (Springfield, Missouri); and Nebraska Artificial Breeders Association (Fremont).[9]

As will be discussed in Appendix A, Midwest Genes, Inc., was later merged with Badger Breeders Cooperative to form the Midwest Breeders Cooperative, in Shawano, Wisconsin.

12.10 1962 Annual Meeting: Cedar Rapids, Iowa

The 1962 annual meeting was held 19–22 August at Cedar Rapids, Iowa.

The meeting was attended by nearly five hundred persons from thirty-five states, Canada, and two foreign countries. Forty-five AI organizations were represented.[10]

At the opening session, President Kellogg stressed three points fundamental to the progress of the AI program: (1) the importance of good public relations; (2) the need for aggressive and determined effort to expand the industry; and, (3) the necessity of operating in accordance with the NAAB code of ethics.

Kenley, chairman of the resolutions committee, presented eight resolutions all of which were adopted during this meeting. These included: (a) an agreement to use the PDCA recommendation that the term *registered* rather than *purebred* be used when referring to animals officially recorded by a breed registry organization; (b) to encourage NAAB members to conduct a progeny test for harmful and lethal genes before bulls are placed in AI service; and (c) to request stronger support for the NAAB coordinated research program.

Having served on the board for two full terms, Kellogg and Heim were ineligible for reelection at this convention. Elected to the board for three-year terms were Vincent Jessup, of Artesia, California, and Reuben Peterson, of Fremont, Nebraska. Lantz Womack was elected president and M. N. Dietrick, of Chicago, vice-president. These two men plus Laurel Behnke, of Clintonville, Wisconsin, and Harry A. Herman, executive secretary, comprised the 1962–1963 NAAB executive committee.

12.11 Fiscal Year 1962–1963

The volume of AI continued to increase slightly during 1963 with 7,673,582 cows serviced.[11] Dairymen were breeding an increasing number of milk cows to beef bulls, totaling to 969,748 such matings for the year. This figure amounted to 13 percent of the total milk cows serviced. For the first time *beef-to-beef* matings were recorded separately by USDA and NAAB, with a total of 235,289 cows for the year (Figure 12.2). It is estimated that some of the milk cows listed as bred to beef bulls were beef cows as some organizations termed it "a beef mating" if beef semen was used and "a dairy mating" if dairy semen was used. Some ranchers were using a dairy bull cross to raise milk yields for nursing. Beef-dairy crosses, the F_1 generation, were demanding a premium in many areas at this time.

With only 6,468,545 dairy-to-dairy inseminations, the smallest number since 1958, NAAB member organizations were beginning to financially suffer because of this decline. In most areas, AI technicians were the ones who were affected the most. It was about this time that the do-it-yourself program, whereby a diaryman was sold semen and trained to inseminate cows, accelerated.

During 1963, the Dairy Herd Improvement Association testing program reached an all-time high with 1,320,423 lactation records reported. According to the USDA, about 17.1 percent of the nation's dairy cow population was on some form of the DHIA test.[12] The percentage of cows on test *having reported records* increased from 43 percent in 1950 to 66 percent in 1963. The more complete reporting of records is largely attributed to the development of central processing of DHIA records. This program made possible a more complete and accurate sire-evaluation program—so vital to the AI program in its expansion of the young-sire-sampling program.

President Kellogg, at one of the 1962 NAAB board meetings, proposed a uniform coding system for bulls in AI use. With the exchange and sale of frozen semen, confusion developed because many organizations had bulls with the same code number. It was agreed that any coding system adopted should be suitable for the electronic processing of records. It was decided to pursue this matter further. Details of the coding system adopted some years later appear in Appendix B.

The executive secretary, in meetings with the USDA Foreign Agricultural Service (FAS), had urged that the sale of semen, as well as breeding stock, be encouraged for export. The FAS at that time was oriented to cattle sales. The NAAB board of directors questioned the wisdom of the NAAB becoming too heavily involved in a program that involved sales. In the earlier days of the NAAB program, the sales of service and semen were largely the functions of each member. Plans, nevertheless, were made to continue working with governmental agencies to explore further possibilities for increasing semen export. These plans involved preparing a circular that would be printed in several foreign languages. This leaflet would give directions for export and import of semen, include health requirements for bulls involved, and provide a list of the breeds (dairy, beef, goats) available from NAAB member bull studs. A general release, to be published in several foreign languages, was prepared by the FAS for use by agricultural attachés. Some frozen semen was exported to developing countries, but the volume was not great. The project was postponed until facilities, necessary for the utilization of frozen semen, became available in the underdeveloped countries.

12.12 Purchase of a Building for NAAB Headquarters

At a meeting of the NAAB executive committee held 30–31 October 1962 in St. Louis, Missouri, the executive secretary advised that the lease on the present headquarters had expired 31 August 1963. He said that "the present building is fast becoming too small for our needs—has inefficient storage space, limited working room, is poorly ventilated and is too cold in the winter. In addition, it will soon need a complete redecorating job."

The committee discussed the problem in detail and concluded that the association either would have to (a) construct a new building; (b) purchase an older building that could be remodeled; or, (c) lease a large building that would eventually be remodeled to meet the needs of NAAB.

It was felt that some money from the NAAB reserve fund, totaling now $50,000, could be used to finance a building.

On motion by Laurel Behnke, seconded by "Matt" Dietrick, it was voted to request the president to appoint a committee that would study the situation and would make the necessary recommendations to the board.

Lantz Womack, the newly chosen NAAB president, designated the following to the NAAB building committee: J. Stanley Earl (chairman), Melvin L. Kenley, Richard Kellogg, Robert E. Elsas (ABS, Chicago, Illinois), Marshall Carpenter (KABA, Louisville, Kentucky), and Kenneth Wallin.

A list of fifteen properties, with description, was forwarded to the building committee. One property, the W. B. Smith Building, 512 Cherry Street, came on the market a few days later at a selling price of $31,500. Realizing that it offered the best site yet presented, the executive secretary placed $700.00 of his own money in escrow and took an option on the building from 28 January to 15 February 1963.

The building committee met in Columbia on 30–31 January 1963. Due to conflict in dates, only Earl, Elsas, and Kenley were able to attend this meeting. They appraised the facilities at 10 N. Ninth Street and "deemed them inadequate and not in keeping with the scope and dignity of our industry," and, therefore, the committee decided to discontinue expending funds on someone else's building.

The committee recommended to the NAAB board that the W. B. Smith property be purchased. Previously, the property had been appraised and found to be moderately priced. Inspection by a builder and real-estate agent found the building structurally sound. By a unanimous vote of the board on 15 February 1963, it was voted that the purchase be completed and the executive secretary be empowered to carry out the details including the necessary remodeling.

The remodeling got underway 1 July and was completed 10 October 1963. The NAAB office was moved into its new facilities. Three downstairs offices were rented augmenting the NAAB income from $1,800.00 to $2,400.00 a year.

12.13 1963 Annual Meeting: Los Angeles, California

The 1963 annual meeting was held 25–28 August 1963 at Los Angeles, California.[13] Over 450 delegates from thirty-three states and Canada attended the convention, representing about forty-seven AI organizations. In his address to the convention group, President Womack lauded the AI industry for its high-caliber leadership, based on free enterprise, during an era fraught with governmental controls.

The resolutions approved by the voting delegates, as presented by Kenley, included: (a) expression of appreciation to the hosts for a most successful convention; (b) expression of thanks to Womack, the retiring NAAB president; and, (c) the renaming of NAAB, which involved deleting the word *artificial* and substituting it with a more appropriate one.

Another matter of discussion centered on the budget. With the new building to be paid for and with increasing costs of operation, the 1963–1964 budget was closely scrutinized. For the fiscal year ending 30 June 1963, the association had an income of $108,996.00 with $82,848.00 from assessments and membership fees. Expenses, not including the cost of remodeling (which had not yet begun), totaled $100,775.00, thus providing an operating margin of $7,221.00. This margin had been accomplished through some "belt tightening" on expenses and added assessment reserves. A total of $28,391.00, with

$15,000.00 from NAAB funds, was allocated for eleven research projects during the year. The net worth of NAAB was reported by the auditor to be $96,144.20.

Elected to the board of directors for three-year terms were W. H. Armstrong (Rocky Mount, Virginia) and Kenley. Behnke was elected as the new NAAB president, and Earl Groff (Strasburg, Pennsylvania) as vice-president.

12.14 Board Action to Aid Industry Problems

Following the 1963 annual meeting, the executive committee met in Chicago on 5 November 1963.

There was discussion regarding the steady decline in dairy cattle numbers, the growing interest in beef AI, and the use of performance-tested sires. It was agreed that, with only 40 percent of the dairy cattle population involved in the AI program and with beef AI beginning to develop, much more effort should be directed toward the expansion of the AI program.

On motion by Groff and seconded by Jessup, it was voted to establish a sales and promotion committee. This standing committee was responsible for promoting the use of AI. Roger Eming, representing the Southeastern Artificial Breeding Cooperative, served as chairman. The other committee members included John Heckman (Armour B.C.I., in Denver, Colorado), Harold Broman (Curtiss Breeding Service, in Cary, Illinois), Lyle Jackson (Southern Illinois Breeding Cooperative, in Breese, Illinois), M. J. (Red) Neverman (Eastern Iowa Breeders, in Cedar Rapids), and Wallace Miller (Minnesota Valley Breeders Cooperative, in New Prague).

At another board meeting in Chicago on 17–18 July 1964, it was agreed to amend the bylaws so that "the one vote, one organization," so prevalent in most farmer-owned cooperatives, would no longer apply. A proposal, which was to be presented at the 1964 annual meeting, suggested the following number of voting delegates for a member:

    5,000 to   49,999 first services, one voting delegate
   50,000 to   149,999 first services, two voting delegates
  150,000 to   349,999 first services, three voting delegates
  350,000 to   699,999 first services, four voting delegates
  700,000 to 1,000,000 first services, five voting delegates
  Over 1,000,000 first services, six voting delegates

The necessity for increasing voting delegates was dictated by the fact that the smaller farmer-owned AI cooperatives were struggling for volume and, in many cases, were merging to prevent financial failure. Meanwhile the privately owned businesses, showing greater capital, more business expertise, and claiming a nationwide operating area, were growing in size. In 1946, and in the early beginning of NAAB, farmer-owned cooperatives were inseminating about 95 percent of the cows on the AI program in the United States. By 1963, cooperatives serviced only 4,776,957 head or 62 percent of the 7,763,382 total.

It also appeared that assessments from members would be lower in 1963 and 1964 than in the previous year. This point worried the board, as did the decline of ads placed in the *A.I. Digest*. The reduction in advertising, which had provided considerable income for the *Digest*, was largely due to the discontinuing use of mechanical freezing equipment, once utilized for frozen semen. Most organizations had, or were, switching to liquid nitrogen as a refrigerant. It was decided by the board that while the *A.I. Digest* served a most useful function, efforts should be made to keep the cost above income, not including not including staff labor, to not over $2,000.00 per year.

The board was also concerned by the suggestion from at least one member that there be a ceiling on assessments. This recommendation came at a time when NAAB activities were expanding and more income was needed. The board decided to propose an amendment to the bylaws to indicate "the first one million services from semen produced by a member organization at one cent each. All first services over one million at ¾ cent each." This amendment was to be a part of the "call of the meeting" for the 1964 annual convention. The executive secretary proposed an assessment on custom freezing of semen, which was steadily growing in volume.

The financial report for the fiscal year ending 30 June 1964 showed total receipts of $97,738.31 and expenditures of $95,867.32, for a difference of $1,870.99 net income. The net worth of the association was listed by the auditor at $98,015.19 as compared to $96,144.20 for 1963. It became increasingly evident that NAAB operations were pretty much on a "break even" basis.

12.15 1964 Annual Meeting: Kennebunkport, Maine

The 1964 convention was held on 8–11 September at Kennebunkport, Maine. More than five hundred people attended, and forty-nine AI organizations were represented.

Considerable time was devoted to discussing problems facing the AI industry and the continuing growth of the NAAB. Another matter of extreme importance involved the changing of the association's name. The use of the word *artificial* in the name "National Association of Artificial Breeders" had been a topic of controversy for many of the NAAB members. In July 1955, the executive secretary offered the following definition:[14]

> Artificial breeding is a commonly used but hardly correct term as any standard dictionary will reveal.
>
> *Artificial,* according to the *Standard College Dictionary,* is an adjective and means (1) produced by art rather than nature; (2) not genuine or natural; affected; (3) not pertaining to the essence of a matter, etc.
>
> *Breeding* (verb), (1) to produce, as offspring; beget; hatch; bear young; (2) to be born; develop; raise; originate.
>
> *Artificial Insemination* is defined as "eutelegenesis" (noun) "In animal husbandry, the method of breeding by artificially impregnating the female, without direct sex contact; artificial insemination."

George Nichols, NAAB assistant secretary, suggested that the word *animal* be substituted for *artificial,* and the abbreviation NAAB could still be properly used. A resolution was prepared and adopted at the 1964 annual meeting to change Article I of the Articles of Incorporation to read "National Association of Animal Breeders, Inc."

Increasing the number of voting delegates was another matter that was resolved at this meeting. The board of directors proposed that the number be based upon the total number of first services made annually, with one slight revision. P. I. Higley, of the American Breeders Service, proposed that no member organization should have more than five voting delegates. The bylaw amendment was changed to state "five voting delegates for 700,000 to 1,000,000 and above first services."

The NAAB board also presented the following resolution to amend Article VIII, Section 2, of the bylaws as follows:

> The dues shall be based upon the number of first inseminations involving semen produced by each member organization, on the basis of one cent per first insemination for the first million inseminations and three-fourths of one cent per first insemination thereafter, and shall in no event exceed one cent per first service except the minimum shall be twenty-five ($25.00) dollars per year.

A motion was made and seconded for adoption. Objections to the amendment were voiced by Higley, who stated that his board of directors felt that the maximum dues in any organization should not exceed $5,000.00 annually. He also felt that his board might support a $7,500.00 maximum. After considerable discussion, Kellogg proposed the following motion to amend Article VIII, Section 2, of the bylaws:

> The dues shall be based upon the number of first service inseminations involving semen produced by each member on the basis of one-cent per

first insemination, except the maximum assessment for the fiscal year for any member organization shall not exceed $10,000.00 for the year of 1964–1965, and $7,500.00 each year thereafter, and in no event shall exceed one-cent per first service except the minimum shall be $25.00 per year.

This motion was seconded and further discussion ensued as to whether or not NAAB could operate on a reduced budget. The proposed change would reduce the annual income by $6,000 to $8,000 for the fiscal year 1965–1966 and thereafter. It was suggested that the NAAB budget eliminate funds for research support. The executive secretary stated that the most important objective is for all AI businesses to be a part of NAAB and all to work toward the same goal of a self-regulated program of cattle improvement. "While some economies can be made," Herman said, "it is going to be difficult to sponsor a growing national program with less funds than we now have."

President Behnke called for a standing vote on the proposed amendment, as presented by Kellogg, and it was passed by a vote of thirty-one delegates for and fourteen against.

The budget for the fiscal year ending 30 June 1965 calling for income of $105,775.00 and expenses of $105,325.00 was presented and adopted.

Max Drake, chairman of the resolutions committee, presented the following resolutions:

1. Commendation of the Holstein-Friesian Association, and American Guernsey Cattle Club for cataloging undesirable recessives and urging all breed organizations to do likewise.

2. Urging all organizations doing custom freezing to strongly recommend that owners of such bulls follow the adopted health standards of the AVMA-NAAB.

3. Requesting the executive secretary of the NAAB to encourage the USDA, Performance Registry International and beef-oriented organizations in general to establish a recording program based on performance with no sire ownership restrictions on the registration of AI progeny.

4. Amending Section 6, NAAB Code of Ethics, to prohibit members, as well as sellers, from quoting the price of bulls purchased at public auction.

5. Appreciation to Earl Groff, retiring director, for his six years of work in behalf of the NAAB.

Elected to the NAAB board was Kenneth Baushke, manager of the Michigan Artificial Breeders Cooperative, in East Lansing. Reelected for three-year terms were Norman Allen, of Schaghticoke, New York, and Laurel Behnke. Behnke was reelected as president. Kenley was elected as vice-president.

12.16 Industry Adjustments Continue, 1964–1965

The years 1964–1965 brought rapid changes to the AI industry. Most NAAB members were operating on an all frozen semen program. The ampule (or unit) of frozen semen was fast becoming a trade commodity and was merchandised widely. Dairy herds were becoming fewer and larger in size. NAAB member bull studs had already merged in many instances.

During 1964, an estimated total of 7,747,955 cows and heifers in the United States were artificially inseminated.[15] Dairy-to-dairy inseminations were totaled approximately at 6,159,599 head. Dairy cows inseminated with beef bull semen totaled 1,117,395 head, or 15.3 percent of the milk cows enrolled on the program, and there were 464,959 beef-to-beef inseminations. It is interesting to note that 2,731 swine and 1,845 milk goats were reported to have been artificially inseminated by bull studs in 1964.

12.16.1 Mergers Widespread

In the spring of 1964, the Southeastern Pennsylvania Artificial Breeding Cooperative (SPABC) in Lancaster and the Western Pennsylvania Artificial Breeding Cooperative in Clarion merged to form the Atlantic Breeders Cooperative, with headquarters at the SPABC facilities in Lancaster.

In April 1964, Evergreen Northwest Breeders in Burlington, Washington; Oregon Breeders Association in Corvallis; Western Montana Dairy Breeders in Missoula; and Century Breeders, Inc., located in Redmond, Washington, all merged to form the All West Breeders. Headquarters were established at Burlington.

In September 1965, the Iowa Breeders Cooperative in Des Moines merged with Consolidated Breeders Cooperative in Anoka, Minnesota. Since Consolidated Breeders Cooperative was a member of Midwest Genes, Inc., the merger made Consolidated Breeders the major stockholder in Midwest Genes. Greg Raps, manager of Iowa Breeders Cooperative, became the general manager of Consolidated Breeders Cooperative, and Russell Seath, then manager of "Consolidated Breeders," nearing retirement age, became operations manager.

In October 1964, four AI cooperatives merged to form the Northeastern Pennsylvania Breeders Association (NEBA). They were Northeastern Breeders Association (Tunkhannock, Pennsylvania); Lehigh Valley Farmers Association (Allentown, Pennsylvania); First Pennsylvania Artificial Breeding Cooperative (Lewisburg); and New Jersey Cooperative Breeders Association (Annandale). A ten-man board, made up of directors from the merging organization, was established to set policy for the new association. Headquarters, including the bull stud, were established at Tunkhannock.

In December 1965, the California Dairy Breeders, Inc., situated in Davis, merged with All West Breeders of Burlington, Washington. Victor Gray, former manager of California Dairy Breeders, was named assistant manager of All West Breeders in charge of the California operations.

These mergers reduced the voting number of NAAB member bull studs by nine. The membership roster for 1966 listed forty-five voting and sixty-seven associate members.

12.16.2 Beef AI Expands: Registration Limited

The use of AI in beef cattle had been gradually expanding since the introduction of frozen semen.

The beef registry organizations, in general, however refused to accept offspring resulting from AI unless the sire and dam were both owned by the applicant. The NAAB continually prevailed upon the

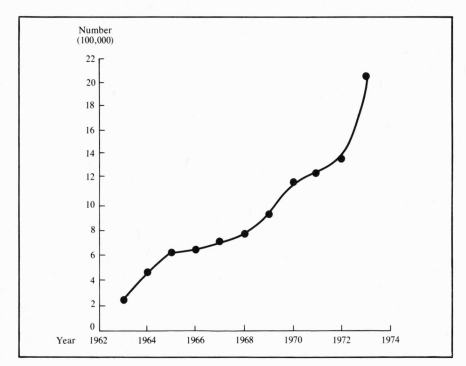

Figure 12.2   Beef cows bred artificially to beef bulls, 1963–1974. Source: *Dairy Herd Improvement Letter* 51:2 (1975), USDA.

beef registry organizations to adopt a more liberal policy such as had existed from the start for registered dairy cattle. These efforts continued to be frustrating. Most permitted joint ownership of sires and also limited the number of calves that could be registered per sire. Some of the beef associations, however, started to require blood typing of sires used for AI. Some purebred beef breeders utilized custom collected semen in their own herds and in other herds where sires were jointly owned at this time. Beef producers, in some cases, were turning to AI to control diseases such as *Vibrio fetus* and trichomoniasis spread by the bull in natural service.

Efforts to obtain liberal rules for the registration of beef cattle were assisted greatly by the NAAB beef AI committee, which was first appointed in 1961. This committee consisted of Ray Woodward, of the American Breeders Service in Wisconsin, chairman; W. H. Armstrong; Glenn Butts, of Ogeechee Farms in Oklahoma; James B. Lingle, of the Wye Plantation in Maryland; and J. Elmo Packer, of the Cache Valley Breeding Association in Utah.

12.16.3 Attempts to Tax AI

Through the years, little effort had been made by state, or federal, agencies to tax semen or AI service. In the spring of 1965, however, the South Dakota Legislative Research Council, Committee on Agriculture and Conservation, proposed a tax on semen sold within the state; a licensing fee for inseminators; and a license for foreign corporations doing AI business in the state. This move was no doubt prompted by beef cattle breeders who were trying to "save the bull market."

The NAAB, with the assistance of representatives from its members doing business in South Dakota, vigorously opposed the proposed tax. Don A. Bierle, an attorney from Yankton, South Dakota, was retained by NAAB. Committee hearings were held at the State Capitol in Pierre. The proponents presented witnesses who claimed "AI reduces the income in South Dakota by reducing bull sales," and some said "The cows don't settle and you lose a calf crop." Other testimony was to the effect "The bulls aren't good enough and you don't know who you are dealing with." After several legislative committee hearings and debate over a period of months, the proposal was tabled.

In New York, artificial insemination and insemination supplies or products were excluded from retail sales tax. "(RCW 82.08.030 [20])." In the State of Washington, semen and insemination equipment was once declared subject to sales tax. However, in June 1965, this ruling was repealed and most of the sales tax collected was refunded.

Generally, semen and artificial service as well as supplies for AI are classified under the heading of "seed" or items involved in agricultural production and are tax exempt. To our knowledge, no state is collecting sales tax on the various facets of the AI program. Several states have an "AI Technician Licensing Act" that requires persons who inseminate cattle for others to be licensed by the state livestock sanitary board.

12.16.4 NAAB Board and Dairy Breed Secretaries Meet

Since the beginning of the AI program in the United States, and with the founding of the NAAB, cooperation with the purebred dairy cattle registry organizations has been excellent. To discuss problems of mutual interest firsthand, the NAAB board of directors and secretaries of the dairy breed associations met together, at Columbus, Ohio, on 18 March 1965.

Items discussed dealt with production, testing programs, registration, type classification, and breed promotion. The PDCA "Requirements Governing Artificial Insemination of Purebred Dairy Cattle" were reviewed and discussed at length. The AI industry leaders voiced their approval of these rules in general but stated that they felt the "purebred industry should not try to make policemen out of AI workers on items that were a purebred herd owner's responsibilities."

12.17 The 1965 Annual Convention

Over 550 delegates and guests attended the eighteenth annual NAAB convention on 22–25 August in Washington, D.C. They represented forty-five AI organizations from the U.S., Canada, Puerto Rico, and Jamaica. There were guests from four foreign countries. Canada's delegation of fifty-five people represented six Canadian studs.[16]

The various NAAB committees submitted reports and outlined their plans for future goals. The executive secretary reported NAAB's progress during the year and emphasized the importance of a uniform coding system for bulls that were used in the AI program. The NAAB board, in cooperation with the Purebred Dairy Cattle Association and the USDA, had adopted a coding system that identified the bull stud, breed, and code number of the bull (see Appendix B).

Article V, Section 7, of the bylaws was changed to read that voting delegates be present when casting a vote. Item 4 of the code of ethics was also amended as follows:

The source and date of all daughter-dam comparisons and/or herdmate comparisons must be shown when published. No daughters with completed normal lactation records of 305 days or less may be omitted. All incomplete records of less than 305 days shall be projected to 305 days by use of appropriate factors for that purpose and shall be used in all calculations unless the cow in question has other completed lactations.

The following resolutions were presented by Max Drake, chairman of the resolutions committee, and approved on motion:

1. Recommend to PDCA that, if living, both the sire and the dam of bulls placed in AI use be blood typed.
2. Acknowledge the leadership of the Dairy Husbandry Branch, USDA, in genetic appraisal of dairy cattle.
3. Encourage members to use the NAAB emblem and slogan as a symbol of quality and prestige.
4. Recommend that all members observe Christmas and Easter as "No Service" holidays.
5. Express appreciation to Laurel Behnke, who tendered his resignation as director and president, for his leadership.
6. Express appreciation to the convention hosts.

Newly elected for a three-year term was Webster Jessup. Keith King, Milking Shorthorn breeder and president of Northern Illinois Breeding Cooperative in Hampshire, was elected to a two-year term, completing the term left vacant by Behnke's resignation. Kenley was elected as president; Reuben Peterson, of Fremont, Nebraska, as vice-president; and Norman Allen was elected to the executive committee.

Chapter 13: The Maturing Years, 1966–1970

13.1 New Programs Develop

Beginning in 1965, there was a steady increase in the number of beef cattle artificially inseminated. It is estimated that in 1965, there were 615,147 beef-to-beef inseminations by AI organizations and services involving semen from ranchers' bulls, an increase of 150,188 head over 1964.[1] Four NAAB members were "all beef" AI operations at this time. There were 449 beef bulls owned by NAAB member studs and all had one or more beef bulls. There were 963,657 milk cows (13.3 percent of the total) bred with beef semen and 6,301,178 milk cows inseminated with dairy bull semen. It was estimated that 43.7 percent of the nation's cows and heifers kept for dairy purposes were artificially inseminated.

Emphasis on Beef. The NAAB program became more positively geared to beef AI during 1965–1966. The beef AI committee, which had served in an advisory capacity, was expanded by formation of the beef development committee. This committee had three subcommittees: beef records and sire evaluation; beef herd management; and publicity and promotion.

Ancel Armstrong, manager of the Kansas Artificial Breeding Service Unit in Manhattan, Kansas, was appointed general chairman of the beef development committee.

In January 1966, an informal conference of AI operators who were interested in promoting beef AI was held in Denver. A similar meeting was held at Chicago in March 1966, and still another in April 1966 at the North Central Conference in Davenport, Iowa. Plans were developed by the beef development committee to organize a formal conference on beef AI, to be held at Denver in January 1967.

13.1.1 First Technical Conference

The first NAAB technical conference, assisted by the technical committee, was held 25–26 February 1966 in Chicago, Illinois. Laboratory managers and other technical workers who felt there was a need for exchange of information and up-to-date reporting of technological developments suggested this conference.

The conference, conducted on a self-supporting basis, was successful. More than 150 persons from twenty-five states and three Canadian provinces attended. Proceedings were published by NAAB and plans were made to continue this conference on an every-other-year basis.

13.2 Symbol of Quality Program

Following board approval in 1965, the NAAB insignia was reworked to include "Symbol of Quality." The purpose was to have a symbol that represented NAAB members and the high standards, integrity, reliability of services, and products offered. This symbol and its indication of integrity, while still in use in the forerunner of the symbol "Know Your Semen Supplier," was adopted by NAAB in 1975.

13.3 Technician Award Program Enlarged

George Nichols reported in March 1966 that over four hundred AI technicians had received the "35,000 Cow Award" and fourteen the "20 Year Award" since the program was initiated in 1958–1959. Many technicians had over twenty years of service and several were approaching, or had exceeded, 100,000 first services. To establish a recognition program for extraordinary performance by technicians, the board approved two new awards: "The NAAB Quarter Century Award" and "The NAAB 100,000 Cow Award."

The "35,000 cow" and "20-year" awards were to be continued on the same basis.

13.4 Frozen Horse Semen

It was reported at the 1965 annual meeting that stallion semen had been successfully frozen by a California group and would soon be on the market. In 1966, a number of NAAB members launched horse-breeding projects utilizing frozen semen. The semen supplier, Horse Breeders' Service in Turlock, California, was approved for associate membership in NAAB. As will be discussed elsewhere, horse AI on a field basis failed due to poor fertility.

13.5 1966 Convention

The 1966 convention, listed as the nineteenth for NAAB, was held 21–24 August at Denver, Colorado.[2] Over 650 delegates and guests attended. Thirty-nine AI businesses, all members of the NAAB, were represented.[3]

Keynote speaker was Peter H. Dominick, U.S. senator from Colorado, who lauded cattlemen for keeping their industry shorn of governmental controls. NAAB President Kenley outlined the association's accomplishments during the past year. He stressed four points for maintaining a strong industry: (1) a favorable existing climate for industry advancement despite declining dairy cow numbers; (2) attracting and maintaining suitable industry personnel and improving their working conditions; (3) establishing pricing formulas consistent with service rendered and existing economic conditions; and (4) preserving good relations with the Purebred Dairy Cattle Association in solving mutual problems.

During the business session, the 1966–1967 budget, calling for $102,925.00 income and expenses of $102,575.00, was adopted. The net worth of the association was $109,633.86.

There were forty-one voting member organizations and sixty-six associate members on the roster 1 July 1966. Resolutions adopted included: (a) encouragement of all NAAB members to use the "Symbol of Quality" in advertising and promotional material; (b) commendation of the NAAB board and the executive secretary for efforts in combating efforts by states to tax semen and AI; and (c) authorization of the NAAB to do all in its power to promote the growth and development of the beef AI program.

Article IV, Section 2(a) [Members], was changed to meet the needs of NAAB member organizations that had consolidated with other members but still operated as semen-producing businesses:

Members of the Association shall consist of any individual operators, corporations, associations, or partnerships which produce semen annually for the insemination of not less than 5,000 first service cows or other farm animals or any corporation or partnership organized as a separate entity by existing members which produces semen either from its own producing livestock or from livestock owned by existing members which meets said minimum annual requirements.

W. Lewis Campbell, manager of the West Virginia Bull Stud in Frederick, Maryland, was elected as a director for a three-year term.

13.6 An Expanding Program

The fiscal year of 1966–1967 was characterized by expanding activities by NAAB and its members.

An estimated total of 7,933,723 cows in the United States were bred artificially in 1966.[4] Most of the gain was in beef inseminations—up by 32,000 head. It was estimated that 47.9 percent of the nation's 15.2 million dairy cows and heifers of breeding age and 1.9 percent of the beef cows and heifers were artificially inseminated.

13.6.1 First NAAB Conference on AI of Beef Cattle

The conference was held 13 January 1967 at Denver, Colorado. The beef development committee and the NAAB staff arranged the program. The purpose of the meeting was to establish the role of AI in beef cattle improvement.

The one-day conference was attended by 115 persons who represented livestock extension specialists, beef ranchers, college administrators, beef breed officials, and AI organization representatives. Delegates were present from thirty states and several Canadian provinces. The conference was considered a success, and it has been continued each year since. During 1977, "Beef AI Conferences" were staged at Louisville, Kentucky, and Denver, Colorado.

13.6.2 Mergers Continue

The merging and consolidations of NAAB member businesses continued at a high level. Mergers of small cooperative organizations, operating under a free-enterprise system, proved to be the logical answer for survival.

In January 1966, Select Sires, Inc., a new AI corporation composed of Central Ohio Breeding Association in Columbus, Kentucky Artificial Breeding Association in Louisville, Northern Illinois Breeding Cooperative in Hampshire, and Southern Illinois Breeding Association in Breese all assumed ownership of *all sires* owned by the four studs.

This merger was described as an amalgamation of sires under common ownership. No property other than bulls was transferred to Select Sires, Inc., at that time. The plan was for each member of the federation to continue to conduct its own inseminating business. As time progressed, additional farmer-owned cooperative bull studs became a part of Select Sires. See Appendix A for details.

By purchasing Eastern Iowa Breeders at Hiawatha in October 1965 and Piper Brothers Breeding Service at Watertown, Wisconsin, in January 1966, the Carnation Farms Breeding Service in Washington, greatly expanded its AI operations.

Eastern Studs Merge. Negotiations began in January 1966 to merge five eastern AI cooperatives with a target date of 1 August 1966 for completion. Organizations involved were New York Artificial Breeders Cooperative, Ithaca; New England Selective Breeding Association (NESBA) at Woodbridge, Connecticut; Maine Breeding Cooperative, Augusta; New Hampshire–Vermont Breeding Association, Concord, New Hampshire; and Central Vermont Breeding Association, Montpelier (Central Vermont Breeding Association, while a part of NESBA, maintained its entity as an organization).

The new organization, Eastern Artificial Insemination Cooperative, Inc. (EAIC), established headquarters in Ithaca, utilizing the former facilities of New York Artificial Breeders Cooperative (NYABC). Charles J. Krumm, manager of NYABC, was appointed general manager. The board of directors of EAIC is made up of representatives from the merged organizations.

Midwest Breeders Cooperative. By a merger, which was effective 1 January 1967, of Badger Breeders Cooperative and Consolidated Breeders Cooperative, a new organization, the Midwest Breeders Cooperative in Shawano, Wisconsin, was formed.

Kenneth E. Wallin, veteran manager of Badger Breeders Cooperative and one of the organizers of NAAB, was selected as the general manager of the newly formed organization. Greg Raps, Doctor of Veterinary Medicine (DVM), former manager of Consolidated Breeders Cooperative and earlier manager of Iowa Breeders Cooperative, was appointed assistant to the general manager.

13.6.3 Cooperative Efforts in Sire Proving

To share in the use of proved sires, which is developed by sampling young sires through the AI program, a number of AI cooperatives combined efforts.

One of the first such programs was Eastern AI Cooperatives, commonly known as EAIC (but not to be confused with Eastern Artificial Insemination Cooperative, Ithaca, New York). The program began in April 1963. The plan called for a unified semen exchange and sire proving by eight bull studs located in New England, New York, New Jersey, and eastern Pennsylvania. William Schaefer, Jr., manager of NEPA Artificial Breeding Cooperative, Tunkhannock, Pennsylvania, served as executive secretary of EAIC. Due to numerous mergers of the cooperating bull studs, the work of EAIC was gradually taken over by the larger organizations.

Frontier Sires. In early 1965, six western and midwestern NAAB member studs banded together to develop a large young-sire-proving project. The organizations involved were Midwest Genes, Minnesota Valley Breeders' Association, Kansas Artificial Breeding Service Unit, Cache Valley Breeding Association, California Dairy Breeders, and the All-West Breeders. Due to the merger of California Dairy Breeders with All-West Breeders, shortly afterward, in 1966, the five remaining studs reorganized the project and adopted the name Frontier Sires.

The plans called for each participating stud to purchase and house a specified number of young bulls. Mergers gradually reduced the participating studs. Those remaining carried on in a modified manner until the bulls involved were phased out.

13.6.4 First Horse AI School

Due to the growing interest in the use of frozen stallion semen, the Graham School, Inc., located in Garnett, Kansas, staged an "AI School for Horsemen" on 20 February 1967. Held at Garnett, the school attracted twenty-three students from sixteen states. Six students were from cattle AI organizations. This AI school was advertised to be the first in the country to feature the use of frozen horse semen.

13.6.5 Beef Improvement Federation Formed

With the growing interest in AI for beef cattle, emphasis for a national beef-testing program was gaining momentum. While some beef registry organizations had adopted some form of performance testing, state "beef cattle improvement associations" were appearing, and while Performance Registry International was quite active, beef producers lacked a national testing program. The opinion among people interested in beef production was that such a program was badly needed.

On 14 January 1967, a "kickoff" meeting was held in Denver, Colorado. Representatives from twenty-six state beef cattle improvement associations, Performance Registry International, beef registry organizations, NAAB, AI organizations, the American National Cattlemen's Association, and USDA were present. F. R. Carpenter, of Hayden, Colorado, a veteran Hereford breeder and a strong promoter for "gain testing," presided over the meeting. At a meeting 12 January 1968 in Denver, the Beef Improvement Federation as an organization was finalized (see Chapter 4).

13.6.6 Minimum Standards for Inseminator Training

There continued to be an expansion of frozen semen sold directly to herd operators. The training of herd operators to do AI brought a demand for AI training schools. As a result, numerous such schools were established. Many were conducted by NAAB member organizations. Some were organized by private individuals. In some cases, the schools lacked skilled instructors and adequate facilities. To provide technical knowledge and instruction necessary for the newer developing program, it became apparent that there was need for collaboration between training schools and the AI industry.

The NAAB established a "Committee on Technician Training Schools" in 1966. The industry was surveyed and standards were established. The first publication, *Recommended Minimum Standards for the Training of Artificial Insemination—Technicians and the Training of Herdsmen—Inseminators,* was published by NAAB in 1967. While revised frequently, it continues to serve as a guide for the industry.

Assessments on Custom Frozen Semen. The custom freezing of bull semen increased steadily as frozen semen became more widely used. The NAAB executive secretary conducted a survey in 1966 that indicated twenty-seven NAAB members, twenty-five individuals or partnerships, eight colleges, and experiment stations were custom freezing semen. Reports from twenty-three members and twelve nonmembers revealed that, in 1965–1966, a total of 819 dairy bulls and 706 beef bulls had been custom collected. It was estimated that at least 1.5 million cows were inseminated with this semen. Since the semen was not produced by member bull studs, no assessments for NAAB were collected.

The NAAB board proposed an addition to the bylaws (Article VIII, Section 2) that an assessment of one-half cent per ampule be paid on all custom-frozen semen obtained from bulls or other farm animals, not owned or under lease by an NAAB member or associate. This bylaw amendment was adopted at the 1967 annual meeting.

13.7 1967 Annual Meeting

The meeting was held 5–8 September at Hershey, Pennsylvania.[5] Approximately 450 delegates and guests from the United States and Canada were registered.

President Kenley mentioned that the fiscal year of 1966–1967 was the first year the full impact of NAAB's structure of dues, as adopted at the 1964 meeting, was felt. The net result was a reduction of $11,000 income as compared to the former rate. He lauded the efforts of the association in balancing the budget even though some activities were curtailed. Attention was also called to the breaking of the price barrier—long overdue—for AI service, semen, and the development of improved marketing and distribution systems for semen.

The newly elected board members included Carl Russell, of Kennebunk, Maine, elected for a three-year term as director; Reuben Peterson, as president; Kenneth Baushke, as vice-president; and W. L. Campbell, of the Maryland-West Virginia Bull Stud in Frederick, as a member of the executive committee.

The 1967–1968 budget, with a deficit of $4,640.00, was adopted. Estimated income was $102,750.00 and expenses $107,390.00. It was becoming increasingly obvious that NAAB could not carry out its program under terms of the fee structure imposed in 1964. Steps were taken at the 1968 annual meeting to improve the situation. The following resolutions were adopted:

1. Express appreciation to the Purebred Dairy Cattle Association for development of the "requirements Governing the Artificial Insemination of Purebred Dairy Cattle" and urge that all revisions be of a nature not to handicap members of NAAB in their efforts to improve both registered and grade dairy cattle.
2. Express appreciation to the Pennsylvania hosts for sponsoring the 1967 convention.
3. Sincere thanks and appreciation to Norman Allen, of Schaghticoke, New York, for his services during two terms as member of the NAAB board of directors.

Don Beck, a technician at Tri-State Breeders in Westby, Wisconsin, received the "100,000 Cow Award." For the past thirteen years, he had averaged over six thousand first service cows per year.

13.8 NAAB Assesses Industry: Develops Plans

Following the 1967 annual meeting, the board of directors and the executive secretary decided to develop long-range plans for NAAB. The activities of the association were steadily expanding. Member bull studs were fewer but doing more business. Beef AI was beginning to grow. The beef conference and the technical conference, sponsored by NAAB, were off to a good start. The financial situation of the *A.I. Digest* was improving, but the number circulating was declining in the United States. This decline was largely due to member organizations selling more semen directly to herd operators. As a result, fewer bull stud technicians were required. Because of insufficient income from member dues and assessments, the program adopted for NAAB was being postponed in some areas.

A five- to ten-year projected program was considered by the board at a meeting in Columbia, Missouri, on 1, 2 November 1967, and these objectives were approved: (1) that NAAB must meet the challenges of a changing livestock industry and the needs of the AI industry in an aggressive manner; (2) adoption of a dues and fees schedule that will adequately support the basic activities of NAAB; (3) added emphasis on the development of a strong AI program for beef. Consideration be given the employment of an additional man to further the efforts to promote beef AI; (4) to actively support and aid in efforts to establish a uniform national beef-testing program; (5) to encourage a pricing system for services and semen sold that will insure a strong and lasting AI industry; (6) adapt the NAAB program to fit the needs of the AI industry resulting from fewer and larger bull studs, direct semen sales, larger herds, and more specialized dairy and beef farm operations; (7) to strengthen research support in collaboration with NAAB member organizations; and, (8) increase efforts to advance the AI program on all fronts including semen sales in foreign countries.

The board on motion by Kenley, which was seconded by W. L. Campbell, adopted a proposal that was to be presented at the 1968 annual meeting. The proposal would increase assessments to 1-1/8 cents per first service on a maximum of 1,000,000 cows per member annually, to be effective 1 July 1968; on 1 July 1970, assessments would be increased to 1-1/4 cents per first service. A maximum dollar amount, not to exceed $10,000.00, would also be allocated to the research fund in any fiscal year.

13.8.1 Violations of Code of Ethics

During 1967 and 1968, several complaints were filed protesting the quoting of the purchase price of bulls by NAAB members. The NAAB board of directors reviewed each complaint. The executive secretary was requested to write the manager and the president of each offending member to inform them of the code that was violated and to request that the practice be discontinued.

Item 6, of the code of ethics, was violated when the proponents of frozen horse semen were claiming to have the method patented. The NAAB executive secretary checked into the matter and found that patent rights had not been granted by the U.S. Patent Office. A letter of admonishment was sent to the offending member.

13.8.2 Status of AI Program: NAAB Membership

A total of 7,847,607 cows and heifers were bred artificially in the United States during the 1967[6] calendar year, a decline of 86,116 head less compared to 1966. There were 6,377,095 cows bred to dairy sires and 807,844 bred to beef bulls. Some 101,641 fewer milk cows were inseminated than in 1966. This figure was not surprising, considering that the number of milk cows was the lowest since 1893. The USDA estimated 14,662,000 milk cows and heifers of breeding age in the United States on 1 January 1967.

There were 662,688 beef cows and heifers inseminated, an increase of 15,000 head over 1966. Three organizations engaged in swine AI reported 7,955 sows bred. There were 73 goats and 1,400 horses reported serviced by AI. There were only thirty-three bull studs in the United States operating during 1967. Two of these, International Beef Breeders, in Denver, Colorado, and Codding Ranch, in Foraker, Oklahoma, were all beef.

Mergers. On 1 March 1968, the Southern Illinois Breeding Cooperative in Breese and Northern Illinois Breeding Cooperative in Hampshire merged to form the Illinois Breeding Cooperative with headquarters established at Hampshire.

Pacific Breeders Cooperative, of Petaluma, California, merged in late 1967 with Genetics, Inc., a new AI organization founded by Harold Schmidt.

Change of Ownership. The two largest NAAB member organizations, both owned by private corporations, exchanged ownership in 1967–1968.

On 31 August 1967 the American Breeders Service, Inc., of De Forest, Wisconsin, was sold to W. R. Grace and Co., through an exchange of stock. P. I. Higley continued to serve as president, and American Breeders Service became a part of Grace's Agricultural Products Division.

The second giant among AI organizations in America, Curtiss Breeding Service, Inc., of Cary, Illinois, was sold to G. D. Searle and Co., of Chicago, on 8 October 1968. It was a stock transaction. Melvin L. Kenley continued as president of Curtiss under the new regime.

13.9 Litigation: Restrictions on AI for Beef Cattle

Throughout the years, the NAAB officers, members of the NAAB beef development committee, managers of AI organizations, and many breeders of registered beef cattle had carried on a continuing plea to beef registry organizations to liberalize their requirements governing the use of AI. These arguments accomplished little, though some beef registry organizations permitted joint ownership of bulls. A summary of the "AI Regulations of Beef Registry Associations" carried in the January 1968 issue of the *A.I. Digest* indicated:

> Among 11 beef registry organizations listed, all but four required that both the dam and sire of progeny from AI to be registered in a same ownership.

Six permitted from two to four owners of a sire. One, Santa Gertrudis, ruled that the sire and the dam must be in the same ownership. Several ruled that semen in storage could not be used, and progeny be eligible for registration, after the sire's death. In some cases, restrictions were placed on the number of AI progeny sired by one bull.

While there were modifications made by some of the smaller and newer beef registry associations, the larger ones held fast to restrictions on the use of AI. One association, Angus, refused to register calves for a herd owner if he sold semen from his bulls for use in other herds, including nonregistered cattle. Needless to say, many beef breeders and AI organizations were angered and frustrated by these restrictions. As a result, legal recourse seemed to be a solution.

United States of America vs. American Angus Association. On 15 January 1968, the U.S. Department of Justice filed a complaint against the American Angus Association, of St. Joseph, Missouri, charging that restricted use of AI is a violation of the Sherman Act, involving restraint of interstate trade and commerce.

The plaintiff charged that (1) the association's members have been restrained in the sale of purebred Angus semen to customers of their own choice; (2) the association's members have been denied the opportunity to register purebred Angus calves that were conceived artificially; and, (3) artificial-breeding companies have been restrained in the sale of purebred Angus semen to the association's members.

The filing of this complaint, it is generally believed, came about when Angus breeders wanted more freedom in the use of AI. Though suspected of being a party to the charges filed by the U.S. Justice Department, the NAAB was not involved. The NAAB had pursued a course of persuasion with all beef registry organizations.

Once the complaint was filed, the NAAB was called upon by the Justice Department for much detailed information on the conduct of the AI program. The NAAB executive committee met in Chicago on 9, 10 February 1968 and decided that the association should intervene in the litigation. As matters developed, the Justice Department preferred to have no intervening parties. The NAAB then played a role in furnishing information and, when requested, the association's attorneys conferred with the Justice Department attorneys.

On 13 July 1970, the Department of Justice agreed to a proposed consent judgment that terminated the antitrust suit filed against the Angus Association. The judgment restricted the Angus Association from imposing limitations on the sale of Angus semen; set up the requirement that 45 calves (maximum) be registered annually from an untested (not progeny tested) bull but not over 10 such calves for any member; and permitted the registration of an unlimited number of calves sired artificially by progeny-tested Angus bulls, though not over 150 annually for any member.

This judgment resulted in some moderation of the AI regulations for all beef registry organizations.

13.10 Proposed Changes in "PDCA Requirements": Dairy Cattle

The six dairy breed secretaries and the NAAB officers and directors met 28 March 1968 in Chicago.

The principal items for discussion related to revisions of the "PDCA Requirements Governing the Artificial Insemination of Purebred Dairy Cattle." Concern was expressed over control of semen that is sold directly and that is resold frequently. There was discussion on the need to revise the PDCA requirements for AI of purebred dairy cattle, which was in effect, with minor changes, since 1966.

13.11 Conferences: National Committees

Beef AI.   The "Second Conference on Beef AI" was held 14 January in Denver, Colorado. Over 170 persons from thirty states and Canada attended. The conference theme was "The Role of AI in Beef Cattle Improvement." Ten speakers and cattle industry breeders appeared on the program.

Technical Conferences.   The "Second Technical Conference on Animal Reproduction and Artificial Insemination" was held 9–10 February 1968 at Chicago and was most successful. Eighteen topics pertaining to animal reproduction, progress in AI technology, research findings, and disease control were presented. About 170 persons from twenty-seven states, Canada, and Europe attended.[7]

Regional Conferences.   The following area conferences, held in 1968, were sponsored by NAAB members with assistance from the association:

North Central Conference. Held 9–10 April at Davenport, Iowa. Attended by 180 persons from thirty states and Canada. Begun in 1948, this conference continues and provides for much informal discussion.

Southeast Invitational Conference. Held 15–16 May at Lexington, Kentucky. While held for several years, this conference was discontinued as AI mergers occurred.

Northeastern Invitational AI Cooperative Conference. Usually held in the late winter, this conference was limited to AI cooperatives. It was discontinued after the present Eastern Artificial Insemination Cooperative, Inc., in Ithaca, New York, was formed.

West Coast Conference. This conference was successful for several years in reviewing problems and progress of AI on the West Coast. It, too, was discontinued as bull studs merged.

13.12 1968 Annual Convention

The convention, listed as the twenty-first annual convention, was held 25–28 August at Knoxville, Tennessee.[8] About 450 persons from thirty-six states and two Canadian provinces attended. Thirty-eight AI businesses were represented.

Kenneth Peissig, chairman of the sire purchase agreement committee, reported that his committee, working with PDCA, had developed a uniform "Purchase Contract for Sires," which outlined conditions of sale and responsibilities of the buyer and seller. The committee recommended adoption and the delegates voted agreement.

Harry A. Herman, chairman of the NAAB research committee, reported that fifteen of the seventeen projects recommended by the committee had received support during the year.[9] Support totaled $23,915.32. Of that amount, $13,915.32 came from coordinated funds supplied by members and $10,000 from NAAB operation funds as approved by the board of directors. One new project, "Improving Reproductive Performance in Beef Cattle," headed by James Wiltbank, of Colorado State University at Fort Collins, was approved. The 1968–1969 research budget was set at $34,200 minimum and $49,260.00 desirable.

13.12.1 NAAB Research and Technical Committee Proposed

R. J. McDonald, of Ontario, a member of the NAAB research committee, presented a proposal prepared by the research committee and the technical committee to streamline efforts and develop a more effective program. This plan called for: (a) combining the research committee and the technical committee into one large committee to be called the research and technical committee. There to be a research subcommittee and a technical subcommittee, respectively. There to be a general chairman and a chairman of each subcommittee. (b) support of research to come largely from "coordinated funds" supplied by members and $10,000.00 annually to be furnished from NAAB funds; (c) the committee to serve as a liaison between research workers and member organizations supporting key projects. The NAAB executive secretary to serve as committee secretary and all policies and project support subject to approval of the NAAB board of directors. The plan was adopted on motion by the voting delegates.

13.12.2 Bylaw Changes: Dues and Assessments

The board of directors recommended that Article VIII, Section 2 (Dues and Finances), be amended as follows:

The dues, effective January 1, 1969, and thereafter, shall be based upon the number of first service inseminations involving semen produced by each member organization, on the basis of 1-1/8 cents per first service insemination except the maximum assessment for the fiscal year for any member organization shall not exceed one million (1,000,000) first services.

This proposal, as expected, was debated. A counter resolution had been received from opponents of the assessment increase affirming their position for a $7,500.00 maximum on NAAB dues; that NAAB discontinue direct support of research from operating funds and serve in a research coordinating role; and that the concept of a graduated increase in assessments per first service be considered. The idea was that smaller organizations should pay higher dues. The NAAB board rejected this resolution as it was not in harmony with budget needs and the association's program.

It was then proposed by opponents of the measure that the 1-1/8 cents per first service be approved but a limit of 750,000 first services be imposed. Strong support was given the NAAB board's original proposal by numerous delegates. The proposal as presented by the board was adopted. This decision marked a turning point for NAAB. It had proved itself as being of service and value, and members were willing to provide reasonable financial support.

Among resolutions adopted were urging NAAB members to complete the annual summary on cows bred and semen sold and report same to USDA before 1 April each year; commending the USDA Dairy Division for its progress in updating sire and cow evaluation methods to fit the changing dairy picture; appreciation to Webster Jessup for his contribution as a director for a three-year term; and to Reuben Peterson serving two, three-year terms as a director and as president in 1967–1968; thanks and appreciation to the Tennessee hosts for the success of the 1968 convention.

Article V, Section 7, of the NAAB bylaws was amended to entitle *six voting delegates* for any member accounting for more than one million first services, with semen produced by the organization, during its fiscal year. Item 4 of the code of ethics was revised to read as follows:

The publication of all sire summaries shall include at least the following: (1) source of information; (2) date of summary; (3) predicted difference; and (4) repeatability factor.

The source and date of all daughter–herd-mate comparisons must be shown when published. No daughters with completed normal lactation records of 305 days or less may be omitted. All incomplete records of less than 305 days shall be projected to 305 days by use of appropriate factors for that purpose and shall be used in all calculations unless the cow in question has other completed lactations. USDA production data and breed association type data, if and when used, shall be the latest information available at the time of publication.

There was debate on these changes. An amendment was proposed to omit the terms *predicted difference* and *repeatability factor*. The question was also raised as to the propriety of details of sire evaluation to be included in the code of ethics. The amendment and the motion failed to pass. The revision as stated above was adopted.

Election: Budget. Elected to three-year terms as members of the NAAB board were Victor Gray, of All West Breeders in Burlington, Washington, and Ronald Hawkes, dairyman from Paul, Idaho, and past president of the Cache Valley Breeding Association at Logan, Utah.

Kenneth Baushke was elected president, and W. L. Campbell vice-president. Keith King, of Oneida, Illinois, and Melvin L. Kenley were elected to the executive committee. The board voted that "when a retiring president continues as a member of the board, he automatically becomes a member of the Executive Committee."

13.13 Revisions in Programs as Industry Matures

The necessity for the NAAB to frequently revamp plans, to meet the needs of the ever-changing AI program, is better understood if one understands the development stages the industry had undergone to date.

The years of 1938 to 1945 were largely a trial-and-error period. This period was one of education, backed by the agricultural extension service at the Land Grant Colleges. There was a lack of technical knowledge. Many managers, and likewise boards of directors, of bull studs were limited in business management experience. The program had to be sold. Many dairy farmers were skeptical about AI succeeding.

The next ten years (1945–1955) resulted in the greatest growth and development the AI program has enjoyed. From 361,000 cows serviced by sixty-seven AI organizations in 1945 to 5,155,240 cows serviced by ninety-three organizations in 1954. There were new organizations formed, thousands of people hired, new buildings were erected, and many new sires added. The demand for service was so great the AI organizations were hard pressed to meet requests. Money flowed fairly free and inflation had not begun to make inroads on purchasing power.

The picture for the AI industry changed greatly from 1955 to 1965. Many AI organizations had financial difficulties. Cow numbers in all but the most populous dairy centers had declined markedly. Competition for business among AI organizations had become widespread. The cost of conversion from liquid to frozen semen and the launching of young-sire-proving programs overtaxed the finances of many small organizations. As related earlier, these factors accounted for the many mergers during this period.

Emphasis was on sales, and management had to struggle to meet the challenge of increasing costs of operation. The NAAB member organizations that survived this critical period became stronger and increased the volume of business, often with added services. When one views the developments occurring from 1955 to date, it is not surprising that NAAB has had to make its share of financial adjustments. Strong, capable, and dedicated leadership from the beginning of the AI program in 1938 continues to be one of the industry's greatest assets.

13.14 Emphasis on Program Expansion, 1968–1969

Following the 1968 annual meeting, increasing efforts were made by the association to expand the use of AI.

During 1968, a total of 7,933,878 cows and heifers were estimated to have been bred by AI in the United States.[10] Of these, 7,133,636 were dairy cows and 795,242 beef cows. This represented an increase of 90,278 dairy cows and 11,423 beef cows over the previous year. For the first time, over one-half (50.5 percent) of the nation's 14,123,000 dairy cows and heifers and about 2.2 of the nation's beef cows were artificially inseminated. There were thirty-one bull studs, two all beef, in operation. A total of 2,380 bulls, 352 beef, were maintained, and 3,334 cows were bred per sire. It was estimated that 413,401 cows were bred with semen sold directly to farmers. There were some 15,000 swine and 442 dairy goats reported for the year. Tri-State Breeders reported over 12,000 sows bred by AI in Wisconsin and the surrounding states. The farrowing rate was 69.7 percent with 9.1 pigs per litter.[11]

Though slight, the increase in total cows and heifers inseminated from 1965 forward was encouraging to the industry.

13.14.1 NAAB Publications and Promotion

Three publications for educational and promotional purposes were authorized by the board of directors in 1968–1969. These publications widely used in the industry were as follows:

1. *How to Make Beef AI Work in Your Herd*. This pamphlet was developed by the beef development committee. It served as a "handout" or

mailing piece stressing the essential points to follow in a successful beef AI program.

2. *Wean More Pounds of Beef.* This sixteen-page bulletin, well illustrated, was prepared by J. N. Wiltbank, of Colorado State University, with the assistance of the NAAB beef development committee. The cost of publishing this booklet was shared by NAAB and the Upjohn Co., in Kalamazoo, Michigan.

3. *Serving the AI Program and the Livestock Industry.* This publication, prepared and issued by the NAAB staff, explained the role of the association in serving the needs of the free-enterprise-operated AI program in the United States.

The pamphlet *Minimum Standards for Technician Training* was revised and five thousand copies were circulated during the year.

These and other publications issued, in addition to the monthly *A.I. Digest*—circulating in every state, Canada, Mexico, and seventy overseas countries—assisted NAAB in becoming an internationally recognized institution.

The association staged an exhibit promoting livestock improvement through AI at the "National Meeting of County Agents," in Louisville, Kentucky, on 12–17 October 1968.

13.14.2 P.R.I. Develops Beef Performance Pedigree

NAAB members and the beef development committee frequently urged that pedigrees of beef cattle carry performance information. While little headway seemingly was being made, no doubt due to a scarcity of "weight-gain" information, Performance Registry International at its 1968 annual meeting on 5–7 September at Sheridan, Wyoming, unveiled its first "Performance Pedigree." This pedigree provided the weaning weight, yearling weight, and weight-gain information useful in selecting breeding stock. It was enthusiastically received.

The NAAB executive secretary, Harry A. Herman, serving his second term as P.R.I. director, was elected first vice-president of the organization at the Sheridan meeting.

13.14.3 NAAB Research and Technical Committee Finalized[12]

Following action at the 1968 annual meeting, the board of directors took steps to combine the technical committee and the research committee.

David Yoder, manager of the Atlantic Breeders at Lancaster, Pennsylvania, was appointed general chairman and Herman as secretary.

A subcommittee for research was appointed. Robert Boese, of Curtiss Breeding Service in Cary, Illinois, was appointed chairman. Other committee members were Harry Rajamannan (University of Minnesota–St. Paul); W. H. Dreher (Midwest-Breeders Cooperative, Shawano, Wisconsin); Harold Hafs (Michigan State University, East Lansing); and R. J. McDonald (Western Ontario Breeders, Woodstock, Ontario).

The technical subcommittee consisted of Howard Kellgren (Louisiana Animal Breeders Cooperative, Baton Rouge), chairman; J. J. Prosser (International Beef Breeders, Denver, Colorado); Roger Ives (Soligenics, Suffield, Connecticut); and Hollis Schwartz (Minnesota Valley Breeders Association, New Prague).

The research and technical committees, one of the association's hardest working, have continued to serve the association and the industry in a commendable manner.

13.14.4 Rules Governing AI: Breed Secretaries and NAAB Board Meet

The NAAB board and officers met with the six dairy breed secretaries—representing PDCA—at Chicago on 2 April 1969 to discuss again proposed changes in requirements governing the registration of artificially sired progeny. The issues dealt largely with changes that had occurred in the AI program—frozen semen, custom freezing, direct semen selling, semen brokers, more within herd inseminations, and less control of semen sold by AI organizations.

The PDCA recommended at this point: (1) the licensing of all bulls used for AI on a fee basis; (2) the approval of semen-producing businesses without a fee; and, (3) the annual registration of individual inseminating technicians for a fee. It was also proposed that semen transfers, between-herd-certificates, and similar records be dispensed with, and no effort be made to follow semen after it left the bull stud. Further, that an enlarged blood-typing program be instituted to check the accuracy of parentage.

The group went on record favoring the establishment, by USDA, of a livestock quarantine station; the strengthening of the DHI[13] program; and encouraging the young-sire-sampling program.

13.15 Exotic Beef Breeds Popular

During 1968–1969, crossbreeding of beef cattle using frozen semen from Simmental, Limousin, Maine Anjou, French Charolais, and others grew in volume. The above breeds, and many other breeds

Figure 13.1.   Simmental-Shorthorn crossbred calf. Born in 1968 in the herd of Austin V. Rugg, of Garnett, Kansas. The sire is the Simmental bull Parisien and the dam a red Shorthorn. (Photograph courtesy of Curtiss Breeding Service.)

to follow, were commonly knows as "exotics," indicating foreign origin. The exotic breeds for most part are dual-purpose cattle in their native land.

Nearly all semen from the exotic breeds came into the United States by way of Canada, where the bulls were stationed. Canada had earlier established quarantine stations and was in a position to handle the importation of cattle from Europe. Due to the prevalence of foot-and-mouth disease in Europe, importation of cattle from that country into the United States was prohibited. The United States did not have quarantine facilities. The only way U.S. cattlemen could obtain service of most exotic bulls was by using frozen semen that was imported from Canada. Emphasis on crossbreeding of beef cattle became the practice for most commercial herds. Hybrid vigor, faster gains, better reproduction, and more profit were the advantages. Cattlemen were enthusiastic about the results from crossbreeding with the exotics. Their only recourse was to use AI. This development paved the way for AI in gaining a foothold in many herds.

Many bull studs, beginning in 1966–1967, introduced frozen semen from exotic bulls. The Curtiss Breeding Service and the American Breeding Service were among the early leaders. Figure 13.1 pictures one of the first exotic crossbred calves to be born in the U.S.

13.16 1969 Annual Convention[14]

The annual convention was held on 17–20 August 1969 at Madison, Wisconsin. Approximately 450 delegates and guests from thirty-three states and several Canadian provinces were registered, representing thirty-one AI organizations, including four from Canada.

As a result of committee action and recommendations, there were several resolutions adopted at this convention that set forth the position of the NAAB on controversial matters:

A Resolution:

Whereas, The National Association of Animal Breeders and its Member organizations recognize their responsibility and obligation to self-regulate the A.I. industry and related services in the interest of insuring the accuracy of all records pertaining to the parentage of resulting offspring, both registered and non-registered, of all species of livestock, and

Whereas, the record indicates that the N.A.A.B. and its Members have demonstrated their desire and willingness to cooperate with all breed registry societies in formulating and complying with regulations necessary to protect and safeguard the accuracy of the pedigrees recorded in the respective Herd Books, and

Whereas, the Purebred Dairy Cattle Association, with whom we have worked closely and harmoniously, have now proposed for our consideration some revolutionary new concepts in our traditional relationship, and

Whereas, these proposals include some provisions which, in principle, are unacceptable to us, and

Whereas, the bull studs in the United States have banded together in the N.A.A.B. to carry out the business of artificial insemination under a free enterprise and self-regulated system and have been compatible with the philosophies and provisions contained in the existing PDCA "Requirements Governing Artificial Insemination of Purebred Dairy Cattle" and wish to see them preserved, except for such revisions as may be necessary to accommodate new developments and practices within the affected industries.

Now, Therefore, Be It Resolved, that the delegate body of this Convention

(1) recognizes the psychological benefits derived from the spot checking program through blood typing, but questions the value of enlarging the program as a means of promoting a greater degree of accuracy in purebred records and opposes any expansion of the program except when it is at the discretion and expense of the particular breed registry association involved.

(2) opposes the principle of licensing either technicians or bulls for a fee

consideration as not being relevant to the primary concern of protecting or safeguarding the accuracy of the breed association records.

Be It Further Resolved, that the Members of the N.A.A.B.

(1) favor the establishment of a joint committee with equal representation from the PDCA and NAAB to consider such revision of the existing "Requirements Governing Artificial Insemination of Purebred Dairy Cattle" as may be necessary and desirable, including consideration of a new system for approving technicians and handling the movement of semen between involved parties and the adoption of reasonable fees designed to offset direct expenses incurred in administering these functions.

(2) recommend that any such proposals as are agreed upon by the joint committee become effective only after ratification by both the PDCA and NAAB.

The above resolution was prepared by the special committee on AI rules. It evolved as a result of differences between NAAB and PDCA regarding responsibility in conduct of the AI program.

Another resolution was proposed by the beef development committee in an effort to encourage uniformity in beef-cattle-performance-testing procedures and reporting of results:

A Resolution:

Whereas, the improvement of economic traits in beef cattle is heavily dependent upon performance testing and the full utilization of progeny tested sires of superior merit through artificial insemination, and

Whereas, there is need for aggressive growth of the performance testing program for meat animals and likewise standardization of testing procedures, terminology, evaluation methods and reporting of records,

Therefore Be It Resolved, that the delegates at the Twenty-Second Annual Convention of The National Association of Animal Breeders urge that all organizations involved in beef cattle improvement aggressively promote performance testing and that steps be taken to bring about uniform testing procedures, uniform evaluation of results and reporting of information,

Furthermore, Let It Be Resolved, that the members of The National Association of Animal Breeders commend the program of The Beef Improvement Federation, and other organizations endeavoring to expand performance testing and develop uniformity in testing procedures and utilization of performance testing data, and pledge their cooperation in forwarding these programs.

David Yoder, chairman of the research and technical committee, reported that thirteen research projects were supported during 1968–1969. Total expenditures were $21,000.00. Receipts were $25,381.27. The committee recommended twenty projects for support in 1969–1970. The committee also presented abstracts of research papers in the field of AI as compiled from scientific journals.[15]

It was recommended by the research subcommittee that the NAAB establish a research award, to be administered through the awards program of the American Dairy Science Association for the recognition and encouragement in researching animal reproduction and AI. The award was to be presented annually and to consist of a $1,000 check and a suitable plaque. The award was given for the first time in 1970 and was based on research accomplishments within a three- to five-year period.

The committee also recommended support for a proposed symposium on "Sex Differentiation in Animals" to be sponsored by the USDA, the Society of Animal Science, and the NAAB, in 1970. It was proposed that $1,000.00 be made available from reserves in the technical committee fund and $2,000.00 to be pledged by NAAB. The symposium was to involve U.S. and foreign scientists who were working on predetermination of sex in animals. The committee report was approved in its entirety.

Robert E. Walton, of De Forest, Wisconsin, was elected to the board of directors for a three-year term; Erton Sipher, of Governeur, New York, was elected to complete the term left vacant by the late Carl Russell. Keith King, of Oneida, Illinois, was elected to the executive committee.

Chapter 14: First Quarter Century: Silver Anniversary, 1970–1971

14.1 Status of the AI Program

The 1969–1970 NAAB fiscal year marked the thirty-first year the organized AI program was in existence in the United States. The NAAB, approaching its twenty-fifth birthday, came to be recognized for its accomplishments in the livestock improvement field.

Dairymen were reaping the benefits of improved inheritance transmitted by the bulls in the AI organizations. In 1940, the average cow on DHI test produced 8,133 pounds of milk containing 331 pounds of butterfat. The 1970 DHI average per cow was 12,750 pounds of milk and 483 pounds of butterfat, an increase of 4,617 pounds milk, 152 pounds of butterfat, in thirty-one years. The average milk cow in the United States with a production of 6,090 pounds milk, 233 pounds butterfat in 1956, was credited with an average of 9,158 pounds milk, 336 pounds butterfat, in 1969.[1] Without the inherited potential for greater production furnished by the AI program, the above gains could not have been realized. Farmers liked the program. There were fewer questions raised concerning its value.

14.1.1 Increase in Cows Bred

An estimated 8,209,444 head of dairy and beef females were artificially bred in 1969.[2] Dairy cattle accounted for 7,285,063 head and beef cattle 924,381 of the total. The USDA estimated 16,209,000 dairy cows and heifers (seventeen months old) and 39,951,000 head of beef cattle of breeding age in the U.S. in 1969. On the basis of total animals eligible, 44.9 percent of all dairy and 2.3 percent of the beef females were artificially inseminated during the year. There were thirty bull studs in operation and an average of 3,501 cows were annually bred per AI sire.

The introduction of the exotic beef breeds was beginning to have an impact on beef AI growth. Favorable beef and milk prices, with grain feeds fairly cheap, were stimuli to both dairymen and cattlemen to invest in better breeding during 1969–1970.

14.1.2 Member Mergers, Affiliations, and Sales: Affiliated Breeders

A new semen exchange organization, NAL Affiliated Breeders, composed of Noba, Inc., Atlantic Breeders, and Louisiana Animal

Breeders Cooperative began operations 1 March 1969. Each member organization maintained its identity. The chief objective of such affiliations is to share the use of bulls.

Select Sires, Inc. On 23 June 1969, Select Sires, Inc., of Columbus, Ohio, federated with six additional AI cooperatives. The new members were Cache Valley Breeding Association, East Tennessee Artificial Breeders, Michigan Animal Breeders Cooperative, Mississippi Animal Breeders Cooperative, Tennessee Artificial Breeding Association, and Virginia Animal Breeding Association.

With the four original members of Select Sires, Inc., which were discussed earlier, this addition of six bull studs represented the largest consolidation of individual organizations to have occurred in the history of NAAB. Each member organization maintained its identity and continued service in its respective area. However, all major administration and semen production come from the central headquarters located in Plain City, Ohio.

International Beef Breeders Sold: On 7 January 1970, Syntex Laboratories, located in Palo Alto, California, acquired the assets of International Beef Breeders (IBB). The plans called for IBB to retain its identity and to operate as a subsidiary of Syntex Laboratories. International Beef Breeders, a leader in beef cattle AI was the first all-beef stud to become a member of NAAB.

14.2 Beef Performance Testing Increases

The Beef Improvement Federation announced in January 1970 that performance testing in beef cattle was expanding with 12 percent more bulls evaluated for growth rate in 1968 than in 1967. The Federal Extension Service reported a total of 749,034 records obtained on U.S. beef cattle in 1968. There were 56,482 bulls performance tested in 1968 by 4,432 breeders compared to 50,482 by 3,710 breeders in 1967. This expansion indicated a growing realization that scientific procedures should be used in evaluating beef bulls.

14.3 Third Technical Conference

Approximately two hundred persons from twenty-five states and Canada registered for the "Third Technical Conference on Artificial Insemination and Animal Reproduction" held 20–21 February 1970 at Chicago. Prominent reproductive research workers and AI personnel discussed fourteen technical topics. They dealt largely with reproduction in cattle, swine, horses, and turkeys.

After a welcome by NAAB President Baushke, the conference got underway with NAAB executive secretary, Harry A. Herman, elaborating on "Challenges in the 1970's."

One of the speakers, Harold Hafs, of Michigan State University, discussed the current NAAB research program. He said that (1) service industries tend to put less money into research and development than product industries; (2) the present NAAB research program, including that of member organizations, is too limited to sustain desirable industry growth and progress; and, (3) there are unused resources both in researchers and potential benefactors that can be of mutual benefit.

14.4 First NAAB Research Award

The first NAAB Research Award was presented 1 July 1970 at the annual meeting of the American Dairy Science Association held at the University of Florida, Gainesville.

Robert H. Foote, professor of Animal Science at Cornell University, was the recipient. The citation was read by R. E. Erb, of Purdue University, who outlined Foote's extensive research in the physiology of reproduction and in AI. The award, consisting of a plaque and a check for $1,000, was presented by NAAB executive secretary, Harry A. Herman.

14.5 Symposium: Sex Control in Farm Animals

The efforts of man to change the sex ratio in animals are recorded in man's earliest history.[3] While most efforts belong in the folklore category, beginning with the turn of the century application of scientific principles were introduced. None has been successful to date. However, with the growth of the AI program, numerous attempts have been made—and continue—to separate the male-producing and the female-producing sperm. Several attempts have been made to sell franchises, involving sexed semen, to bull studs.

To clear the atmosphere on claims being made, and to bring the latest information available on "sex control" to the livestock industry, a symposium, entitled "Sex Ratio at Birth: Prospects for Control," was arranged. The sponsors were the American Society of Animal Science, NAAB and the USDA Agricultural Research Service.

The two-day symposium was held 31 July–1 August 1970 at Pennsylvania State University, University Park. The leading investigators from the United States, England, West Germany, Israel, and Scotland were invited to speak on the subject of sex differentiation.

Eleven papers were presented by as many authorities. Each presented a somewhat different viewpoint and approach. Every suggested method, based on a scientific approach, for producing sex differentiation, known to date, was reviewed. It was agreed that none of the methods attempted for changing sex ratio had been effective. While some pessimism prevailed, the investigators were in agreement that new approaches should be continued. A proceedings of the symposium was published.[4]

14.6 The 1970 Convention: Utah

The 1970 annual convention was held 15–18 August at Salt Lake City and Logan, Utah.[5] Over six hundred persons from thirty-six states, Canada, and several foreign countries were registered.

David Yoder reported on the activities of the NAAB research and technical committee. As of 30 June 1970, eight research projects had received financial support under the revised program. Total support amounted to $20,147.68 with all but $9,750 derived from funds supplied by NAAB members. He also presented the 1970–1971 research budget, which called for expenditures of $27,705. Howard Kellgren, of Select Sires, reported for the technical subcommittee.

Eight resolutions were presented and adopted at this meeting. They included: (a) commendation and support for the National DHI program and the National DHI Association; and, (b) a strong plea for all concerned agencies to combine forces to expand beef performance testing and to adopt uniform procedures and standards.

In April 1969, President Baushke appointed a bylaws committee consisting of Kenley, chairman, Keith King, and Herman to review the present bylaws and to suggest any necessary revisions. The committee completed its work, recommending several changes that were approved in April 1970 by the board of directors. The proposed changes were circulated to the membership and were approved at the 1970 business meeting. Several statements had been clarified upon advice of the NAAB attorney, Robert C. Smith, of Columbia, Missouri.

Changes presented and adopted by the voting delegates included:

1. Establishment of four membership classifications:
   a. **Regular members.** Those individuals or organizations located on the North American Continent and operating a bull stud producing over 10,000 service units of semen annually.
   b. **Affiliate members.** Non-voting membership for those individuals or organizations on the North American Continent engaged in the artificial

insemination of livestock; custom freezing; and, storage of semen and selling of semen.

c. Associate members. Any individual or organization not eligible for regular, affiliate or Canadian membership, but engaged in, or having an interest in supporting the AI program. Non-voting.

d. Canadian members. Individuals or organizations located in Canada engaged in the operation of a semen producing business, not enrolled as a Regular member. Non-voting.

2. Set forth provisions for reclassification of membership. (This step became necessary because of the many mergers and consolidations of bull studs.)

3. Broadened eligibility for voting delegates by including bona fide employees, director or owner of a Regular membership organization.

4. Provided for the NAAB nominating committee to nominate twice as many candidates as there are vacancies to be filled on the board.

The revised bylaws became effective 1 January 1971.

Newly elected to the NAAB board were Alton Dale Block, manager of Tri-State Breeders Cooperative in Baraboo, Wisconsin, and David Yoder, manager of the Atlantic Breeders Cooperative in Lancaster, Pennsylvania. Erton Sipher was reelected to a three-year term after filling the vacancy created by the death of Carl Russell. W. Lewis Campbell, was elected president, and Robert Walton, vice-president. Sipher was elected to serve on the executive committee.

14.7 The Silver Anniversary Year, 1971

By action of the NAAB board of directors, the year of 1971 was declared to be the "Silver Anniversary" of the association. The 1971 convention was also designated as the twenty-fifth.

While purely happenstance, some confusion developed in the numbering of the earlier, and present, NAAB annual meetings. This discrepancy also occurs in the numbering of the earlier annual convention proceedings. The "Managers' Meeting" of 1946, which was held at Tiffin, Ohio, might well be considered the first meeting. The 1947 annual meeting at Cedar Rapids, Iowa, was instead publicized as the "First Annual Convention—National Association of Artificial Breeding Managers." In 1948, when the annual meeting was held in New York, the publicity and reports stated "First Annual Meeting of the National Association of Animal Breeders." The first-printed proceedings of the association's annual meeting appeared following the meeting held on 17–19 August 1949 at Dyersburg, Tennessee. The program was entitled the "Second Annual Convention of the National Association of Artificial Breeders." The proceedings were similarly worded, but interestingly the cover page reads, "Proceedings National Association of Artificial Breeders of Dairy Cattle."

With the Silver Anniversary Year properly proclaimed, most NAAB events during the year made some cognizance of "25 years old."

14.7.1 More Cows Bred in 1970

During 1970, an estimated 8,578,778 cows and heifers—7,344,420 dairy and 1,234,358 beef—were artificially inseminated in the United States.[6] This was 15.0 percent of the total cattle of breeding age, based on USDA estimated totals of 15,861,000 dairy and 41,289,000 beef females. The figures quoted account for 46 percent of the dairy and 3.0 percent of the beef cattle of breeding age.

Some inseminations with dairy bull semen were to beef cows as ranchers used dairy bulls to increase milk production in the F_1 females. Beef bulls bred 1,234,358 beef cows and 615,322 dairy cows for a total of 1,849,680. The total beef cows inseminated in 1970 are largely those reported by NAAB member bull studs. United States beef breeders were privately importing, by way of Canada, consider-

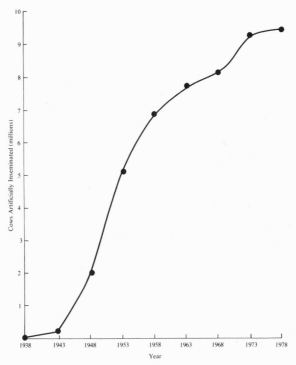

Figure 14.1.   Estimated total cattle (dairy and beef) in the United States bred artificially since 1938. Years of 1972 through 1978 were computed on units of semen —1.7 units = 1 service dairy; 1.3 units = 1 semen beef.

able semen of the Simmental, Limousin, French Charolais, and Maine-Anjou breeds during this period. The number of beef cattle bred with semen resulting from private transactions is not known.

From 1965 to 1971, the trend in dairy cattle inseminations took a healthy upswing (see Figure 14.1). Some of the financial problems the industry encountered during the decline in dairy AI from 1960 to 1965 were easing. NAAB members, as well as the association, had weathered this first serious adjustment in operations. A more optimistic attitude prevailed throughout the AI industry in 1970–1971 than had been true for the past half-dozen years.

14.7.2 Fifth Conference on Beef AI

The "Fifth Annual Beef AI Conference" held 16 January 1971 at Denver, Colorado, was most successful.[7] The conference was dominated by beef producers. Many of the ranchers were involved with the new exotic breeds, two of which (Simmental and Limousin) held their introductory breed sales at the National Western Sales Pavilion with rousing successes. Also well represented were colleges and universities with livestock specialists, researchers, instructors, veterinarians, and students. The registration indicated 438 persons attended, an increase of more than 100 from 1970. Registrants came from thirty-five states. There were visitors from five foreign countries.

Thirteen speakers, including panelists, covered a wide array of subjects dealing with reproduction and beef production in an ever-changing agriculture.

14.8 The Silver Anniversary Convention

The most significant event of the year was the association's "Silver Anniversary Convention." Held 29 August–1 September 1971 at Columbus and Tiffin, Ohio, this convention can best be described as "a series of highlights." About 750 people from thirty-five states and three foreign countries attended. Canada sent a delegation of thirty people. Fourteen manufacturers and suppliers of AI equipment had exhibits.

Clifford Hardin, then U.S. secretary of agriculture, was the keynote speaker. Hardin, who grew up on a farm in Indiana, lauded the role of agriculture in supplying consumers a plentiful supply of food at a moderate price. He said, "The price at which the public buys its food and fiber has risen only 7.8 percent in the past 20 years, while a recent labor contract called for a 42 percent increase over the next 42 months."

NAAB President Campbell reviewed the developments in the AI program the past twenty-five years. He noted that it had been a period of constant change and one of rapid developing technology. Campbell stated, "A Silver Anniversary is one to be proud of and one that recognizes the organization *has fulfilled a need and performed a function sufficiently well* to keep the organization together."

Reports were also presented by the various NAAB committees. The report of the credentials committee was made by Don Pelkey, chairman. He advised that seventy-three official voting delegates were registered. The report of the executive secretary emphasized the accomplishments of three decades of AI in cattle improvement; covered the activities of the association during the fiscal year ending 30 June; the growth of the AI program in the U.S. under a free-enterprise system of self-regulation; and urged that NAAB members continue to stand together on issues of mutual concern. On behalf of the DHI relations committee, Chairman Sipher reported on the working relationships between NAAB and the total DHI program. He also advised that according to a survey made by the NAAB executive committee that NAAB members were directly and indirectly furnishing support of over $300,000 a year to encourage DHI testing. According to Robert Boese, chairman of the research and technical committee, seven projects were supported during the year; the 1971 NAAB Research Award was presented to Harold D. Hafs, of Michigan; and the "Technical Conference" was held. Research receipts totaled $24,043.75 and disbursements $24,285.72. The committee recommended three additional projects for support in 1971–1972 and presented a budget requesting $40,854.00 for research.

During the business session of the convention, reports were presented by the sire health and management committee, Dr. J. B. Herrick, the chairman. He urged positive action on the part of the AI industry on the interstate movement of semen. The 1971–1972 budget, based on an estimated income of $127,087.50 and $126,665 expenses, was adopted. The association had a net gain of $4,825 in 1970–1971.

Eight resolutions were adopted, including:

1. Urge the U.S. Department of Agriculture to negotiate with countries that ban the import of breeding cattle and semen from the U.S. so as to develop ways and means to exchange cattle and semen *on a reciprocal basis*.

2. Petition Congress and the Office of Management and Budget to provide stronger financial support for dairy cattle research at the federal level.

3. Commend the American Dairy Association, the National Dairy Council and Dairy Research Inc. (UDIA) for initiation of research aimed at demonstrating the facts regarding the efficiency of milk and animal products in the human diet.

4. Urge that the U.S. Department of Agriculture be maintained at "full Cabinet status in the U.S. Government."

5. Attempt to amend the Meat Import Act of 1964, which regulates imports of fresh, chilled, frozen, canned, and cured beef so that all types of beef are considered by the U.S. government under beef import quotas.

Three Wisconsin AI technicians were presented the "NAAB 100,000 Cow Award" by George Nichols, the NAAB assistant secretary. The technicians were Anders Spahr, of East Central Breeders in Waupun, Alan Welzien, and Wendell Henke, both of the Tri-State Breeders in Baraboo.

On the last day of the convention, guests gathered in the President's Ballroom to attend a luncheon in honor of Harry A. Herman.[8]

14.9 The NAAB Dedication Ceremony

On 31 August, the men at the convention boarded buses for Noba, Inc., at Tiffin, Ohio, for the NAAB founding dedication ceremony. In cooperation with NAAB, the Ohio hosts had arranged for a large granite stone to which a bronze plaque, bearing the names of the twenty-nine managers who attended the first "Managers' Meeting" at Tiffin, was to be mounted. The granite stone and plaque facing west is within a few yards of the building where the first "Managers' Meeting" was held and the idea for a national association was initiated.

Enos J. Perry, professor emeritus at Rutgers University, and organizer of the first AI cooperative in the United States, presided at the dedication ceremonies.

In his opening address, Perry said, "This organization has put order and continuity and science into the field of animal breeding. Long may it live." He then unveiled the plaque saying, "Now we dedicate this plaque. We dedicate this site, and ourselves, to the continued service and to the genetic improvement of the animal industry throughout the world. Furthermore, we are here to pay heartfelt tribute to those 29 men who were courageous enough and far-sighted enough to meet in those early days of 1946."

Glenn Salisbury, director of the University of Illinois Agricultural Experiment Station and associate dean of agriculture, was introduced next. Salisbury had presented the topic "Looking Ahead in Artificial Breeding" at the 1946 meeting. He recalled "how it all began" including the group meeting in Noba's incompleted building. He said, "The heat, or warmth for that December day was contributed first by a pot-bellied stove, and by the red-hot enthusiasm this group (the 29 managers) held for a new development which would contribute to livestock improvement." He then reads the names of the managers who were in attendance at the first meeting.

Eleven of the twenty-nine managers whose names appear on the plaque were present to participate in the ceremony. As his name was read, each stepped forward and with a wrench gave the bolts holding the plaque in place a symbolic turn.

Preface to the names on the plaque is this message:

> The National Association of Animal Breeders held its founding meeting on this site: December 2–3–4, 1946. This site is hereby dedicated to a quarter century of service by the NAAB and to the honor and memory of the following managers who attended the first meeting.

Lewis Campbell, NAAB president, accepted the plaque on behalf of the association. He then introduced Arthur Loschky, president of Noba, Inc., with the statement that the plaque was entrusted to Noba, Inc., for safekeeping. Loschky accepted the responsibility stating, "We hope to keep this honored memorial in good condition as a reminder, to the agricultural world of tomorrow, of these pioneers and their role as an organization to breed better cattle for better living."

Chapter 15: New Challenges and Goals

15.1 The Second Quarter Century

The fiscal year of 1971–1972 marked the beginning of the second twenty-five years for NAAB. From the standpoint of continuing radical changes in agriculture and livestock technology, in addition to unsettled economic conditions throughout the world, there was little reason to be suspect that the course of the AI program would be smooth. It was now thirty-six years since the organized AI program in the United States had been operating.

In the passing years, revolutionary changes had taken place with respect to: semen evaluation, processing, and distribution; the use of frozen semen had transformed the industry from largely one of service to a sales program; equipment had been vastly improved; sire-evaluation methods had made more progress in the past twenty-five years than in all the centuries preceding; there were fewer AI organizations and more cows were bred per sire and per organization; international trade in semen had developed; and operating under a free-enterprise system there was stern competition on the domestic and international market among all AI organizations.

The NAAB members were strongly united with respect to common industry problems. Since its beginning, not a single member had withdrawn from NAAB because of dissatisfaction with the job the association was attempting to do. Neither had a member been expelled. The task of producing "Better Cattle for Better Living" moved forward under the guidance of dedicated people.

15.2 Status of the Industry, 1971–1972

The number of dairy cows bred to dairy bulls by AI, in spite of a declining cow population, continued a slow upward trend. Crossbreeding in beef cattle steadily increased. The introduction of the exotic breeds from Europe was proving to be a great catalyst in stimulating AI activity in beef herds.

During 1971, an estimated 8,643,089 head of cattle, 7,285,171 dairy, and 1,357,918 beef were bred artificially.[1] This was 15.7 percent of the cattle of breeding age, based on estimates of 14,976 dairy females and 40,123,000 beef females, in the United States. The numbers quoted comprise 48.6 percent of the dairy and 3.4 percent of the beef cattle of breeding age. An average of 3,620 cows were inseminated per AI sire.

Membership. The NAAB membership, in 1971, was the largest to date. There were thirty-seven regular members (voting), thirty-six in the United States; twelve Canadian members (four largely beef); twenty-eight affiliate members, all in the United States; and sixty associate members, thirty-one in the United States, and twenty-nine in fourteen different overseas countries. All beef studs in the U.S. were NAAB members.

15.3 William M. Durfey, Jr., Named Executive Secretary

Since Harry A. Herman, the first executive secretary of the NAAB, had passed the established retirement age of sixty-five by some three years, the NAAB board had been searching for a replacement since 1970. In April 1972, William M. Durfey, Jr., of Houston, Texas, was hired. He assumed responsibility as the NAAB executive secretary 1 September.

Since May 1968, Durfey served as director of performance and research for the American International Charolais Association in Houston, Texas. He worked closely with university and research stations carrying on animal research while coordinating the association's research activities with cooperating institutions. He was responsible for designing, developing, and promoting the Charolais Herd Improvement Program (CHIP), a complete record-processing program for all phases of beef cattle production, and also for coordinating the association's programs with the various state beef cattle improvement associations.

His wide travels have brought him in close contact with all segments of the beef industry and national beef cattle organizations.

15.3.1 J. Rockefeller Prentice

J. Rockefeller Prentice, founder of American Breeders Service, Inc., died on 13 June 1972, at Phoenix, Arizona. While at times because of his strong belief in competition and the free-enterprise system Prentice was a controversial figure, his contributions to the AI industry and the NAAB were immense. More details concerning his role in the AI program will be found in Appendix A.

15.4 NAAB Research Award 1972

E. F. Graham, professor of dairy science at the University of Minnesota, St. Paul, was the recipient of the third NAAB Research Award. Graham, a prolific research worker in AI and animal reproduction, was chosen in part for his studies involving the chemistry of bull and boar semen. The award, consisting of a plaque and a $1,000 check, was presented to him on 29 July at the annual meeting of the American Dairy Science Association in Blacksburg, Virginia.

15.5 Twenty-Sixth Annual Meeting

The twenty-sixth annual meeting of NAAB was held 20–23 August 1972 at Arlington, Illinois.[2] Over five hundred persons registered, from thirty-five states, Canada, Mexico, and the West Indies. There were seventeen exhibitors with AI equipment and products, the most numerous in recent years.

The voting delegates and member organizations approved several bylaw changes, amended the code of ethics, elected directors, adopted the 1972–1973 budget, and passed nine resolutions during this meeting.

The bylaw changes, proposed by the board and adopted by the delegates, were as follows:

Article VI, Section 3—Term. Decreed that any director elected to fill a vacancy for the remainder of an unexpired term did not become ineligible for re-election until the completion of two full three-year terms.

Article VIII, Section 2—Dues and Fees (a) Regular Member. Increased dues to 1-1/4 cents per first service for a maximum of 750,000 dairy and 750,000 beef, and 0.60 cents per ampule or other single service units of semen custom frozen.

(b) *Affiliate Members.* Established dues of 1-1/4 cents per first service insemination on semen sold or used, excluding that semen purchased from other NAAB members, and 0.60 cents per single unit of semen custom frozen.

Section 3. Empowered board of directors to establish conversion factors to convert semen sold or unaccounted for to a first-service basis (after a survey of the membership, a conversion factor of 1.7 units equal one first dairy and 1.2 units equal one first service beef was adopted). (The factor of 1.2 units chosen for beef is due to the fact that the breeding season for beef cattle is short. Most beef cows that do not settle on the first service are turned back to the range and a follow-up bull—usually of another breed—is used).

These revisions of the bylaws became effective 1 July 1972.

The adopted amendments of the code of ethics were approved as follows: Item 4, publication of all sire summaries shall include at least (1) source of information; (2) date of summary; (3) evaluation factor (either PD of sire superiority); and, (4) reliability factor (either repeatability or confidence range).

The resolutions adopted included:

1. Commend the USDA for establishing a maximum security quarantine station at Fleming Key, Florida, and urge the U.S. Congress to make funds available for construction.
2. Thanks to W. Lewis Campbell, outgoing NAAB president, who served six years on the board, two as president, for his contributions to NAAB and the AI industry.
3. A resolution of appreciation to Harry A. Herman for his years of service as the first NAAB executive secretary, and conferring upon him the lifetime title of NAAB executive secretary emeritus.
4. Expression of appreciation to the co-hosts of the twenty-sixth NAAB convention.

During the election of new board members and officers, Wilfred Broussard, of Natchitoches, Louisiana, was elected to a three-year term. Robert E. Walton was elected president, and David Yoder as vice-president. Elected to the executive committee were William Durfey, the NAAB secretary, and Erton Sipher.

15.6 NAAB AI Industry Exhibit and Museum

Located at "The National Agricultural Hall of Fame," at Bonner Springs, Kansas, the NAAB-sponsored AI industry exhibit and museum was erected in 1971–1972. This site was chosen because it is virtually at "the crossroads of the nation." The "National Agricultural Hall of Fame," so designated by the USDA, is a federally chartered nonprofit educational institution.

The NAAB Museum Committee. The desire for an AI industry exhibit and museum had been expressed by many NAAB members over the years. As technological changes came about and equipment became outmoded, there was the feeling that, for historical purposes, the story of the AI program in America should be portrayed. It was further suggested that the NAAB should undertake the task.

Accordingly, president Melvin Kenley, in 1967, appointed the following "AI Museum Committee": Enos J. Perry, New Jersey; Jay Harris, Washington; Wallace Miller, Minnesota; Roy Snyder, Ontario, Canada; Harry A. Herman, Missouri; and Max Drake, Ohio, chairman.

Funds were raised by a special assessment for all NAAB members. The response was nearly 100 percent. The NAAB board authorized support of $2,000. Construction began in 1970 and the exhibit was practically completed during the summer of 1971.

The Spangler Printing Co., in Kansas City, Missouri, and the George Eschbaugh Advertising Firm, in Wilson, Kansas, were engaged to draw up plans and construct the exhibit.

The Exhibit. The purpose of the exhibit is to show the advantages of livestock improvement utilizing AI. The growth of the AI program for both dairy and beef cattle is portrayed through graphs and pictures. There is a display of both old and modern equipment used in AI.

15.7 Developments, 1972–1973

The fiscal year of 1972–1973 found NAAB, under new leadership, appraising the past and gearing programs to fit the changing AI picture. During the year, revisions were made in the bylaws and the code of ethics. A change was made in Columbia headquarters and an international marketing program was initiated.

15.7.1 The Industry Picture

The NAAB membership roster indicated thirty-five regular (voting) members; twenty-nine affiliate members; twelve Canadian; and fifty-eight associate members (thirty-two in the U.S. and twenty-six foreign).

The number of dairy cows remained at about the same level, but beef cattle inseminated showed a continuing gain.

15.7.2 AI Use Estimated on Semen Units

A departure from the earlier practice of estimating the number of cattle artificially bred from reports furnished by AI organizations and custom semen freezing businesses was made in 1972–1973. The USDA and the NAAB that had cooperated for many years in compil-

Table 15.1. Status of Artificial Breeding in the United States: Units of Semen, 1970–1973

	Units Sold		Units Custom-Frozen			
Year	Dairy	Beef	Dairy	Beef	Total Dairy	Total Beef
1970	11,312,269	1,926,930	422,560	979,635	11,734,829	2,906,565
1971	10,926,992	2,078,632	439,468	1,029,529	11,366,460	3,108,161
1972	11,883,075	2,390,505	484,601	1,091,297	12,367,676	3,481,802
1973	11,393,806	2,683,417	570,618	1,146,332	11,964,424	3,829,749

Source: *Dairy Herd Improvement Letter* 51:2 (1975), USDA.

ing estimates on cattle artificially bred decided to base tabulations on units of semen sold and units of semen custom frozen.

This change, while made reluctantly, came about by necessity. In 1971, and beginning earlier, it became obvious that many AI organizations were unable to assess accurately the number of first services resulting from semen they distributed. Each year, more semen was being sold directly to herd operators. Custom freezing of semen from privately owned bulls, particularly in beef, was growing steadily. Some breeders would buy semen, or have it custom frozen and stored for future use. There also came into the picture an increasing number of "semen dealers" who purchased frozen semen and merchandised it. It was impossible to know how much semen was used and how much remained stored in liquid nitrogen "jugs" throughout the country each year.

The obvious disadvantage of the unit system of reporting is that the number of cows artificially bred cannot be accurately estimated in our present AI program. To minimize the disadvantages of being unable to compare the number of cows artificially bred with preceding years, the USDA and NAAB coordinated the collection and analysis of data obtained from 1970 to 1973. These estimates on the basis of semen sales and custom freezing are shown in Table 15.1.[3]

Some idea of the cows actually artificially inseminated can be gained by converting semen units to first services. Neither the USDA nor NAAB have recommended a conversion factor to be used with the semen unit plan. However, in August 1972, the NAAB board approved 1.7 units for dairy and 1.2 units for beef in figuring assessments. While purely an empirical approach, the writer found that by assuming the semen units sold plus one-half the units custom frozen approximated the total semen units utilized. Dividing the total units for dairy by 1.7 and the total beef units by 1.3, the estimates closely agree with those published for 1970 and 1971 for cows bred artificially (see Table 9.1 and Figure 14.6). It is not suggested that these conversion factors would be suitable for all situations.

15.7.3 Carnation-Genetics Affiliation

On 1 November 1973, the Carnation Company, in Los Angeles, California, acquired the assets of Genetics, Inc., in Hughson, California. The name of the firm became Carnation Company, Genetics Division, with Harold Schmidt, founder of Genetics, Inc., serving as division manager. Clarke A. Nelson, vice-president, Carnation Co., was named an officer and adviser for the new division that included bull studs at Hughson, California; Watertown, Wisconsin; and Carnation, Washington.

15.8 Preliminary Sire Summaries

During the 1972 annual convention, Dean Plowman and Frank Dickinson, both of the USDA Dairy Division, met with the NAAB board and proposed a plan to speed up the proving of young sires; namely, the arranging for the DHI record-processing centers to make early summaries, thus providing preliminary proof on young sires rather than waiting for the regular or semiannual summary. The cost of this extra service was to be borne by the AI organizations. However, the advantages of early proof would result in considerable savings in sire proving and increase the number of bulls that could be sampled. The cost was estimated to be about $22,000 annually.

The board viewed the proposal favorably and requested Durfey to make a survey of NAAB member bull studs to determine if they were willing to pay 0.3 cents per usable record for the service. This survey found bull studs willing to finance the plan, and it was put into effect.

15.9 Foreign Marketing Committee Appointed

With increased activity developing on semen exports and international trade, President Walton and the NAAB board established a foreign marketing committee in June 1973. The following were appointed to the committee: Donald Ausman (Midwest Breeders Cooperative), chairman; Robert Boese (Curtiss Breeding Service); Richard Chichester (Select Sires); Richard Cotta (Carnation–Genetics); and Robert Pelkey (American Breeders Service).

The name of this committee was later changed to international marketing committee.

15.10 Seventh Conference on Beef AI[4]

The 1973 NAAB-sponsored conference on "Artificial Insemination of Beef Cattle" was held 13 January at Denver, Colorado.

The general conference theme, "Options for Profit Using AI," was developed by beef industry leaders including ranchers, AI personnel, breed association representatives, and college workers.

The leadoff speaker was Lloyd Miller, the executive secretary of the American Angus Association. In outlining progressive stages in the breed's history, he said that liberalizing AI rules in 1972 by the Angus Association board was an important forward step, one that coincided with implementing the National Angus Sire Evaluation Program. Opening up AI regulations, he said, increased AI service certificates issued by the association from 3,789 in 1971 to 10,971 in 1972. AI in Angus herds, he said, will increase, along with the number of Angus bulls in AI service. Robert Long, vice-president of the Ankony Angus Corporation, emphasized cow-herd fertility and daughters of sires doing a good job in selecting herd replacement heifers.

Ancel Armstrong, president of New Breeds Industries in Manhattan, Kansas, outlined the profit-making potential involved in using the new beef breeds making their American debut.

15.11 A New Headquarters Building

Due to an extremely dry summer in 1972, the ground along side the NAAB headquarters building had settled and this was aggravated, as was found later, by a blocked drain from a downspout.

The board of directors debated the matter of (a) repairing the building; (b) moving NAAB headquarters to another city; or, (c) erecting a new building in Columbia. A committee consisting of President Walton, Vice-President Sipher, David Yoder, and Executive Secretary Durfey was established to study the matter and make recommendations. This committee moved rapidly. After checking the cost of real estate and of operational expenses in other cities, the decision was reached (1) to sell the building in its present state for $50,000, and (2) to erect a one-story building meeting the needs of the NAAB in a Columbia location.

There were several bidders for the building. On 12 February 1973, George Fuchs, a practicing Columbia physician who had opened his first office in the building, purchased it.

The board then purchased a lot at 401 Bernadette Drive in a newly developing business area and contracted for the construction of a single-story steel building measuring 40 X 60 feet. The new building is a steel-framed structure with a brick-facade front. It was completed, sufficiently for occupancy, by 1 August 1973, and the furnishings and equipment from 512 Cherry Street were moved into the new location on 3 August.

The building was dedicated at the time of the NAAB board meeting on 30 October 1973 in Columbia.

The new headquarters has ample parking space. The interior is carpeted throughout and has about 1,400 square feet of secretarial and office space. There is about 800 square feet for working space, equipment, and storage at the rear of the office section.

15.12 Mrs. Dorothy R. Schnering, 1893–1973

Dorothy R. Schnering, former president of Curtiss Breeding Service, died 17 February 1973.

Dorothy was the only woman to date in the United States to serve as president of an AI business. She was the widow of Otto Y. Schnering, founder of Curtiss Candy Company Farms. The Curtiss AI program, launched in 1949, grew from the Curtiss purebred herds. (Further details on Curtiss Breeding Service will be discussed in Appendix A.)

15.13 NAAB Research Award to William E. Hansel

Recipient of the 1973 Research Award was William E. Hansel, professor of animal physiology at Cornell University. The fourth winner of this award, Hansel was chosen for his research in animal reproduction and AI. He has been engaged in research on estrus regulation of cattle for many years.

The award was presented 26 June 1973, at the sixty-eighth annual meeting of the American Dairy Science Association held at Washington State University in Pullman.

15.14 International Trade Activities

The export of frozen semen to overseas countries and developing countries elsewhere by U.S. bull studs steadily increased. The NAAB in cooperation with the USDA and the Foreign Agricultural Service began participation in a seven-year research program with Poland. The project called for the mating of twelve thousand Polish-Friesian cows annually with semen from young, unproved, Friesian sires from nine participating countries. Each country, including the U.S. is to furnish semen from forty young sires annually. The objective is to determine which crosses work best and to improve Poland's national dairy herd of about eleven million cows.

During the year, NAAB also sponsored an exhibit booth, in cooperation with several AI organizations doing overseas business, at the Cremona Trade Show in Italy. The availability of frozen semen from U.S.-owned proved sires was emphasized.

15.15 The 1973 Convention

The 1973 NAAB convention was held on 12–15 August at New Orleans, Louisiana.[5] From the standpoint of attendance, program features, and entertainment, this was one of the more successful NAAB meetings. Almost seven hundred persons registered. They came from thirty-eight states, Puerto Rico, and Canada. Forty-one AI organizations were represented. During the three days, there were many speakers, special-interest sessions, panel discussions, and the association's annual business meeting.

NAAB President Walton, in his address to the delegation, said that 1972–1973 had been a dynamic year for the association, a year in which "basic issues have been faced and positive action initiated." Walton also paid tribute to Durfey, serving his first year as executive secretary, for picking up the reins of responsibility in this critical time of industry changes and "changing of the guard."

Durfey, the NAAB executive secretary, advised that during 1972 some 12.8 million units of dairy semen were sold for an increase of 7.6 percent over the preceding year. Beef domestic semen sales totaled 2.4 million units, up 18.2 percent over 1970.

Reports were presented by the nominating committee and the treasurer. Resolutions and proposed bylaw changes were presented, briefly discussed, and adopted during the business session. The 1972–1973 budget was also adopted.

Due, in part, to the association converting from a first service basis to a semen unit basis for reporting the volume of business done by members, portions of the bylaws were amended as follows:

1. *Regular members* were defined as those firms on the North American Continent, engaged in the operation of an AI business, and producing 10,000 service units of semen annually.
2. Established the number of voting delegates permitted Regular members on the basis of one voting delegate for the first 99,999 service units of semen to as many as seven voting delegates for those businesses producing 1,250,000 or more service units of semen annually.
3. Established annual dues for Regular members; Canadian members; and, Affiliate members on the basis of semen units.
4. Added an amendment enabling changes in the Code of Ethics, as well as the bylaws, to be made at any called Annual Meeting of the Association.

The amendments of the code of ethics involved beef progeny testing; preliminary dairy sire proofs; and inseminator-training programs.

Newly elected to a three-year term was Wallace E. Erickson, of Waupun, Wisconsin.

Three AI servicemen were presented the "NAAB 100,000 Cow Award." They were Phil Foth and Roy Honer, both of Tri-State Breeders, and Melford Sorensen of East Central Breeders.

Chapter 16: Thirty-Five Years of Service: Markets Expand

16.1 Domestic and International Semen Sales Improve

The year of 1973 marked the thirty-fifth year the organized AI program had been in operation in the United States. Due to the dedicated efforts of thousands of workers in the AI field, cattle breeders, breed registry organizations, agricultural college research and extension personnel, and the USDA, continuous progress had resulted.

With the reporting of AI volume in terms of semen units sold and custom frozen, the status of the dairy program for 1973 as compared with 1971 and 1972 is shown in Table 16.1.[1]

Dairy cattle numbers had declined slightly since 1973 and the AI volume for dairy had likewise been at a fairly constant level. Custom freezing of dairy bull semen increased 25.2 percent in 1972 over 1971 and in 1973 was 27 percent higher than in 1972. An increasing number of privately owned dairy bulls were custom collected, and, in many cases, the semen entered trade channels.

The picture for beef AI in 1973 and as compared to the two preceding years is shown in Table 16.2.[2] A gain of about 14 percent on domestic semen sales was recorded for both 1972 and 1973. During both 1972 and 1973, international sales of semen showed considerable gain. Dairy bull semen sales led the way and are indicative of the

Table 16.1. Dairy Semen Sold and Custom Frozen

Year	Domestic	International	Total Sales	Custom	Grand Total
1973	11,338,980	682,307	12,021,287	534,972	12,556,259
	(−4.05%)	(4.3%)	(−3.6%)	(27.0%)	(−2.6%)
1972	11,817,867	654,162	12,472,029	421,060	12,893,089
	(8.65%)	(39.8%)	(9.9%)	(25.2%)	(10.4%)
1971	10,876,840	467,989	11,344,829	336,314	11,681,143

Source: Data from NAAB files.

Table 16.2. Beef Semen Sold and Custom Frozen

Year	Domestic	International	Total Sales	Custom	Grand Total
1973	2,719,052	348,733	3,067,785	1,092,334	4,160,119
	(14.1%)	(63.3%)	(18.15%)	(−4.6%)	(11.2%)
1972	2,382,955	213,590	2,596,545	1,144,903	3,741,448
	(14.7%)	(34.2%)	(16.1%)	(8.5%)	(13.7%)
1971	2,076,778	159,121	2,235,899	1,055,409	3,291,308

Source: Data from NAAB files.

esteem accorded the dairy sires that are available in America. International sales of beef semen also showed a healthy gain, up 135,000 units for a 63 percent gain. However, dairy semen exports were nearly double those for beef.

The increase in international semen sales can be credited to a forward step by NAAB in arranging for a marketing agreement with the USDA and the Foreign Agriculture Service in 1973. The export of frozen semen has continued to grow and is an important source of income for many semen-producing businesses in the United States. During 1979, semen exports totaled 2,156,000 units, an increase of 21 percent over 1978.

16.2 Product Liability Disclaimers

The NAAB developed statements relative to product liability for the information of patrons. In view of the normal biological variations in semen quality and expected conception rates, this step became logical.

It must be realized that with much of the frozen semen being sold directly to and utilized by herd operators, the semen-producing business has little control over the handling and utilization of the semen once the sales transaction is made.

16.3 Requirements Governing AI of Purebred Dairy Cattle

The long-delayed revisions in "Requirements Governing Artificial Insemination of Purebred Dairy Cattle" became effective 1 July 1974. The changes adopted were made necessary by the progress of the AI program—frozen semen, direct semen sales, and semen merchandisers. The product of a joint committee representing NAAB and the PDCA, the revisions were put into effect following negotiations that were begun in 1968. The dairy breed associations are by virtue of their articles of incorporation charged with the enforcement of requirements pertaining to animals recorded in their respective herd books.

16.4 The Eighth Conference on AI of Beef Cattle: "How to Make AI Work"

The "Eighth Conference on AI of Beef Cattle" was held 12 January 1974 at Denver.[3] Over 550 persons attended this one-day conference. While beef cattle ranchers predominated in numbers, there were twenty-four AI organizations from the U.S. and seven from Canada represented.

The theme for the conference was "Procedures to Maximize Conception." As an interesting sidelight, a survey, concerning the use of AI in beef herds, was conducted. Eighty-eight ranchers responded. They represented ranching operations in twenty-one states and Canada. Their breeding-cow herds averaged 622 head. Over 80 percent of the ranchers used direct herd service with either the herd owner and/or hired help doing the inseminating. The remainder relied largely on commercial AI technicians.[4]

Dan Laster, of the U.S. Meat Animal Research Center at Clay Center, Nebraska, stated that good management was the key to high conception. Factors he listed as affecting the conception rate included nutrition, state of lactation, calving interval, abnormal heat cycles, failure to show estrus and breed effects. He said that during the past ten years conception rates had been improved largely by better management and by educational programs conducted by AI organizations. E. J. Carroll, veterinarian from Colorado State University, cautioned ranchers to use only semen from bulls that have been under veterinary surveillance to minimize disease. Also, if teaser bulls were used for heat detection, they should originate within the herd.

Azel Lewis, of Lake City, Florida, and former manager of Duda Ranches, explained how their program grew into one of the country's most extensive beef AI operations, with a conception rate reaching 90.3 percent on ten thousand head. He credited the program's success with good management, improved nutrition, and dividing the breeding herd into six groups to stagger the insemination schedule.

16.5 Fifth Technical Conference

The "Fifth Technical Conference on Artificial Insemination and Reproduction" was held 14–16 February 1974 at Chicago.[5]

Several of the topics discussed included freeze-drying of bovine spermatozoa. E. F. Graham, of the University of Minnesota, emphasized the critical points of using this procedure and said that the presence of glycerol was a likely reason for earlier failures in freeze-drying bovine spermatozoa, and that buffer systems have been developed that permit successful freeze-drying of semen without the use of glycerol. In reporting on the status of ova transplants, Graham said that a survey showed more than six hundred pregnancies were recorded by seven organizations using the technique. He predicted that, even with the problems involved, it would become an important tool for rapid genetic improvement. R. G. Saacke, of Virginia

Polytechnic Institute, discussed effects of freezing, thawing, and extenders on semen packed in straws. Optimum results were achieved with increased glycerol levels and rapid thawing. W. E. Erickson, manager of the East Central Breeders in Waupun, Wisconsin, reported on AI in swine. With frozen semen, farrowing rates have averaged about 50 percent, ranging from 30 to 90 percent. He said that farrowing rates are acceptable if the sows are inseminated on the second and preferably the third day of heat. B. W. Pickett, of Colorado State University, reported varied results with heterospermic fertility involving cattle and rabbits. His studies showed higher fertility when semen from three bulls was mixed. In another session, Pickett reported that successful results from frozen horse semen depend on developing better extenders. The economic aspects of banking bovine semen were discussed by Glenn Coulter, of Cornell University. He outlined the precautions to be followed in sampling bulls to make semen banking economical.

16.6 Semen Shipped to Poland for Research: International Trade Fairs

Early in 1974, following plans that were begun in 1972, the first lot of semen made available by NAAB members was shipped to Poland as part of a study of Holstein-Friesian cattle from nine different countries. The project is carried out in cooperation with FAO and is partially financed by PL 480 funds.

Continuing efforts in developing international trade in semen, the NAAB international marketing committee staged exhibits at trade fairs in Cremona and Verona, Italy. A new brochure entitled "The Artificial Insemination Industry of the USA" was prepared and published in English and Italian, as well as in other languages for worldwide distribution.

16.7 The 1974 Convention

The 1974 NAAB convention was held 11–14 August at Toronto, Canada. This annual meeting proved to be one of the largest ones held to date. There were about 750 persons registered from thirty-five states and ten Canadian provinces. Twenty commercial exhibitors displayed AI equipment, supplies, and breed literature.

In his address to the delegation, NAAB President Walton outlined the numerous NAAB programs in process. He said, "The future of AI, whether dairy or beef, is as big as we dare prepare for. The only limitation is our ability to perceive, plan and implement for the future needs of a great and growing industry.''

During the business session of the convention, Bob Boese, vice-president of the Curtiss Breeding Service in Cary, Illinois, and Brainard Palmer-Ball, dairy leader from Louisville, Kentucky, were elected to serve on the seven-man NAAB board, replacing Victor Gray, Sedro Woolley, and Ronald Hawkes. The new officers were David Yoder, elected as president; Erton Sipher, as vice-president. Wallace E. Erickson was named to the executive committee.

In addition to election of directors, delegates at the business session also voted several bylaw amendments, adopted resolutions, and established assessments to support the NAAB research program. The bylaw amendments adopted were (1) member, which enabled federated members of an AI business, custom-freezing business, and bull studs to become eligible for regular membership if the necessary requirements were met; (2) voting, which increased the number of voting delegates in proportion to dues paid; (3) board of directors, which became a board of nine directors rather than seven as provided in the original bylaws; and, (4) dues, fees, and assessments, which provided for support of the NAAB coordinated research program by voting three-tenths of one-cent assessment on each service unit of frozen semen, either sold or custom frozen, up to a maximum of 1,250,000 units from dairy sires and 1,250,000 units from beef sires.

The resolutions that were approved included: (a) to encourage all NAAB members to sample as many young sires each year as economically feasible, and, (b) to express appreciation to the Ontario hosts for a delightful and informative convention.

Special recognition was given to AI servicemen who, during the past year, had received the "NAAB 100,000 Cow Award." They were Francis Moon, of Tri-State Breeders; Marcel D. Showers, of East Central Breeders; and Harvey Nichols and Oliver Voigt, both of Midwest Breeders.

Lloyd Dale Van Vleck, professor of animal science at Cornell University, was introduced as the 1974 NAAB Research Award recipient. His contributions in developing methods for the application of quantitative genetic information in sire evaluation are numerous. He is the author of over 130 scientific publications.

16.8 Decline in Beef Prices, 1974–1975

During 1974–1975, the price of commercial beef cattle declined some 30 to 50 percent. Cattle were in plentiful supply with about 132 million head reported by the USDA on 1 January 1975. This situation

Table 16.3. Semen Sales and Custom Freezing in U.S., 1974–1975 [1]

	Dairy Units	Beef Units
Domestic Sales	10,887,613	2,869,962
International Sales	761,756	385,153
Custom Freezing	594,804	1,673,993

[1] Preliminary NAAB figures.

had a momentarily depressing effect on artificially inseminated beef cattle. It also served to reduce some of the enthusiasm for crosses to the exotic breeds. The result was a decline in the units of beef semen that were sold. In August 1975, the NAAB reported on semen units sold and custom frozen (see Table 16.3).[6]

It is believed that ranchers and dairymen who had a reserve of semen in storage tended to use it and avoid cash outlay in 1974–1975. There was also a decided increase in the number of beef bulls custom collected during this period, another indication of reducing cash expenditures by cattlemen. Historically, slumps in milk prices have had a depressing effect on the volume of AI. It is reasonable to assume that the same situation applies to beef AI.

16.9 International Semen Sales Expanded

One of the highlights in 1974–1975 was the continued growth in international sales of semen. The international marketing committee and the NAAB executive secretary maintained a steady effort to develop overseas sales (see Tables 16.1, 16.2, and 16.3). Many contacts were made and several "technical teams," coordinated by NAAB and the USDA Foreign Agricultural Service, visited many foreign countries to promote superior germ plasm furnished by bulls in the United States.

During May 1975, a five-man team of USSR research and technical specialists in crop and livestock production visited the United States. Their trip was coordinated by the USDA and NAAB. In a nineteen-day tour, they visited agricultural experiment stations and several universities to review projects on reproductive physiology. They also visited nine NAAB member organizations and became acquainted with the conduct of the program for AI as conducted in the United States.

For the second year, NAAB sponsored an exhibit booth at the Verona International Fair (Italy), in cooperation with American

Breeders Service, Carnation-Genetics, World Wide Sires, and USDA Foreign Agricultural Service.

These activities are promoting a steady increase in semen sales to overseas countries. Some of the health restrictions imposed by other countries on U.S.-produced semen (in many instances serving as a trade embargo) are being overcome.

16.10 "Know Your Semen Supplier" Campaign

In late 1974, the dairy and AI industries in North America were distressed by the first-known fraudulent merchandising of a counterfeit product that was represented as being frozen semen from prominent registered Holstein-Friesian bulls. Also, a large quantity of frozen semen was stolen from one of the AI organizations in Canada, and semen had been smuggled back and forth across the U.S. and Canadian borders violating customs regulations. Through the vigilance of members of the Ontario Association of Animal Breeders and the Ontario Canada Provincial Police, the distributor, and his accomplice, was apprehended.

The honesty and integrity of cattle breeders, likewise producers and sellers of semen, are paramount in establishing and maintaining the confidence of patrons. As a result, AI organizations proceeded to keep stored semen under lock and key; inventories were regularly made, and sales and shipments were more carefully recorded. The NAAB immediately launched a program to emphasize to semen purchasers the importance of, "Know your semen supplier and the prod-

Figure 16.1. Insignia and slogan: "Know Your Semen Supplier." Used by NAAB members in conjunction with the association's emblem.

uct he sells." This program stresses the wisdom of dealing with known and legitimate AI organizations. Commonly called the "Know your semen supplier program," it has been adopted as part of the advertising slogan of most NAAB members.

The auditing program of Certified Semen Services, Inc. (see Chapter 17) now assists in protecting the integrity of semen supplied by its members.

16.11 Swine AI Developments: U.S.

Artificial insemination in swine, as an adjunct to cattle insemination programs, was attempted by several NAAB members beginning in 1966–1967. Liquid semen was used. Most organizations found it difficult to make the program pay its way. One organization that persisted and made reasonable progress was the East Central Breeders in Wisconsin.

In 1974, as a result of technical developments by E. F. Graham and Bo Crabo, both of the University of Minnesota, frozen swine semen in pelleted form became commercially available. Efforts were made to develop a large swine AI cooperative in Wisconsin that would involve several AI organizations, pork producers, and meat-packing establishments. This move was encouraged by East Central Breeders. The plans for a large cooperative, however, never fully materialized, and in November 1975 East Central Breeders Cooperative sold the swine AI division to United Suppliers, Inc., in Iowa. Jack Eichel, swine AI supervisor at East Central, was hired by the buyers to manage the project.

The swine AI program operated by United Suppliers, Inc., has grown steadily.[7] Further refinements have been made in swine semen freezing, storage, and shipping. Insemination techniques have been improved and field operations have been made more practical. At last reports, United Suppliers, Inc., is furnishing frozen semen from outstanding boars of several breeds to hog operations in several states. Swine semen is being exported. The conception rate ranges from 50 to 80 percent on a first-service basis and litter size averages about eleven pigs.

Swine AI is practiced by several private operators in Indiana, Illinois, and several other states on a limited scale. Developments on swine AI in Canada are reported in Appendix A.

These developments represent a continuing effort on the part of U.S. swine producers to utilize AI. The cost of labor has been looked upon as the strongest deterrent to the program. The use of frozen semen offers a means to make the program more practical under U.S. conditions.

16.12 NAAB Research Award to Freeman

The recipient of the 1975 NAAB Research Award is Dr. A. E. Freeman, professor of animal breeding and genetics at the Iowa State University in Ames. Dr. Freeman's research efforts have emphasized sire evaluation, cow evaluation, and genetic parameters. His work has added to the accuracy of estimation of bull and cow-breeding values. His counsel has been of great value to AI organizations throughout the nation, and he has participated in many NAAB-sponsored conferences.

16.13 "Pete" Schmitz Establishes Record

Too much credit cannot be given the inseminator-technician. As emphasized in the section on NAAB technician awards, the men who serve as inseminators for the various AI organizations are a dedicated lot. One of these, "Pete" Schmitz, a serviceman (inseminator) for Tri-State Breeders, established an all-time high record of 6,811 first services during 1974. While several inseminator technicians have exceeded the 6,000 cow barrier for a year, the records indicate Schmitz was first on the list for 1974.

16.14 Twenty-ninth Annual Convention: Dallas, Texas

This NAAB annual convention was the first to be held in the State of Texas. On 17–20 August 1975, approximately 575 delegates, representing the U.S., Canada, and four foreign countries, attended this meeting. This convention was historic in the annals of NAAB, in addition to a fine program, for two reasons: (a) the NAAB board was increased to nine directors; and, (b) the association voted to adopt a program of "self-regulation." Executive Secretary Durfey got the meeting underway by reviewing the industry's economic status and reported that during the fiscal year (1974–1975) over 12 million units of dairy semen and 4.9 million units of beef semen, including domestic, custom, and international, were produced and sold. Dairy AI volume remained steady but there was considerable decline in beef AI (see Table 16.3).

As previously mentioned, the board was increased from seven to nine members. Robert E. Walton, having served two consecutive three-year terms, was not eligible for reelection. Wilfred Broussard completing his first term as a director was reelected, but for a two-year term. Three additional directors were elected: Thomas Lyon, representing Midwest Breeders Cooperative in Shawano, Wisconsin;

Charles Michaels, Kansas Artificial Breeding Service Unit in Manhattan; and Bryce Weiker, Noba, Inc., in Ohio.

The resolutions that were adopted during this meeting included:

1. Commend the Ontario Canada Provincial Police for an alert and thorough investigation in the first-known fraudulent merchandising of frozen semen;

2. Urge the veterinary services of the Animal and Plant Health Inspection Service of the USDA to enact regulations governing the health of bulls producing frozen semen for interstate shipment;

3. Authorize the NAAB board of directors to develop and implement a program of self-regulation for members and nonmembers alike if such a program is in the best interests of the AI industry.

Special recognition was given to six AI technicians who had qualified for the "NAAB 100,000 Cow Award": William T. Williams (ABS, Wisconsin); Donald Paulson (East Central Breeders); Alfred Harlow (Midwest Breeders, Wisconsin); Dan Schultz (Midwest Breeders); Goodwin Anderson (Tri-State Breeders); and Keith Wolfe (Tri-State Breeders).

Chapter 17: A Matured Program: New Horizons, 1975–1978

17.1 Broad Program of Activities Outlined

The year 1976 marked the thirtieth year of existence of the NAAB. Now a matured and respected force in livestock production circles, the association was active in nearly every important program involving genetic improvement in dairy and beef cattle as well as other farm animals.

Following the twenty-ninth annual convention at Dallas, the board and officers gave attention to an expanding program of activity for the association.[1] Major considerations included:

1. Appointment of a steering committee, composed of the executive committee and an ad hoc committee, to study self-regulation and to carry out the legal steps to form a subsidiary organization, wholly owned by NAAB to provide services for the AI industry;

2. To revise the association bylaws to meet legal requirements growing out of interpretations associated with a trade organization to provide flexibility in meeting current conditions;

3. Implementation of a program to list bulls in AI use reported to be a carrier of genetic defects. (This type of a program had been a part of NAAB activities since 1953 and involved the reporting and withdrawal of such bulls on a voluntary basis by member bull studs.);

4. To emphasize the development of international markets for frozen semen inculding: (a) representation at the "Eighth International Symposium on Animal Reproduction and Artificial Insemination," in Krakow, Poland; (b) visit state farms in Poland where the research project in which Holstein-Friesian bulls from the U.S. are being compared with sires from eight other countries is underway; (c) to continue sending technical teams to Europe, including the USSR, and to participate in trade fairs; (d) to produce a brochure setting forth improvement in milk production resulting from the use of AI proved sires;

5. To develop standards for semen quality that might have general use;

6. To encourage and support research and programs aimed at improved evaluation methods for dairy and beef sires;

7. To strive for adoption of federal regulations governing the interstate movement of semen and to cooperate with USDA officials in efforts to secure uniform sire health regulations;

8. To evaluate the role of the *A.I. Digest* with respect to editorial content, advertising potential, readership, and costs, and to seriously consider an industry advertising program.

17.2 Certified Semen Services, Inc.

In February 1976, NAAB representatives attending the annual meeting of the Purebred Dairy Cattle Association at Columbus, Ohio, announced the formation of a subsidiary organization, the Certified Semen Services, Inc. (CSS).[2]

This action resulted from the resolution calling for "self-regulation," which was adopted at the twenty-ninth annual convention. It was explained that CSS would endeavor to provide: product (semen) protection; technical assistance for accurate identification of sires and semen; aid in sampling young sires; and to create confidence among all users, beef and dairy, of frozen semen. In essence, the CSS program is designed to provide "accurate and positive identification of bovine semen from production to the point of sale." It is considered a "semen identification auditing service."[3]

The first CSS officers consisted of: president, Charles Krumm (manager of Eastern AI Cooperative in Ithaca, New York); vice-president, Robert E. Walton (president of American Breeders Service in De Forest, Wisconsin); treasurer, David Yoder (NAAB president and manager of Atlantic Breeders Cooperative in Lancaster, Pennsylvania); secretary and executive vice-president, William M. Durfey (executive secretary of NAAB in Columbia, Missouri); and board member, Thomas Lyon (Midwest Breeders Cooperative in Shawano, Wisconsin).

Certified Semen Services was established as a self-supporting NAAB subsidiary with fees paid by participating AI organizations. The NAAB bylaws were revised in 1976 requiring all NAAB members to belong and support the CSS program. However, NAAB membership is not a prerequisite for participation in the CSS program. The outline of proposed activities as presented to the dairy breed registry organizations in part supplanted some of the "PDCA Requirements Governing the Artificial Insemination of Purebred Dairy Cattle" that were adopted in 1943.

The "PDCA Requirements" as originally developed were established by the Purebred Dairy Cattle Association, Inc., and the dairy cattle breeding committees of the American Dairy Science Association. The guidance of the AI program in America at this time was largely through efforts of dairy extension workers at the various agricultural colleges and dairy breed organizations (PDCA). The NAAB was not organized until 1946–1947, and prior to 1953, the AI organizations had no formal spokesmen in the formulation of the rules pertaining to the use of AI in purebred cattle.

The "PDCA Requirements" were revised numerous times as the AI program developed, as frozen semen gained popularity, and as direct semen sales increased. The NAAB as early as 1953 had representatives on the PDCA committees charged with developing rules and regulations for the AI program. The PDCA regulations for AI served a most useful purpose in guiding the program. The philosophy that integrity and accuracy in conduct of the AI program for grade cattle are just as important as for purebreds prevails.

The formation of CSS in 1976 may be viewed as a further maturing of NAAB and the AI program in America, which is conducted by self-imposed regulations. With frozen semen becoming a trade commodity—bought and sold by various individuals, custom collected, stored, often sold, and with identification entirely dependent upon the integrity of the seller—it became increasingly apparent that the responsibilities of the AI industry should be enlarged. The goals for CSS were summarized by NAAB President David Yoder as follows: (a) our basic moral and ethical obligation to protect the user of our product; (b) to prevent the need for formal government intervention or control; and (c) to enhance the overall image of and confidence in the AI industry and thus, hopefully, increase the use of herd-improving sires.[4]

Further objectives of CSS considered the future standards of semen quality, sire health, and semen identification.

On 1 August 1976, Gordon A. Doak, of Columbia, Missouri, was named service director of CSS.[5] One of his first assignments with CSS was to initiate a national program as it relates to CSS members. He was also named technical director of NAAB.

17.2.1 Each Dairy Breed Establishes AI Requirements

There was some disagreement from 1968 forward among the six dairy breed organizations in establishing a uniform set of rules governing AI. There are responsibilities and obligations, however, that no breed registry organization can delegate. These include blood typing and listing of bulls used for artificial insemination. All breeds were agreed on these matters. The general philosophy prevailed that the burden of proof regarding information furnished on an application for registration is the responsibility of the owner of the animal. The position of NAAB members, and likewise CSS, is that a bull stud, or semen-producing business, can be responsible only to the extent that the semen unit sold by it is bona fide with respect to the sire indicated. A part of the function of CSS is to carry out a surveillance program among its participants to insure that each unit of semen is accurately identified.

In late 1976, each of the six dairy breed registry organizations decided to establish its own requirements governing the registration of purebreds that resulted from AI. The uniform "Requirements

Governing Artificial Insemination of Purebred Dairy Cattle'' and the joint PDCA-NAAB committee on AI were abolished.

The AI requirements of five of the dairy breed organizations are uniform at the present time, and all six have similar objectives.[6] The role of CSS is being well accepted throughout the cattle-breeding industry. It serves as the service arm of NAAB.

17.3 J. O. Almquist Receives 1976 NAAB Research Award[7]

The 1976 NAAB Research Award was awarded to J. O. Almquist, professor of dairy physiology at Pennsylvania State University. The presentation was made at the seventy-first annual meeting of the American Dairy Science Association, which was held in Raleigh, North Carolina.

Almquist, veteran research worker in the field of animal reproduction and AI, has made many beneficial contributions to the AI program. His research in semen extension and preservation; the value of antibiotics on controlling bacterial organisms in diluted semen; and techniques in securing maximum sperm production are most noteworthy.

17.4 Thirtieth Annual Convention

One of the most significant annual meetings of the NAAB, from the standpoint of action and policies approved by the delegate body, was held 16–18 August 1976 at Seattle, Washington. At the start of the meeting, NAAB President Yoder reviewed the progress of NAAB over the years, changes in policy proposed to meet current developments, and urged cooperation among all segments of the AI industry. The executive secretary reported on NAAB activities during the year, including staff changes, USDA Sire Summaries, international semen sales, and a summary of semen sales. Durfey pointed out the sharp decline in beef semen sales during 1975, which represented a 10 percent drop in dairy semen on the domestic front, but an increase of 28 percent in international dairy semen sales (see Table 17.1).

Action taken by the voting delegates at the business session included:

Election of Directors. Albert Fox, of New York, and Henry Roth, of Pennsylvania, were elected to the board for the first time. Two directors—David Yoder and Erton Sipher—retired, having each served two, three-year terms. Robert Boese, Curtiss Breeding Ser-

Table 17.1. Sales of Beef and Dairy Semen and Units of Custom Freezing, 1974–1975

Year	Domestic Sales	International Sales	Custom Freezing
Dairy Semen Sales and Custom Freezing			
1974	10,887,456	761,756	616,104
1975	9,760,054	979,850	613,164
	−10.36%	+28.6%	−0.5%
Beef Semen Sales and Custom Freezing			
1974	2,869,659	385,153	1,675,843
1975	1,712,264	167,938	1,425,958
	−40.3%	−56.4%	−15%

Source: *Proceedings of the Thirtieth Annual Convention of the National Association of Animal Breeders* (1976), NAAB, Columbia, Mo.

vice, was elected president; W. E. Erickson, Select Sires, vice-president; William M. Durfey, executive vice-president; and Bryce Weiker, Noba, Inc., to the executive committee.

Amendments of Bylaws. In order to better conform with the association's expanding sales program, to industry developments, and to CSS, several bylaws were rather extensively revised:

Article V, Section 2, was changed to enable the NAAB president to call a special membership meeting only upon a majority vote of the board.
Article VI, Section 3, was amended to require a lapse of three years for a director who had served two consecutive terms to become eligible for election.
Article VIII, Section 4, "Special assessments may be adopted by a majority vote of the voting delegates at the annual membership meeting called for that purpose."

The articles of incorporation were also amended to conform with the bylaw changes:

The statement relative to principal business and purpose of the corporation was amended to include functions carried out by CSS and as outlined in 17.2.
Membership Classifications. The number of membership classifications was reduced from four to two. The former classes of Regular, Associate, Affiliate, and Canadian were realigned into Regular and Associate members only.
Regular Members (*Voting*). Eligibility applies to any individual or business entity whose principal office is located in the United States or in the Commonwealth of Puerto Rico who is engaged in the collection, processing, and freezing of semen for the artificial insemination of livestock or holds equity in a member organization in a ratio approximating the proportionate usage of semen produced.... And, Regular members agree to

comply with such standards and requirements established by Certified Semen Services, Inc.

Associate Members. Any individual or business entity in any country is eligible to a nonvoting associate membership if engaged in a business or occupation related to the artificial insemination industry or has an interest in cooperating with or assisting the artificial insemination program.

Voting Privileges. Only Regular members are entitled to vote. Number of voting delegates ranges from one for the first part of 99,999 units of semen to eleven voting delegates for 2,250,000 or more service units.

Another revision stated that the position now known as "Executive Secretary" of the NAAB will be entitled "Executive Vice President."

The resolutions that were passed were as follows:

1. Utilize the word "Recommended" instead of "Code" on sire health standards so the document reads: "Recommended Minimum Standards for the Health of Bulls Producing Semen for Artificial Insemination";

2. Authorize the board of directors of the NAAB, Inc., to study the desirability of expanding CSS services into the areas of semen quality, sire health, inseminator training, and other appropriate areas. Also, that the NAAB board be empowered to direct CSS to develop and implement such programs;

3. A fourth resolution expressed appreciation to the hosts for a successful convention.

At the close of the convention, three technicians who had qualified for the "NAAB 100,000 Cow Award" were presented plaques by President Yoder: Vernal Dunham (Tri-State Breeders Cooperative); Norbert Schutz (Tri-State Breeders Cooperative); and Robert Oswald (Midwest Breeders Cooperative). Also honored at the traditional banquet was George Nichols, NAAB assistant secretary, who had earlier announced his retirement to become effective 1 September 1976. Nichols had served fifteen years as assistant secretary and editor of the *A.I. Digest.* During those years, he made many contributions to the AI program and to the livestock industry.

17.5 New Programs Initiated, 1976–1978

During the last half of 1976 and throughout 1977, the NAAB was involved in several new programs including:

1. A "Beef AI Conference" to be held at Louisville, Kentucky, as well as Denver, Colorado;

2. Agreed to support the concept of estrus synchronization but not to become involved in promoting individual products;

3. Entered into an agreement with the Utah DHI Computing Center for processing data relative to the "NAAB Sire Summary for Calving Ease";

4. Began discussions with dairy breed registry organizations to develop a modified type of evaluation program that can be used in sire evaluation summaries and better meet the needs of the AI industry;

5. In February 1976, Patty Poage joined the NAAB staff as communications director, replacing Charles Pattyson, who resigned;

6. The technical committee initiated research on a method of semen packaging that might identify counterfeit semen;

7. Efforts were continued to boost semen sales both on the domestic and international markets. The NAAB committees, the board of directors and the NAAB staff at Columbia, under the leadership of executive vice-president Durfey, continued to advance the development of new and expanding programs.

17.6 AI Digest Succeeded by the Advanced Animal Breeder

The official publication of the NAAB, the *A.I. Digest,* was renamed the *Advanced Animal Breeder.* The first issue under the new title appeared in May 1977.

The *Advanced Animal Breeder* succeeds the *A.I. Digest* that has been NAAB's formal publication since January 1957. Prior to that, beginning in September 1953, the *NAAB News* was published bimonthly as a service to its members and the industry. NAAB management explained that the *Advanced Animal Breeder* "will continue to serve as an internal communications magazine for NAAB members and the growing AI industry."

Editor of the *Advanced Animal Breeder* is Patty Poage, who assumed the position of NAAB communications director in February 1977.

17.7 Sire Summary for Ease of Calving[8]

A national program for recording ease of calving in Holstein cattle was developed by NAAB during 1976–1977. The program compiles data relative to the ease or difficulty of calving by a sire's daughters. Thus, another measure of a sire's ability to transmit desirable characteristics is available.

Data are collected from herds cooperating with participating AI organizations and processed at the DHI Computing Center, in Provo, Utah. The first summary, compiled on a within-stud basis and involving over seventy-five thousand calvings and thirteen hundred bulls was reported in July 1977.[9] The results of this first summary and others to follow will be coordinated with production data in the USDA Sire Summary. This information is of value to cattlemen in choosing sires for mating to first-calf heifers and small cows where the incidence of difficult calving is usually the highest.

17.8 NAAB Research Program, 1977–1978

From the beginning of the AI program in America, research has played a prominent role in developing the modern technology in practice today. Before NAAB was formed in 1946, individual bull studs were supporting research at their respective agricultural colleges. In 1953, a formal research program dubbed the "Coordinated Research Program" was initiated by NAAB. At that time, an assessment of one-fourth of a cent per first service cow was earmarked for research. Throughout the years, the research program has supported at least one hundred or more projects during the twenty-five-year span.

The present program is supported by a special assessment for research, applicable to regular NAAB members, at the rate of three-tenths of a cent per service unit of semen used or sold, up to a specified maximum, which is payable quarterly.

For the year 1977–1978, twelve projects were approved for a total of $56,400.00. The projects are listed to illustrate the basic and in-depth work underway.

Continuing Projects

"Optimizing Conception Rates and Calving Interval in Artificial Breeding" (second year)—Robert H. Foote, Cornell University; "Investigation of Sire Evaluation Procedures for Dystocia in Dairy Cattle and Description of Sire Evaluation Program" (third year)—Freeman and Berger, Iowa State University; "Study of a Sperm Defect: Proposed Viral Etiology and Relation to Subfertility" (second year)—J. W. Costerton, University of Calgary; "Bovine Blood Serum Induced Head-to-Head Agglutination of Motile Bovine Spermatozoa" (third year)—P. L. Senger, Washington State University; "Biochemical Assessment of Fertilizing Capacity" (second year)—Duane L. Garner, Oklahoma State University; and "Nature, Cause and Effect of Congenital Defects in Cattle" (second year)—H. W. Leipold, Kansas State University.

New Projects

"Early Breeding of Heifers and Suckled Beef Cows"—Harold D. Hafs, Michigan State University; "Bovine Embryo-Specific Antigens: Hemagglutination Test, A Basis for Pregnancy Diagnosis in Cattle"—J. J. Reeves and R. W. Wright, Jr., Washington State University; "Determining Conception Rate on Artificial Insemination Sires Using Dairy Herd Improvement Data"—K. R. Butcher, N. C. State University; "Fate of Spermatozoa in the Female-Phagocytosis"—R. G. Saacke, VPI and State University; "Objective Evaluation of Bull Sperm Motility by Computers"—R. P. Amann, Penn State University; "The Use of X-Ray

Fluorescence Spectrometry for Monitoring Tagged Inks in Semen Packages''—Jack Henion, Cornell University.

17.9 NAAB Research Award to Henderson

Charles Henderson, professor emeritus of animal breeding at Cornell University, received the 1977 NAAB Research Award. The award was presented at the seventy-second annual meeting of the American Dairy Science Association, which was held on 28 June at Iowa State University in Ames. Henderson is well known for his long and distinguished career in applying statistical measures to cattle-breeding methods. He developed theoretical methods for partitioning genetic variations into components that have found practical application in evaluating the genetic worth of dairy cattle. He was an early, and continuing, leader in developing and promoting the young-sire-selection program as a means of speeding genetic improvement for milk production. This program has resulted in more rapid improvement in production for AI-serviced herds.

17.10 Thirty-first Annual Meeting and Convention

The thirty-first annual convention of NAAB was held 24–26 August 1977 at Louisville, Kentucky.

Over six hundred persons attended the convention. They came from thirty-three states and Canada.

NAAB President Boese reviewed the association's activities that were underway and emphasized that NAAB members were accomplishing many things as an association that could not be achieved by an individual member. In his annual report, NAAB Executive Vice-President Durfey noted that sales of dairy bull semen had increased about 10 percent during 1976 and international sales by 42 percent. Beef semen sales, as a result of depressed beef prices, had dropped over 50 percent in the past two years (see Table 17.2).

Robert E. Walton, vice-president of CSS, on behalf of President Charles Krumm, who was recovering from surgery, gave the CSS report. He praised the efforts of the CSS officers in getting the association into operation. Thirty-seven CSS member AI organizations, in forty-four locations, were visited by Gordon A. Doak, service director. Cooperation among the dairy breed registry associations in revising their AI rules was also praised by Walton.

Action taken by the voting delegates at the business session included the election of officers and the adoption of new resolutions. Robert Boese was reelected to a three-year term as a director, as were

Table 17.2. Units of Dairy and Beef Semen Sold and Custom Frozen

Year	Domestic	International	Total Sales	Custom	Grand Total
			Dairy		
1975	9,760,054	979,850	10,739,904	613,164	11,353,068
	(−10.4%)	(28.6%)	(−7.8%)	(−.48%)	(−7.4%)
1976	10,753,149	1,389,921	12,143,070	545,841	12,688,911
	(10.2%)	(41.8%)	(13.1%)	(−10.9%)	(11.8%)
			Beef		
1975	1,712,264	167,938	1,880,202	1,425,958	3,306,160
	(40.3%)	(−36.3%)	(−42.2%)	(−14.9%)	(−32.9%)
1976	1,368,618	179,110	1,547,728	1,270,646	2,818,374
	(−20.1%)	(7%)	(−18%)	(−11%)	(−15%)

Source: Data supplied by NAAB office, Columbia, Mo.

Brainard Palmer-Ball and Harry Roth, operations manager and sire analyst of Atlantic Breeders Cooperative. Thomas Lyon, of Midwest Breeders Cooperative, was elected to the executive committee. Retiring from the board was Wilfred Broussard, who had served two terms. The resolutions adopted included: (a) to support efforts of the National Dairy Herd Improvement Association to maintain dairymen input into DHI policies and to urge NAAB members to support the DHI program at the local level; and (b) to express appreciation to hosts for the thirty-first annual convention.

The following eight technicians qualified for the "NAAB 100,000 Cow Award": Roger Stone (Tri-State Breeders Cooperative); Victor Ganser (Tri-State Breeders Cooperative); Erling Larson (Tri-State Breeders Cooperative); Elroy Nelsen (Midwest Breeders Cooperative); Albert Ullmer (Midwest Breeders Cooperative); Alvin Fasen (Minnesota Valley Breeders Association); John Caspers (Minnesota Valley Breeders Association); and Marty Leonard (Eastern AI Cooperative).

17.11 Seventh Technical Conference

The "Conference on Artificial Insemination and Animal Reproduction," sponsored every other year by NAAB and the NAAB technical committee, has enjoyed a steady growth in interest and attendance. As the conference name indicates it provides an opportunity for presentation and discussion of the latest technical developments, as well as the research that is underway, in animal reproduction and AI.

The seventh technical conference was held 14–15 April 1978 at Madison, Wisconsin. There were 269 registrants, representing twenty-eight states, four Canadian provinces, Bulgaria, France, Norway, USSR, West Germany, and the West Indies. Also represented were thirteen colleges and universities.

The two-day program included presentations by twenty of the nation's leading research workers in the physiology of reproduction and AI. The topics were primarily designed for personnel involved in the collection and processing of semen; the management of bulls; and the utilization of techniques to obtain a high AI conception rate in the field. Two workshops, one dealing with laboratory procedures and the other with bull management in AI studs, were held.

There was emphasis on the production of high-quality semen, its identification from collection until used, and the advantages of different packaging methods.[10]

17.12 The 1978 North Central Conference

This conference sponsored by a committee composed of AI organization personnel, extension dairymen, and with assistance from the NAAB has been held every year since 1948. Beginning in 1977, the NAAB marketing and communications committee was assigned the responsibility for the organization of the conference. The general tone of the conference had been toward marketing for some years since many technical phases of the AI program were covered in the NAAB technical conference and the NAAB beef AI conference.

The 1978 North Central Animal Breeder's Conference was held on 2, 3 May at Milwaukee, Wisconsin. The theme was "Building a Solid Customer." Attendance was the highest to date with 310 persons from twenty-two states and four Canadian provinces registered. There were fifteen exhibits shown by equipment manufacturers.

Sixteen speakers presented topics such as "keeping the bull off the farm," herd-mating programs, progeny identification, heat synchronization, and calving ease.[11]

17.13 NAAB Research Award to Ralph Erb

Ralph Erb, professor of animal physiology at Purdue University in W. Lafayette, Indiana, was awarded the 1978 NAAB Research Award. For over thirty years, Erb has been a leader in research involving the physiology of animal reproduction. Included are studies on infertility in cattle; hormone levels in blood, milk, and urine with respect to fertility and infertility; the feeding of urea; and a series of experiments on inducing lactation by the use of exogenous hormones. He is also the author or co-author of over 160 scientific papers and numerous other publications.

The award was presented at the seventy-third annual meeting of the American Dairy Science Association, which was held 11 July at Michigan State University.

17.14 The Thirty-second Annual NAAB Convention

The thirty-second annual convention was held in Boston on 8–11 August 1978. There were 670 persons, including representatives of seventy-three U.S., Canadian, and Israeli AI organizations, registered. NAAB President Boese outlined the activities of the association and said, "The policies developed over the years have been by consensus, reached by compromise and have had the ultimate objective of providing the greatest benefit we (NAAB) can for our farmer and rancher customers." Executive Vice-President Durfey discussed the sale of semen and custom freezing of semen for the past year. There was an increase in domestic and international sales as well as custom freezing in 1977 (see Table 17.3 for summary).

Export of frozen semen from dairy bulls in the United States has steadily increased, and in 1977 (Table 17.3) nearly 1.5 million units, valued at $7.3 million, were sold abroad. Europe and South America are the principal importers of dairy bull semen from the United States. This demand results from the high milk production and good conformation of U.S. dairy cattle.

The domestic sales of beef semen have been lower since 1972 when the demand for the exotic breeds was high. However, during 1977, over 1.1 million units were sold to the domestic market and 161,721 units were exported. South America and Central America are the principal markets for U.S. beef bull semen (Table 17.3).

More beef bull than dairy bull semen is frozen and has been so since tabulations have been made. It will be noted in Table 17.3 that the units of beef semen custom frozen in 1977 are nearly double that of dairy.

Reports were also presented by the various NAAB committees. It was reported that NAAB member businesses are now paying about $1.3 million annually to dairymen on DHI test for cooperating in the proving of young dairy sires. This amount is in the form of cash payments, and discounts are allowed on semen and services.

CSS President Walton reported that thirty-eight AI businesses had signed CSS contracts for 1978 and that a budget for advertising and promoting CSS had been approved.

The treasurer's report indicated an operating balance of $37,515 for

Table 17.3. Domestic and Export Sales of Semen and Custom Freezing, 1977

Dairy		Beef	
Domestic Sales		**Domestic Sales**	
Holstein	9,770,967	Angus	454,200
Jersey	455,515	Simmental	202,868
Guernsey	350,663	Polled Hereford	130,372
Brown Swiss	188,609	Hereford	96,775
Ayrshire	70,094	Brahman	38,596
Milking Shorthorn	17,766	Limousin	37,898
Other Dairy	53,813	Charolais	36,262
Total	10,907,427	Maine Anjou	24,850
		Red Angus	16,581
		Chianina	12,502
		Other Beef	58,261
		Total	1,109,165

Export Sales			Export Sales		
	Units	Dollar Value		Units	Dollar Value
Europe	583,429	4,138,732	South America	98,071	315,751
South America	314,426	1,127,346	Central America	25,874	71,943
North America	262,788	.934,376	Asia	14,015	37,652
Central America	143,558	396,460	North America	9,421	29,795
Middle East	112,854	512,176	Africa	7,955	34,072
Asia	38,671	101,306	Europe	6,385	23,396
Africa	11,947	89,104	Total	161,721	512,609
Total	1,467,673	7,299,503			

Custom Frozen		Custom Frozen	
Holstein	435,727	Angus	246,242
Jersey	48,886	Hereford	130,940
Guernsey	24,326	Simmental	123,428
Ayrshire	19,915	Polled Hereford	78,086
Brown Swiss	15,905	Brangus	67,927
Milking Shorthorn	4,163	Charolais	59,993
Other Dairy	24,685	Limousin	53,362
Total	573,607	Brahman	26,082
		Santa Gertrudis	18,821
		Chianina	17,989
		Other Beef	191,145
		Total	1,014,015

Source: *Proceedings of the Thirty-second Annual Convention of the National Association of Animal Breeders* (1978), NAAB, Columbia, Mo.

the year, showing NAAB to be in a good financial position with assets of $229,930.

During the business sessions of the convention, Thomas Lyon, Charles Michaels and Bryce Weiker were reelected for second terms on the board of directors.

Wallace E. Erickson was elected as president; Thomas Lyon as vice-president; and Bryce Weiker, to the executive committee.

The wording of bylaws one through five was slightly revised to correct the language and to include omissions in the amendments that were adopted in 1976. Article VIII was also amended to stipulate that dues are based on a calendar-year basis and to be paid quarterly.

The resolutions that were adopted included: (a) to confirm the continuation of payment to Dairy Record Processing Centers for transmission of records in progress to the USDA; (b) to continue an educational program and to encourage members to discontinue using bulls that are known to be carriers of inherited defects; (c) to continue to publish information on ease of calving; (d) to establish that if a pedigree estimate of a young sire's transmitting ability for production is used that it be identified as a Pedigree Index (PI) and defined as one-half (1/2) the sire's PD plus one-fourth (1/4) of the maternal grandsire's PD, based on the latest USDA Sire Summary for Production; and (e) to authorize the development of a sire-health-auditing program by CSS.

The "NAAB 100,000 Cow Award" was presented to seven technicians: Maurice Proulx (Centre d' Insemination Artificielle du Quebec) [the first Canadian to receive this award]; W. D. Walker (Louisiana Animal Breeders Coop.); Willard Wesley (Midwest Breeders Cooperative); Michael McLaughlin (Midwest Breeders Coop.); Robert Scott (Minnesota Valley Breeders Association); Albert De Salles (U.S. Genes, in Modesto, California); and Homer Noble (U.S. Genes).

The "Quarter Century Award" plaque was presented to Robert Boese.

NAAB Executive Vice-President Durfey announced a new award—"The NAAB Distinguished Service Award"—to be presented for the first time in 1979.[12] This award is to be given to a person who has made a significant contribution to the AI industry during the most recent five years, and who is currently active in his field of work at the time of nomination.

17.15 Forty Years of Evolutionary Change

In my presentation of developments in the AI program in the United States I have frequently emphasized the growth of the NAAB.

Due to developing technology, the experiences gained as the program grew, and economic factors that were constantly involved, the morphology of the AI industry has undergone many changes in reaching its present status. These developments have influenced the destiny and procedures of NAAB in a pronounced manner. The development of the AI program in America over the past forty years can be best described as consisting of four stages.

The first category is the experimental stage, the early developments. As is well known, the early use of AI by horse breeders and a few cattlemen involved the collection of semen from the vagina following service and transferring it to other animals in estrus. Beginning about 1936–1938, the first attempts to practice modern-day AI methods were largely experimental in the United States. Universities and agriculture experiment station workers led the way in developing techniques for the collection, evaluation, processing, and shipping of liquid semen. Storage of semen was relatively short, depending entirely on ice and refrigerators for the maintenance of low temperatures.

These early efforts, spurred on by the success of the Europeans in developing cattle AI programs, resulted in a limited application to farm herds. The Land-Grant Colleges, through research, educational, and extension programs, made major contributions to the development of the early AI program. Dairy extension workers led the way in organizing the first "AI Associations" and, with research people assisting, supplied the best technical information that was then available. The results were sufficiently effective to demonstrate that AI could eliminate keeping a bull; provide service to a better bull than most dairymen could afford to own; and help control venereal disease in the herd.

The second stage of development, from 1939 to 1950, was characterized by the formation of many cooperative breeding associations to provide AI service to dairymen. By 1950, there were ninety-seven bull studs, an all-time high, in business in the United States. Of this number, eighty-three were farmer-owned cooperatives; two were operated by universities; and twelve were privately owned or proprietary businesses. Farmer-owned cooperatives inseminated 80 percent of the 2.6 million dairy cows reportedly bred by AI in 1950.

During this period, the AI program became less of an educational and extension project of the agricultural colleges. Farmers formed organizations (cooperatives) to provide cattle-breeding services for their herds. The agriculture colleges continued to play a leading role in education and assisted in establishing and operating the AI associations. In many cases, the early associations were small, lacked capital, business volume, and sound management. For those reasons, many failed.

The surviving organizations made progress by expanding volume, adequate financing, good management, employing capable personnel, utilized the best sires available, and carried on effective member relations programs. Membership loyalty was then, and continues to be, a strong asset among farmer-owned cooperatives.

The AI industry was developing and capitalizing on the improvement made in herds where AI was properly utilized. During these first two stages, the AI industry was relatively noncompetitive. The chief competition was "the bull on the farm." Cooperative AI organizations operated in fairly well-defined service areas. The cooperatives, at this time, respected each others' boundary lines and rarely extended service into another's area except by mutual agreement. Toward the end of the second stage, strong competition had developed between the farmer-owned cooperatives and the proprietary or privately financed businesses. Competition was destined to increase and to accelerate rapidly with the introduction of frozen semen in 1954–1955.

The third stage is signified by expanding competition due to the growth of the AI cooperatives and the entrance of proprietary firms into the field. This was a natural development as the successful cooperatives presented attractive balance sheets and assets during the first twenty-five years of the AI program. The opportunities for profit appealed to independent operators and an increasing number entered the field.

The large numbers of dairy cattle in the country and the farmers owning these cattle constituted a market that dramatically increased in both cooperative and proprietary firms from 1946 (537,376 cows enrolled) to 1965 (7,879,682 cows). During these years, the use of AI became an integral part of the dairy-cattle-breeding program. Beginning about 1955, the AI industry became a segment of the free-enterprise-competitive system. Under this system, we now have individuals, partnerships, proprietary corporations and cooperative corporations competing with each other for business.

The perfection of frozen semen had a tremendous effect on the competitive nature of the business. The ampule of semen became a trade commodity. Frozen semen can be shipped worldwide. It can be stored for many years with fertility maintained. Whereas the use of liquid semen limited the operational area to within a few hundred miles of the bull stud at best, frozen semen requires no boundaries. The proprietary organizations soon became nationwide operators. They provided strong competition for all local cooperatives as well as among themselves.

The larger proprietary organizations started to discontinue the use of technicians to provide a complete AI service to dairymen and began to sell semen directly to herd owners. This gave rise to the "do it yourself'' system of AI that now accounts for well over 65 percent of the business. Herd operators are trained by the semen-producing businesses to inseminate cows. Training schools of three days to two weeks—some only one or two days—became part of the AI program.

Improved sire-evaluation methods came into the picture during this stage. The AI-proved sire with many tested progeny, PD, and ''repeatability'' of the proof came into demand. The organizations with a battery of ''High Plus'' proved sires commanded increasing business. This factor forced every organization that survived to maintain top transmitting sires. It accounted in no small part for many of the mergers and consolidation of cooperatives. Also, many cooperatives set up semen-exchange programs and joint-sire ownership as a means of increasing ''bull power''.

The declining milk cow population—from 27 million head in 1946 to about 15 million cows and heifers of breeding age in 1975—greatly increased the problem of financial survival and the competition for business. The number of AI organizations, largely as a result of mergers, dropped from ninety-seven in 1950 to thirty in 1970.

In the fourth and current stage of the AI industry, competition is increasingly keen. However, reorganization and adjustments, some beginning in stage three, have been made. There are fewer and larger bull studs. In 1971, there were twenty-six bull studs operating and only twenty-one by 1975. This drastic reduction in the number of studs came about by mergers and consolidations, largely among the cooperatives, as a means of survival. The decline in dairy cattle numbers, sales volume leveling out at about 55 percent of the milk cow population, a slower growth in beef AI than anticipated, and the steadily rising costs of operation all forced consolidations or concentration of semen-producing businesses. A good example is Select Sires, Inc., a cooperative at Plain City, Ohio. That organization consists of a federation of eleven cooperative bull studs. Eastern AI Cooperative, which serves the State of New York and the entire New England area, is another good example.

Continuing is the trend toward a semen sales, rather than a complete AI service program. Frozen semen is shipped from the east coast to the west coast, from the west to the east, and across the country, and cooperatives generally compete with each other. Several of the larger farmer-owned cooperatives still maintain a staff of skilled technicians and provide a complete AI service rendered by professionals. However, semen is in nationwide supply for direct herd service and inseminator training is available. If he so desires, a herd operator in remote areas can obtain frozen semen from a number of AI businesses.

Currently, through purchase of proprietary AI firms, large conglomerate corporations are active as AI organization operators. Three of the largest in the nation are in this category. These independent firms offer a genuine challenge to all cooperatives and smaller proprietary organizations by virtue of the capital and manpower available. However, no organization has a monopoly. The genetic pool of superior germ plasm is fairly well dispersed among the AI businesses.

The current picture finds the custom collection and freezing of bull semen at an all-time high. During 1976, about 0.6 million units of dairy bull semen and 1.3 million units of beef bull semen were custom frozen. While some of the dairy bulls involved in breeders' herds are quite outstanding, much of the semen comes from young bulls, under joint or syndicate ownership, being sampled for proving. In case of the beef bulls, we find many bulls in the owner's herd being used artificially by this means. The immediate question is whether or not maximum genetic improvement is achieved by the use of such bulls.

While growth in an organization does not guarantee success, it becomes evident in this day that only the larger, stronger, AI businesses can provide the sire-service programs and the manpower the market demands.

Advances in technology and research, along with effective marketing programs, will continue to bring financial success to aggressive AI businesses in this competitive field.

Competition is desirable if it results in efficiency of operations, requires use of the best sires available, improves service, and lowers costs to the livestock producer. However, duplication in service areas and increased costs to cattle breeders can also result. It must be said, however, that the free enterprise, competitive system has given America a unique and outstanding AI industry.

Appendix A: Development of the AI Program by States and Organizations in the United States, Puerto Rico, and Canada

Introduction

The early growth and development of the AI program in the United States involved many people. Among the pioneers were cattle breeders, farm leaders, extension dairymen, agricultural college teachers, research workers, county agents, vocational agriculture teachers, state and federal agencies, and bankers. It is beyond the scope of this study to pay tribute to all. We will, therefore, briefly narrate early developments in establishing AI programs and bull studs in the respective states and provinces and their progress through the ensuing years. In so doing, I will strive to mention some of the key leaders whose foresight and dedication made possible the AI program.

Custom-Freezing Businesses. One of the important developments, following the introduction of frozen semen in 1953 and on, has been the establishment of many custom-semen-freezing businesses throughout North America and other countries. The custom freezing of semen not only involves collection on the farm, or at facilities to maintain privately owned bulls, but freezing, storage, and distribution. As frequently mentioned in earlier chapters, most bull studs do custom freezing. There are, however, many privately owned or

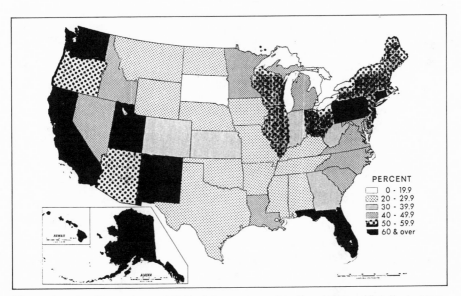

Figure App.1.   Percentage of U.S. dairy cows and heifers bred artificially in 1971.
Source: U.S. Department of Agriculture.

269

corporate entities involved. Most operate in areas involving several states. Beef cattle breeders use custom-frozen semen to a high degree. Since the number of custom-semen-freezing businesses varies from time to time and no accurate cataloging of them is available, we have not attempted to identify many by name. Their role in the AI program is important, and their contributions are many.

Since the AI program has developed, the most rapidly in the more populous dairy areas, they will receive special mention. Figure App.1 indicates by states the percentage of dairy cows and heifers bred artificially in the United States during 1971.

The activity by states in terms of units of semen sold and custom frozen for 1973 is indicated in Table App.1.

Alabama

No formal state organization was formed. There were seventeen county AI associations established during 1947 under the leadership of J. C. Grimes, a professor at Alabama Polytechnic Institute in Auburn. A bull stud with eleven bulls of the Jersey, Guernsey, and Holstein breeds was established at Auburn. The name Alabama AI Association was adopted, and W. E. Alston, Jr., was hired as manager.

The first year, 5,976 cows were inseminated. In 1954, with frozen semen being utilized, the bull stud was discontinued and semen was purchased for the local units from American Breeders Service. Alabama is now serviced by out-of-state AI organizations.

Alaska

Under the guidance of William J. Sweetman, of the Bureau of Dairy Industry, USDA, and stationed at the Alaska Experiment Station in Palmer, plans for an AI organization in Matanuska Valley were laid in 1947. Dr. O. E. Reed, chief of the Bureau of Dairy Industry, USDA, met with local dairymen. There were only about one thousand cows in milk and about half were indicated as ready to enroll in the program.

The Bureau of Dairy Industry agreed to furnish the bulls. The farmers and the Agricultural Experiment Station at Palmer agreed to furnish transportation.

The Matanuska Valley Breeders Association began operations at Palmer in 1948. Financing came from cooperative membership, the Alaska Agricultural Experiment Station, and territorial aid. The membership is direct, and there are six directors. The experiment station continues under the general supervision of the board of directors. Holstein, Red Dane, and Guernsey bulls were maintained at the beginning, but, at present, only Holsteins are kept.

In 1965 A. L. Brundage, a professor of animal science (Palmer), assumed guidance of the AI cooperative. Most of the semen presently used is imported from state-side organizations, thus providing a wider choice of high-plus-proven bulls. During 1971, a total of 1,816 dairy cows, 93 percent of the total, were inseminated. The program in Alaska provides an excellent oppor-

Table App.1. Units of Semen Sold and Custom Frozen in 1973 by States [1]

State	Units Sold Dairy	Units Sold Beef	Units Custom-Frozen Dairy	Units Custom-Frozen Beef	Total Dairy Units	Total Beef Units
Alabama	59,829	27,894	2,741	1,787	62,570	29,681
Alaska	2,441	242	0	0	2,441	242
Arizona	63,366	7,139	0	1,412	63,366	8,551
Arkansas	36,790	34,216	1,383	4,780	38,173	38,996
California	1,072,499	64,947	107,404	42,325	1,179,903	107,272
Colorado	69,091	57,521	6,691	87,560	75,782	145,081
Connecticut	89,715	3,871	30,968	1,170	120,683	5,041
Delaware	8,673	1,118	0	0	8,673	1,118
Florida	275,679	89,478	1,778	27,279	277,457	116,757
Georgia	87,650	28,127	1,596	7,564	89,246	35,691
Hawaii	22,955	657	0	330	22,955	987
Idaho	129,367	29,545	375	11,108	129,742	40,653
Illinois	323,390	48,294	19,802	24,859	343,192	73,153
Indiana	192,552	48,730	17,311	24,234	209,863	72,964
Iowa	263,458	123,503	6,214	31,964	269,672	155,467
Kansas	137,040	83,513	11,036	37,847	148,076	121,360
Kentucky	124,169	78,671	3,092	24,250	127,261	102,921
Louisiana	127,601	51,234	1,149	8,136	128,750	59,370
Maine	80,248	1,050	2,775	0	83,023	1,050
Maryland	187,387	18,381	26,715	11,140	214,102	29,521
Massachusetts	68,937	2,782	96,063	3,620	165,000	6,402
Michigan	326,525	51,441	19,630	40,368	346,155	91,809
Minnesota	918,540	171,735	11,974	11,057	930,514	182,792
Mississippi	48,365	41,555	24,074	13,175	72,439	54,730
Missouri	164,197	99,248	3,415	72,652	167,612	171,900
Montana	19,263	153,112	674	17,546	19,937	170,658
Nebraska	71,009	80,939	2,502	24,681	73,511	105,620
Nevada	15,674	3,943	0	1,796	15,674	5,739
New Hampshire	44,538	1,359	1,486	0	46,024	1,359
New Jersey	48,219	7,366	2,352	1,129	50,571	8,495
New Mexico	34,645	10,110	1,377	1,539	36,022	11,649
New York	1,167,334	26,292	48,441	14,302	1,215,775	40,594
North Carolina	136,832	26,860	2,259	1,287	139,091	28,147
North Dakota	52,138	77,929	3,189	11,452	55,327	89,381
Ohio	470,094	166,280	41,866	44,003	511,960	210,283
Oklahoma	55,277	50,715	2,328	100,138	57,605	150,853
Oregon	113,897	39,882	8,743	13,055	122,640	52,937
Pennsylvania	595,905	114,291	63,540	13,368	659,445	127,659
Puerto Rico	121,635	92	0	0	121,635	92
Rhode Island	5,090	145	0	0	5,090	145
South Carolina	50,249	14,683	2,099	964	52,348	15,647
South Dakota	60,238	117,614	3,224	13,949	63,462	131,563
Tennessee	126,035	50,602	5,797	21,480	131,832	72,082
Texas	149,774	170,809	17,787	181,313	167,561	352,122
Utah	95,374	24,505	7,512	1,483	102,886	25,988
Vermont	230,667	6,628	5,183	0	235,850	6,628
Virginia	183,458	30,889	9,496	31,168	192,954	62,057
Washington	746,207	61,679	14,212	27,023	760,419	88,702
West Virginia	37,348	9,025	3,420	2,292	40,768	11,317
Wisconsin	1,307,903	257,278	41,363	4,324	1,349,266	261,602
Wyoming	16,450	36,019	6,270	85,781	22,720	121,800
Unnamed States	174,722	118,966	0	40,429	174,722	159,395

Source: *Dairy Herd Improvement Letter* 51:2 (1975), USDA

[1]Zeros indicate no activity.

271

tunity for constructive breeding in an area where fresh milk is scarce and high in price.

Arizona

Arizona Cattle Breeding Association. In 1945, the Arizona Cattle Breeding Association, a cooperative, was formed by dairymen in the Mesa area. W. R. Van Sant, poultry and dairy specialist at the University of Arizona in Tucson, was leader of the project. The first year, 854 cows were bred. Due to privately financed AI businesses starting up, and semen from adjoining states becoming available, the association was disbanded in 1951.

Eads Cattle Breeding Service. In 1942, Seth Eads, a successful dairyman in Phoenix, established the privately owned Eads Cattle Breeding Service. Holstein bulls in his dairy herd were used to provide semen. By 1949 three inseminators and one fieldman were employed and service was furnished for Guernseys, Holsteins, and Jerseys. In 1954, about six thousand cows were serviced. During 1960, Eads began distributing frozen semen for American Breeders Service. In 1966, the organization discontinued operation.

Herschede Ranch. Herschede Ranch, a purebred Hereford establishment and owned by W. Foy Herschede, of Hereford, began artificial breeding in 1954. About one hundred Hereford cows were bred annually using frozen semen from Herschede-owned bulls at the ranch and two bulls jointly owned. The use of frozen semen enabled joint use of these bulls, described as being "dwarf free."

New's Artificial Breeding Service. Gay D. New, who served as manager of the Arizona Artificial Breeding Association, started his own AI business in 1950 at Phoenix. Soon after beginning operations, he became affiliated with American Breeders Service as a distributor and ceased his personally owned business in 1961.

Arizona Bull Ranch. This business started in 1974 by Earnest St. John and Vernon St. John, of Glendale, in connection with the St. John Dairy Operations. The St. John Brown Swiss herd is known worldwide for its many "excellents" and world-record milk producers. Semen from bulls in use in the St. John's herd is marketed—some abroad. A custom-collecting business is also operated under the management of J. N. Farnham.

Arizona Breeders. Paul Prince, of Mesa, moved to Arizona in 1958 and established a semen-distribution and custom-collecting business. This business has distributed semen from several bull studs. At the present time, most of the semen used by Arizona cattle breeders comes from out of state. A total of 34,866 cows and heifers, 30,238 dairy, and 5,528 beef were artificially serviced during 1971.

Arkansas

Arkansas Artificial Breeders Association. This cooperative had its beginning in 1946 under the direction of Harry Goforth, of Fayetteville, and Marion Wasson, of Conway. In 1947, the state legislature appropriated $20,000 to finance a barn, which was to be built on the University of Arkansas land in Fayetteville. The name *Arkansas Dairy Breeders Association* was adopted but later was changed to the *Arkansas Artificial Breeders Association*. The association began business in 1948 with Harry Goforth serving as president of the seven-man board of directors. Twenty bulls—twelve Jerseys, four Guernseys, and four Holsteins (which were purchased mostly from public donations)—were available the first year. Later, Milking Shorthorn and Angus bulls were added.

In 1961, the association began purchasing frozen semen from Noba, Inc., in Tiffin, Ohio, and the bull stud was phased out. In 1962, the association's program was taken over by Arkansas Milk Producers Association.

California

The dairy herds in California are large, averaging three hundred cows with some over four thousand head. Drylot management is followed. The artificial breeding program in California has followed a somewhat different trend than that of the midwestern and eastern states. Small AI organizations, in terms of herds serviced, were formed by individuals in many areas. With large herds, and about 40 percent of the cows replaced annually, maintaining a large battery of superior bulls is costly. Most heifers are raised a considerable distance from the milkshed area. California dairymen demand high-production potential.

Tulare County Breeders Cooperative. A group of dairymen organized the Tulare County Breeders Cooperative, Inc., in February 1944. Financing was done by an assessment to the membership of $2.00 per cow enrolled. The first sign-up was for about fifteen hundred cows. Local veterinarians Arch Inman and Carl Lambert did the inseminating. George Barnes was the first president. However, in 1942, a group of dairymen had organized the Sacramento Dairy Breeders, which was carrying on artificial breeding. This Tulare group as well as several other county associations formed the nucleus for California Dairy Breeders. The volume of business was small, with about one thousand cows the first year and not over three thousand the best years. In late 1947, the cooperative expanded to include Kings and Fresno counties.

The Tulare Breeders Cooperative, Inc., merged with the California Dairy Breeders in 1951. At that time, all property was moved to the stud owned by California Dairy Breeders.

San Pasqual Breeders Association, Inc. The San Pasqual Breeders Association, Inc., a cooperative, began operations in 1944. The membership was limited, consisting of dairymen Stanley Trussel, Charles Judson, Trussel Brothers, Henry Fenton, and Julit Judson. The capital was furnished entirely by the members.

The stud was established near Escondido. Service was provided for only the above named dairymen and about twelve hundred cows were inseminated annually. V. C. Bunker, DVM, was the manager and technician. By 1953, services were extended to Chula Vista, San Diego, Chino, Hemet, and

Victorville. In December 1954, the San Pasqual Breeders Association was sold to Roger Jessup Farms Artificial Breeding Service in Artesia.

Santa Clara Breeding Association. The association was organized in 1945, at San Jose. M. S. Beckley, farm adviser in San Jose, was active in promoting this cooperative. It had a short life, being taken over by American Breeders Service after two years of operation. While already active in the area, American Breeders Service established a new bull stud at Palo Alto in 1952.

California Dairy Breeders. The organization was established originally as the Sacramento Dairy Breeders. It was formed by a group of six dairymen in Sacramento County, with the assistance of Ray Gieberger, extension agent, and the Animal Science Department of the University of California in Davis cooperating. Cantrell Castello, William Sims, and Carl Herzog were leading supporters. It was a direct membership cooperative, financed by the members.

As early as 1942, steps were taken toward organization in an exploratory way. The University Animal Science Department agreed to cooperate with interested dairymen for one year and, if agreeable, to expand over a twenty-year period. The breeding fee was set as $7.00, with $4.00 allocated to the inseminator, $2.00 to the university, and $1.00 to the association. The members agreed to purchase revolving fund certificates at $3.00 per cow for the number they decided to breed the first year. Additional cows were enrolled at the same fee.

In August 1943, the corporation was formed, but actual operations did not begin until January 1945. The bulls were housed at the university's dairy barns. The University of California in Davis, through efforts of S. W. "Tom" Mead, provided laboratory and office space and technical assistance. The first year, fifteen hundred cows were serviced. Ten years later, the volume was thirty-five thousand. Bulls of all dairy breeds, except Ayrshire, were available. Angus bulls were also kept for the benefit of western dairymen who preferred to breed first-calf heifers to beef bulls.

Robert Sherwood served as manager of California Dairy Breeders for several years. Later, the association purchased land and moved its headquarters. Victor Gray became the manager and served until 1964 when California Dairy Breeders merged with All West Breeders, in Burlington, Washington. He continued as manager of the California area and assistant manager of All West.

Roger Jessup Farms Artificial Breeding Service. The Roger Jessup Farms Artificial Breeding Service was organized in 1948, as an individually owned company, with Roger Jessup, R. Vincent Jessup, DVM, and Webster M. Jessup as owners. Kenneth B. Baushke was employed as manager of the company and served until 1963. The bull stud was located at Artesia in conjunction with the Jessups' farm dairy. In 1955, the Jessups purchased Golden West Breeders in Modesto, California. This was a small breeding business set up by purebred Holstein breeders in the area to extend the use of their bulls. In 1966, Robert E. Carroll, DVM, of Anaheim, California, who had started a small AI service some years earlier, sold his operation to the

Jessups. In 1954, they had bought San Pasqual Breeders Association. Cows bred by AI, including semen sales, by the Jessups reached 110,000 head in 1966.

In late 1967, the assets of Jessup Farms Artificial Breeding Service were sold to Genetics, Inc., a bull stud, and semen sales business initiated the same year by Harold J. Schmidt, DVM, of Hughson, California. The Jessups continued to prove sires and while, under terms of the sale, barred from doing business in the United States, carried on a small semen-export business to Mexico.

Excelsior Farms Artificial Breeding Service. Excelsior Farms Artificial Breeding Service, in Corona, is owned by Excelsior Farms. The Excelsior Farms Dairy was formerly located at Garden Grove, California, and the AI business was started in 1948. Due to real-estate developments, the operation was moved to its present location in Corona.

C. L. Ranney, DVM; Walter Ranney, his brother; and Harry Ashen were the chief promoters of the early business. It is financed entirely by private capital. In 1967, Excelsior affiliated with Genetics, Inc., as a cooperating bull stud. "Excelsior," at the present time, serves as a distributor for semen from its privately owned bulls, Carnation-Genetics Division, and several other bull studs including Eastern Artificial Insemination Cooperative, which is located in Ithaca, New York. Semen and services are supplied for about forty to fifty thousand cows. A custom-freezing business is also operated.

Walter Ranney, one of the owners, is general manager of the business, with Robert Sherwood serving as operations manager.

Excelsior Farms Breeding Service is the only early-day-founded AI organization in California, nearly thirty years later, operating under its original ownership and management.

Pacific Breeders Cooperative. Located in Petaluma, this organization was originally called the Sonoma-Marin Dairy Breeding Cooperative, which was organized in 1949. A group of local dairymen including Bill Bianchini and Ray Gambonini, who served as president and secretary, respectively, were responsible. The cooperative was named after the two counties it originally served, Sonoma and Marin. Abe Solomon served as the manager.

The name *Pacific Breeders* was adopted in the 1950s to fit more appropriately the expanding area it serviced. The first year, less than one thousand cows were bred, but by 1966 the total was about forty-two thousand head.

In 1967–1968, Pacific Breeders Cooperative affiliated with Genetics, Inc., at Hughson, California.

Adohr Farms. Adohr Farms' AI program was established in 1950 to provide service for their own herd, of some twenty-two hundred cows in milk, and for producers of the Adohr Dairy. The Adohr Farm setup was established in 1916 by Merritt H. Adamson in the San Fernando Valley. In 1942, operations were moved to Camarillo. Arthur Knight, one of the outstanding large-scale dairy managers in the country, who has been with Adohr since 1929 and farm manager since 1942, headed the AI program. In 1966, about ten thousand cows were serviced. Like many, Adohr Farms is being taken over for real-estate development and the dairy and breeding service has been discontinued.

Genetics, Inc. (Carnation-Genetics Division). Genetics, Inc., was started the fall of 1966 by Harold J. Schmidt, a veterinarian and well-known Holstein breeder. It is a corporate firm, established to make available top-proved sires and to guide their use in a constructive manner. In addition to Genetics, Inc., acquired sires, the bull stud was enhanced by the acquisition in 1967 of Pacific Breeders Coop., and the Jessup Breeding Service.

The organization became a nationwide operation in addition to serving California herds. A working arrangement was set up with Excelsior Farms, as well as Carnation Farms, Inc., in Washington. Clarke Nelson, vice-president of Carnation Co., and Walter Ranney, part owner of Excelsior Farms, were members of the first board. In addition to its own production and sale of semen, Genetics became a distributor for Tri-State Breeders, Noba, Inc., and others. The business grew, with 319,215 cows serviced in 1971.

On 1 November 1972, Carnation Co., and Genetics, Inc., merged. On that date, the Carnation Co., in Los Angeles, acquired the assets of Genetics, Inc. The breeding operations of the two firms were gradually merged. The bull studs in Watertown, Wisconsin, and in Hughson, California, were continued. The new firm is known as Carnation Company—Genetics Division. W. R. (Skip) Pringle is the division manager with headquarters at Hughson. Clarke Nelson, of Carnation Co., serves as the chief officer and adviser.

The Bull Bank—Golden State Breeders. In 1968, Melvin W. Lauriton and Roy Selover started operations at Escalon, California, with a small bull stud and a custom-freezing business named The Bull Bank. One of the functions was to collect and process semen from young bulls that were owned by private breeders for sire-proving purposes. Selover discontinued as a partner in 1969.

The business continued to be operated as a bull stud, semen-distribution center, and custom freezing, with AI supplies sold, by Mr. and Mrs. Melvin Lauriton and their son Paul. In 1977, The Bull Bank was sold to Faria Dairy, Inc. Operations have been expanded, and the business now operates at the same address under the name Golden State Breeders.

American Breeders Service, Inc. While its principal office is in Wisconsin, American Breeders Service (ABS) started distributing semen in California in 1950. In 1952, a bull stud was established at Palo Alto, California, in the "Old Stanford Barn" at Stanford University.

In 1960, the West Coast offices of ABS were moved to a new location in Palo Alto, and a frozen-semen-distribution center was established. The program grew under the leadership of Don Pelkey, Harlan Koch, and other long-time ABS employees.

Curtiss Breeding Service. Prior to 1950, Curtiss Breeding Service had been shipping semen direct for use in a few California herds. In 1950, a Curtiss distributorship was set up at Turlock with veterinarians R. L. Collinson and Ike Bohlender serving the entire state. Known as Western Improved Stud

Service, it continued until 1954 when James Kuhtz became the Curtiss district manager. With frozen semen being fully used in 1959, the West Coast operations extended to additional states. (The operations of the Curtiss Breeding Service will be discussed in more detail under Illinois.)

In summary, the AI program in California is unique in that there have been many organizations involved. Some of the locals, most of which failed, and the various semen distributorships have not been mentioned. There have been many of these; some have succeeded, and some have failed.

California has about 875,000 milk cows including many prominent purebred herds. Production per cow is high, averaging over fourteen thousand pounds of milk per lactation. The dairymen insist on cows that "wear well." Herd analyses and careful sire selection are emphasized. At present, about 70 percent of the state's cows are bred artificially.

Colorado

The early work in AI began at Colorado State University in Ft. Collins, when senior veterinary students, in the early forties received instruction in artificial insemination. A small bull stud was maintained in connection with the college dairy herd. V. D. Stauffer was the original veterinarian in charge of the AI unit. Upon Stauffer's entry into private practice, H. J. Hill, DVM, a graduate of Colorado A&M College took over management in 1947. About five hundred cows were bred artificially that year.

The veterinary college insisted that all inseminations be done by veterinarians. This decision worked against the expansion of the program. Out-of-state competition moved in and serviced the herds in the major dairy areas.

Research work in reproduction and fertility testing of bulls received early attention from the Colorado State University Veterinary and Animal Science staffs. Aided by private grants and a state appropriation, the Colorado State Animal Reproduction Laboratory was established. This laboratory at present is supervised by B. W. Pickett and associates. Studies are conducted on all phases of reproduction in cattle and horses as well as the preservation and handling of frozen semen. An insemination service is not maintained. Extensive fertility testing of beef bulls, many under range conditions, and a semen-custom-freezing business are conducted. At the present time, about 35 percent of the dairy cows and 3 percent of the beef cows in the state are bred artificially.

International Beef Breeders. Located at Denver, Colorado, International Beef Breeders (IBB) was founded in 1960. It had the distinction of being the first *all beef* AI organization in the United States.

The idea for IBB was conceived in 1946 when two of the founders, L. M. Cropsey, DVM, and Rand van Dervoort, were employees of Curtiss Candy Farms, located in Cary, Illinois. After 1950, both men devoted all their time to beef cattle AI and related beef-cattle-breeding projects. In 1953, they organized a commercial beef AI project at the Jennings Brothers Ranch, in Highmore, South Dakota. This largest continuous beef AI project in North America has inseminated between two thousand to five thousand head of cattle every year since 1953. The success of this first project led to the establishment of IBB. The founders were L. M. Cropsey, president; Rand van Dervoort, secretary-treasurer; and field service manager, William Brittain, and his manager Ray Smith, of Mahogany Farms in Michigan; and Jennings Brothers, of Highmore, South Dakota. A corporation with forty stockholders, all cattlemen, was formed.

Denver was selected as the site for a headquarters because of its location as the geographical center of the beef cattle industry. A farm a few miles north of Denver was purchased. Barns, laboratories, and offices were built. Housing units for bulls provided each with a grazing and exercise area. The major part of the physical plant was completed in 1963.

IBB soon developed beef AI customers throughout the United States, Canada, and Mexico. As the exotic beef breeds came into the picture, bulls of practically all of the British breeds and the exotics were located at IBB. Large quantities of frozen semen were collected and stored for future use and sale.

In 1970, IBB was sold to the Syntex Corporation, a pharmaceutical firm, in Palo Alto, California. William Brown, DVM, was named vice-president and general manager. The IBB program was carried on under his guidance.

In 1977, the Syntex Corporation sold IBB to Bov Imports, Inc., a beef AI business originally based in Quebec, Canada. The Bov Import, Inc., controlled bulls are now housed at IBB. The business continues under the leadership of Richard E. McCormick, of Denver, Colorado, who is a veteran in the AI field.

Premier Breeding Center. Shortly after the sale of IBB to the Syntex Corporation, L. M. Cropsey and several beef breeders established the Premier Breeding Center, in Bennett. Located about twenty miles east of Denver, Premier has new and modern facilities for handling bulls and semen production.

The general manager, Carman F. Hawkins, DVM, moved from IBB to Premier and has guided developments. The collection, processing, storage, and sale of beef semen is the principal business.

In 1973, Premier introduced the American Beef Friesian to U.S. cattlemen. To establish the new breed, five Friesian bulls, sixteen Friesian cows, and four heifers were imported from Ireland. In the British Isles and Ireland, Friesians have been bred for milk and meat for many years. The Beef Friesian bulls are mated to Angus, Hereford, Charolais, and crossbred females. The Friesian-cross dams excel in milk yield, and weaning weights of calves often exceed six hundred pounds.

Connecticut

Artificial insemination of dairy cattle started in Connecticut in 1939. Three small bull studs began operations that year. They were Tolland County ABA located at the University of Connecticut, Storrs; Litchfield County ABA located at Litchfield; and Fairfield—New Haven County ABA—Southbury.

In 1941, breeding associations were formed among dairymen in New London and Windham counties. Bull studs were not established, and these two groups purchased semen from other studs.

In 1945, the three bull studs and the two county groups united to form the Connecticut Artificial Breeding Association (CABA), which was a federated cooperative, with headquarters at Woodbridge. The president was Lester Reynolds, who served for ten years.

Paul Heller became manager in 1947 and is credited with developing an efficient organization. During 1954, over forty-one thousand cows were serviced.

In 1959, due to increasing costs, competition, and the availability of frozen semen from out of state, CABA merged with Massachusetts Selective Breeding Association, Rhode Island Breeders Cooperative, and Central Vermont Breeding Association, under the name of New England Selective Breeding Association. Gerald Hayes was president of CABA at the time. Paul Heller became general manager and served until 1960. He was succeeded by Henry Soborg.

In January 1962, all of the New England Selective Breeding Association members; Vermont–New Hampshire Breeding Association; Maine Breeding Coop; and Rhode Island Breeders Coop., merged with the New York Artificial Breeders Cooperative in Ithaca. The result was the formation of Eastern Artificial Insemination Cooperative, Inc., with headquarters at Ithaca, and with Charles J. Krumm, general manager. In the consolidation of cooperatives, each merging organization is represented on the board. The bull stud at Ithaca provides service for patrons throughout the consolidated area.

CABA is believed to be the first AI organization to demonstrate the use of streptomycin as an antibiotic to be routinely used in extended bovine semen. It was also the first to demonstrate that streptomycin when used alone is superior to its use with penicillin and other antibiotics.

Herdsman. In 1968, Lucien La Liberty founded the Herdsman Corporation to produce and freeze semen in Putnam, Connecticut. The first few years, semen was distributed for other bull studs and from a few bulls under contract in breeders' herds. About 1972, a bull stud of some thirty sires was established at Verona, New York. Fred Hendricks was employed as manager of Herdsman, and the trade territory increased. In 1971, about sixteen thousand cows were serviced. Herdsman also sold AI supplies.

In 1976, Herdsman disposed of the Verona, New York, facility and its sires. A program of working with breeders in sampling and developing young dairy sires that would be offered to domestic and foreign semen markets was announced.

Agricultural Company of Pan America, Inc. Located at Guilford, Connecticut, the Agricultural Company of Pan America (Agripan) in 1960 began acting as an exporter (really a forwarding company) for frozen semen. The company was organized shortly after World War II by R. W. Bishop, to expedite the global shipment of chicks, hatching eggs, and poultry.

Lindsay Brown joined the company as vice-president in the 1950s and initiated the shipping of frozen semen, supplied by U.S. bull studs, to overseas countries. By 1962, over thirty thousand units of semen were being exported annually, and Agripan was acting as exporter for a dozen bull studs.

Due to the death of Lindsay Brown in 1965, Paul E. Rebner became general manager. In 1969, a separate corporation, Agripuire International

Ltd., was formed to handle semen exports. AI equipment is also merchandised on a global basis.

Delaware

Artificial breeding started in Delaware about 1940 at the University of Delaware Farm, in Newark. In November 1941, a group of about seventy progressive dairymen met at the Glasgow Grange Hall, near Newark, and organized the New Castle County Cooperative Breeding Association.

E. L. Symington, a veterinarian in Sussex, New Jersey, was employed to manage the association and to do the inseminating. Within six months, a second veterinarian, Douglas C. Reid, a graduate of Ontario Veterinary College, was hired. There were 1,484 cows bred the first year. The number did not increase in succeeding years, however.

In 1946, the breeding service was reorganized to form a "State Cooperative." A bull stud was established at the Delaware Agricultural Experiment Station and sires were centralized. In 1948, William L. Crossan, Jr., became manager. By 1952, about six thousand cows were being bred annually. The program received strong support from the staff at the University of Delaware with special credit due to George M. Worrilow, extension dairyman and later vice-president of the university, and Delmar Young, an extension dairyman.

Because of business difficulties, on 1 December 1953, the Delaware Association merged with Southeastern Pennsylvania Artificial Breeding Cooperative in Lancaster. Since that time, the area in Delaware has been known as The Delaware District and has a director on the SEPABC (now Atlantic Breeders) board. At present, about 69 percent of the state's total dairy cows are bred artificially.

Florida

In 1948, seventeen local AI cooperatives were formed by C. W. Reeves, an extension dairyman at the University of Florida. Liquid semen was purchased from out-of-state organizations including American Breeders Service, Noba, Inc., and Curtiss Breeding Service. In 1957, frozen semen was put into use.

In 1960, most of the Florida dairymen and beef breeders using AI went on a "do-it-yourself program."

In 1949, a total of 7,102 cows were serviced. In 1971, 165,000 dairy cows (77 percent of the milk cow population) and 20,441 beef (2.1 percent of the total) were serviced by AI. By avoiding the costs of bulls, facilities, and personnel to operate its own bull stud, the Florida people escaped some of the expense and tribulations encountered by other states.

The dairy herds in Florida average 245 cows per herd. Some including Velda Farms, Tallahassee, began as early as 1954, to have semen custom collected and used in their own and affiliated herds. Charles S. Coble was among the early leaders.

Noba—Lake Worth. In the late 1960s, Noba, Inc., established a beef unit and semen sales unit at Lake Worth, Florida. James Christian was the first manager. Its operational area is the southeastern United States.

Georgia

Beginning in 1948, the extension dairymen in Georgia took the lead to promote the use of AI in dairy herds. No bull stud was established. Local cooperative associations, farmer controlled, were organized. In the beginning, semen was purchased from the American Breeders Service and the Southeastern Artificial Breeding Association stud, located at Asheville, North Carolina. At present, the local associations and private breeders buy frozen semen from many NAAB member bull studs.

Leaders in the early Georgia program were Frank W. Fitch, extension dairyman, University of Georgia; J. O. Chandler, county agent in Monticello, who helped organize the first local association; and Tyron Reynolds, of Millen, who served as an AI technician in 1949.

Noba—GA. In December 1974, Noba, Inc., unveiled a new facility at Elko, Georgia. Its function is to house young Noba bulls being sampled for progeny test and those awaiting proof and to provide a custom-collection service. Plans call for the eventual housing of 175 to 200 bulls. Millard Jones was the first manager of the new facility. He was succeeded by Larry Guthrie in December 1975.

Hawaii

There are no commercial bull studs in Hawaii. The cattle population in 1975 consisted of about 127,000 head designated as beef and 18,000 as dairy.[1]

Artificial insemination has been carried on since the early 1950s with semen shipped from the U.S. mainland. One of the early importers of semen was the University of Hawaii Agricultural Experiment Station. Finding the herd was infected with trichomoniasis all of the bulls were slaughtered, and, as a result, semen was imported. The bull battery has now been reestablished and AI is used in the herd.

The larger dairy herds, managed similarly to those in California, are bred artificially. The large beef herds and the dairy herds have used AI for many years. Several AI organizations on the U.S. mainland ship frozen semen, both beef and dairy, to Hawaii. Over 60 percent of the dairy cows are serviced by AI.

Harvey M. Volrath, extension animal husbandman, and the University of Hawaii have greatly supported the program.

Idaho

Idaho was later in starting an AI program than some of the other western states. This was because dairymen and the Agricultural Extension Service, in cooperation with the University of Idaho, under the direction of Ivan H. Loughary, had established a number of "natural service bull studs" throughout the state. In 1942, three cooperative bull associations, with 19 bulls, servicing 692 cows for the year were reported.[2] There were probably more bull associations as only those with three or more bulls were listed. The bulls were transported, by a small trailer, to the dairyman's farm, the cow was serviced, then the bull was loaded up and transported to the next farm. The bulls used in the natural-service–bull-stud program were privately owned, but they were selected by the University of Idaho Dairy Extension staff. This program did not result in a large number of cows being bred to superior bulls. Meanwhile, the AI program was demonstrating its success, and Idaho dairymen began taking a dim view of the natural-service studs.

Bluffview Artificial Insemination Service. Lloyd E. Johnson, New Plymouth, Idaho, was manager of the Payette County Stud when natural service was used. Due to the reluctance of the extension service to concede the shortcomings of the natural-service studs, Johnson resigned in 1944 and formed his own bull stud and AI business. By 1946, his Bluffview AI Service had seven bulls and inseminated about one thousand cows annually. In 1954, the volume reached 18,658 first services. Bluffview was purchasing some semen from the Curtiss Breeding Service and also from the Jessup Farms Artificial Breeding Service during this time.

When frozen semen became available beginning 1955 and on, many out-of-state AI organizations invaded Idaho. In 1959, over 98,000 cows, representing 40 percent of the total, were bred artificially. In December 1959, Bluffview AI Service merged with the Jessup Farms Artificial Breeding Service, and the bull stud was moved there. Johnson continued to operate an inseminating service using frozen semen from the Jessup organization.

Shortly after the first AI organization started in Idaho, the legislature passed a licensing act placing all AI operators under the jurisdiction of the State Department of Animal Industry. Service is now supplied by out-of-state AI businesses.

Illinois

Illinois has been the birthplace of four prominent AI organizations:

Northern Illinois Breeding Cooperative. Through the efforts of J. Rockefeller Prentice in 1939, and C. S. (Dusty) Rhode, an extension dairyman at the University of Illinois, the Northern Illinois Holstein Artificial Breeding Association was formed. Service began in April 1940, and 1,054 cows were serviced.

In 1946, a total of 16,229 cows, representing nearly 50 percent of those registered, were artificially bred. The association became recognized throughout the country for its outstanding Holstein bulls. Rhode served as adviser on sires purchased, and he was ably assisted by Wilbur D. Goeke.

In 1949, necessitated by the addition of other breeds of cattle, the name *Northern Illinois Breeding Cooperative* (NIBCO) was adopted. The volume had grown with 117,285 cows bred in 1955. Forty counties were being serviced with seventy technicians in the field.

In February 1968, Northern Illinois Breeding Cooperative and Southern Illinois Breeding Association in Breese merged to form the Illinois Breeding Cooperative. This move provided more efficient service and economy in operations.

Southern Illinois Breeding Association. Southern Illinois Breeding Association (SIBA), a cooperative, was incorporated in June 1945 and began service the following September.

C. S. Rhode led in the organization work. He was aided by milk-marketing cooperatives, DHIA groups, and county breed organizations.

Finances were obtained by patrons paying one-half their annual herd breeding fees in advance, and a loan from a milk-marketing cooperative, the Sanitary Milk Producers, of St. Louis, Missouri. F. G. Stevenson was employed as manager. He organized the staff, engineering the construction of a modern bull barn, laboratories, and office facilities at Breese. Howard Boland, of Paris, Illinois, served as president in the 1960s.

SIBA serviced 5,483 cows during the first year and by 1965 was averaging about 67,000 annually. The decline in dairy cattle numbers in the southern Illinois area from 1950 on reduced the volume severely. The fall of 1967, the boards of directors of Northern Illinois Breeding Cooperative and of Southern Illinois Breeding Association voted to merge the two organizations. This merger resulted in the formation of the Illinois Breeding Cooperative (IBC).

Illinois Breeding Cooperative (IBC). The amalgamation of NIBCO and SIBA became official on 1 March 1968. Over 99 percent of the patrons voted in favor of the merger. Headquarters was established at Hampshire. Wilbur D. Goeke continued as general manager. Upon his retirement a few years later, Jess Barnes became manager.

The first board of directors was composed of representatives of the merging organizations as follows: Richard Muehling, president; Norman Henke, vice-president; J. A. Hickey, secretary; and Howard Boland, treasurer.

During 1965, both SIBA and NIBCO affiliated with Select Sires, Inc., which was then headquartered at Columbus, Ohio. In the growth and development of Select Sires, the bull stud at IBC was gradually integrated. Illinois Breeding Cooperative then received most of its semen through Select Sires.

However, some of the IBC patrons and officers, who were long accustomed in selecting their own bulls and in having a major voice in policymaking, were unhappy with the Select Sires, Inc., affiliation. On 1 August 1973, IBC terminated its membership in Select Sires and made contractual arrangements to purchase semen from its nearby neighbor, the Curtiss Breeding Service in Cary. Under this arrangement, IBC continues to provide service throughout the state and also operates a custom-semen-freezing business. A few of the IBC patrons chose to remain with Select Sires, Inc., and receive service through that organization's Illinois-Wisconsin distribution center. This group later became known as Prairie State-Select Sires.

The American Dairy Guernsey Associates of Northern Illinois (American Breeders Service). When the Northern Illinois Holstein Artificial Breeding Association, in Dundee, failed to use only proved bulls and refused to add Guernsey bulls, J. Rockefeller Prentice withdrew and established the

American Dairy Guernsey Associates of Northern Illinois on 28 February 1941. It was a nonprofit corporation. Three high-indexed-proved Guernsey bulls were assembled in the same barn formerly utilized by the Northern Illinois Holstein ABA. P. I. Higley, later ABS president, was fieldman and general manager of the unit. As a result of this corporation, the American Breeders Service was organized, which will be discussed in detail in the section on Wisconsin.

Curtiss Breeding Service. Curtiss Breeding Service started business on 21 March 1949, under the name of Curtiss Candy Company Farms Improved Stud Service. That was the day the first cow outside the Curtiss Farm herds was inseminated. The use of AI, confined entirely to the farm's own cattle, had begun seven years earlier. Resident Curtiss veterinarians handled semen collection and inseminating details.

Entrance into a commercial AI business started as a project of the Curtiss Candy Farms Division, a large-scale farming operation, which was started in 1942 by Otto Y. Schnering, founder of the Curtiss Candy Company. An extensive livestock operation, with outstanding herds of all six dairy breeds, three beef breeds, hogs, and sheep, was established. There was a poultry operation of nine thousand laying hens, and as a "sideline" ten thousand broilers produced weekly; a mink farm; and a trout-rearing farm.

Production testing was carried on in the dairy herds and show strings of all breeds were exhibited at fairs. The farm and its livestock were publicized nationwide. Delbert Kingston, former secretary of the Wisconsin Guernsey Breeders Association, was the livestock superintendent.

Curtiss entered the commercial AI field at a time that competition among the organizations, already entrenched, was becoming intensified. There was some criticism from agricultural extension workers who felt that AI should be carried on by farmer-owned cooperatives.

The business grew slowly at first. Only 7,136 cows were bred in 1949 and 45,472 in 1950. Curtiss had a strong following among many purebred breeders and received good support in that area.

Two developments aided in the growth of the Curtiss Improved Stud Service. First, services were offered to beef bulls beginning in 1949, and that same year technician-training schools were started. Trainees, carefully screened, and often herd managers, were brought to Cary for a three-week training course at what was dubbed "Otto Schnering College." The training course was shortened to two weeks and eventually to one week. In 1963, a mobile training school was developed. It moved about the country providing five-day sessions for training dairymen and ranchers in AI techniques. Some "on the farm" training of inseminators was also begun. In 1968, the name of the course was changed to Curtiss Technician Training School.

The technician-training courses not only trained inseminators, but they developed customers for Curtiss semen. Some trainees became technicians or distributors for Curtiss, and some of them inseminated cows in their own herds and other herds in their area.

In 1950, Curtiss established its first "direct service" herd by selling semen to a large Florida herd. Others were to follow. Some reports indicate that the term *direct service* was first coined at Curtiss. The earlier practice was to sell semen directly to technicians. In 1954, this practice was discontinued and distributorships were established. By 1972, Curtiss employed seventeen hundred distributors, technicians, dealers, and district sales managers in the field. Over 1.2 million first service cows were reported for the year.

In 1950, foreign semen sales were initiated. These first endeavors were greatly publicized. Vials of semen were attached to the legs of carrier pigeons at the Tokyo Japan Airport. The pigeons were released, and they flew to farms where cows were to be serviced. By 1972, Curtiss had thirty-seven distributorships in foreign countries.

On 10 January 1953, Otto Schnering died. His wife, Dorothy R. Schnering, who had helped him in his business and who was always deeply interested in the livestock and breeding farm projects, became the president of Curtiss Improved Stud Service in 1954. Following Schnering's death, there were rumors that the Curtiss Improved Stud Service, along with all the cattle and farms, would be sold. Prentice, owner of American Breeders Service, advised Dorothy to sell out at ABS's price or he would take over and run Curtiss out of the AI business by competitive means. Dorothy vowed she would not let this happen and proceeded to take matters into her own hands.

Dorothy Schnering requested the company's board of directors to let her run the "bull company," which was her late husband's beloved project, for as long as she could show a profit. Thus, she became the first woman to head a large AI business, the second in the United States in volume of business. The business grew under her guidance. In 1960, the name of the business was changed to Curtiss Breeding Service. It was under her guidance that frozen semen, top-ranking dairy and beef bulls, export trade, and an expanded field staff came into the picture. An office building and a new bull barn were constructed at Cary.

On 17 February 1973, Dorothy R. Schnering died at the age of eighty.

Melvin L. Kenley: General Manager. William E. Hunter, manager of the Curtiss Farm Division, had returned to the candy company after the herds and farmland were sold. Herman Adams succeeded him as manager of the AI business until 1959. That year, a general reorganization occurred. Melvin L. Kenley was named general manager. Kenley had come to Curtiss Farm as a junior accountant in 1947. He worked successively as assistant office manager, general manager, vice-president, executive vice-president, and in 1963, at the request of Dorothy R. Schnering, succeeded her as president.

Finances continued to plague the candy company, and in 1964 it was sold to Standard Brands, Inc., a large food conglomerate with offices in New York City. Curtiss Breeding Service, Inc., was sold as part of the package. Dorothy, as a condition of selling her bloc of stock, obtained a six-month's option to buy back the "bull company." She then gave the employees of Curtiss Breeding Service an opportunity to buy part of the stock that she was entitled to acquire. Many took advantage of the opportunity. Once again, Curtiss Breeding Service had survived and became in part an employee-owned-stock company. Growth continued. Several outstanding beef bulls including the early "exotics" were added. Curtiss obtained exclusive semen rights to the famous Simmental bull, Parisien, located in Alberta, Canada. By special arrangement, semen from twenty-seven Simmental and one Brown Swiss bull in Switzerland was imported. Arrangements were also made to distribute semen from Sam #951, the renowned Charolais bull, at Litton Ranch, in Chillicothe, Missouri.

Curtiss Breeding Service was not to continue under the same ownership too long. In October 1968, the AI business was sold to G. D. Searle and Co., a large pharmaceutical firm at Skokie, Illinois. In May 1972, Searle Agriculture, Inc., was formed with the Curtiss Breeding Service becoming a division. John O. Quinn served as president of Searle Agriculture, Inc. Kenley served as president and chief executive officer of Curtiss Breeding Service. To accommodate its increasing number of bulls and growing sales volume, a new production center was established near Elburn, Illinois.

In 1975, Kenley resigned to set up his own business. Glenn W. Pirrong was named president, and Robert Boese, vice-president of the breeding service. Operations were carried on with an expanding export trade developed.

Change in the ownership of Curtiss Breeding Service came about again in late 1978 when Idgara Ltd., a Colorado-based syndicate, purchased the assets of Searle Agriculture, Inc. The breeding service continued to be operated, but the name of Searle Agriculture, Inc., was changed to Curtiss Breeding Industries, Inc. In the new regime, Thomas B. Neff, president of Idgara, was elected president of Curtiss Breeding Industries; Pirrong became executive vice-president; Robert D. Long, vice-president of marketing; and Boese, vice-president of international operations. In 1979, the Cary plant was phased out and all operations moved to Elburn.

Armour and Company Beef Cattle Improvement Research (Armour BCI). Recognizing the need and the opportunity for improvement in beef cattle, the Armour Company, a Chicago-Illinois-based–meat-packing firm, entered the AI field in the late 1950s. The goal called for a program of using the best beef sires available to provide AI service for cooperating herds. Beef-bull-performance-testing and progeny-evaluation programs were initiated.

Harold Hill, a veterinarian on the staff of Colorado State University, directed the project. Headquarters were established on a farm near Denver. A laboratory and semen-freezing facilities were set up. George Norwood was hired as the production manager and in charge of the bulls and semen processing. An outstanding battery of bulls was assembled, and several field representatives were employed to sign up and work with cooperating herds in the Midwest and Rocky Mountain areas. An agreement was made with owners of the Codding Ranch, in Foraker, Oklahoma, to erect the necessary facilities and to carry on an extensive performance-testing–sire-proving program. The Codding Ranch was selected because of its longtime experience in performance testing and the outstanding herd of cattle that had been developed.

Unfortunately, AI in beef cattle was not widely accepted and beef breeders were not too well versed on the advantages of performance testing. The program grew slowly. Thousands of units of frozen semen, from outstanding beef bulls of all breeds, piled up in the Denver storage. It was then decided to sell frozen semen rather than depend entirely on cooperating herds for utilization. The time was right for this move. Very few established AI organizations had a strong lineup of beef bulls. The Armour BCI officials decided to sell semen to existing AI organizations for their beef demands. It seemed to be a "natural." The AI organizations could merchandise frozen semen from many outstanding bulls available and yet be spared the capital investment. This program started off strong and for several years many NAAB members utilized Armour BCI semen. Troubles in pricing developed, and as a result the Armour Company had numerous policy changes. The project had proved to be costly.

Hill was replaced by Harold White as general manager. A 620-acre farm, with excellent barns and facilities, near Morris, Illinois, was procured. Personnel, bulls, and equipment at the Denver location were moved to Morris. Norwood continued as production manager. Emphasis on progeny-proved sires continued. Some eighty sires, representing all of the major beef breeds, and many already proved were available. Semen sales were growing, but not enough to alleviate the financial burden. Sales managers, John Heckman and Charles Calkins, toured the country, working with AI organizations and beef breeders in efforts to boost semen demands. Due to changes in semen pricing, which left most cooperating bull studs in an embarrassing position, sales dwindled. Most AI organizations began establishing their own beef bull studs. The Codding Ranch, disappointed in the revenue from semen sales and the progeny-testing program, filed suit against Armour BCI. As a result, the Armour Beef Cattle Improvement research program was phased out. The Coddings did not win their lawsuit, but did inherit the array of buildings Armour had erected on their property. These facilities were later put to use as a custom-freezing and beef unit by Noba, Inc., and continues so.

The Armour project was planned to meet a need that still exists. Its goals were sound and represented the first effort to date for a major packing company to attempt to improve beef cattle by modern methods.

Indiana

The first AI efforts in Indiana began in 1939 at Purdue University. An Angus bull unable to work naturally was used by the AI method.

Between 1940 and 1945, several small farmer-owned cooperatives were formed by E. A. Gannon and Frank De La Croix, both extension dairymen at Purdue University. There were some seventeen small associations, some owning a few bulls and others buying semen, in operation. One of the leaders in setting up a bull stud was Leroy Coplen, of Plymouth, Indiana. He was then employed by the Indianapolis Dairymen's Cooperative Association, Inc. Later, he became a fieldman for ABS. The stud was located at Carmel and later became the Indiana Artificial Breeding Association. ABS bulls were used. P. I. Higley was the manager and Leroy Coplen, fieldman. By 1945, about twenty-two thousand dairy cows were serviced.

Late in 1946, ABS purchased the Carmel Association and organized the Indiana ABA. Cooperative breeding associations were organized in forty counties. These local cooperatives used semen from ABS. Twenty-two proven bulls were now located in the Carmel stud. By 1949, the Indiana ABA (American Breeders Service) was breeding cows in eighty-five of the ninety-two counties in Indiana. In 1972, over 90,000 milk cows were bred to dairy bulls, 6,034 dairy cows to beef bulls, and there were 19,763 beef to beef matings. About 34 percent of the state's dairy cows are now artificially bred. The Carmel stud was later closed out and merged into its central operations by ABS.

The principal bull stud was that of ABS because the Purdue extension dairymen favored a program utilizing "plus proven" bulls. However, AI organizations from several states service many cows in Indiana.

Iowa

Artificial insemination of dairy cattle began in Iowa in 1942. The Linn County Cooperative Artificial Breeding Association at Cedar Rapids and the North Iowa Breeders' Cooperative at Kanawka (Wright County) organized about the same time. Three thousand cows were artificially bred in 1942. Other organizations developed soon after, including bull studs at Humboldt, Emmetsburg, Mason City, Dennison, Iowa City, Des Moines, and around the borders of the state at Trenton, Missouri; Fremont, Nebraska; and New Prague, Minnesota. There were forty-seven local associations in Iowa by 1947.

Small in size, organized without experience as to the problems of maintaining a bull stud, and finding the volume of business too small to support the bulls needed for a liquid semen program, these locals soon merged, sold out, or disbanded. However, 53,180 cows were artificially bred in Iowa during 1946.

The AI organizations that grew into prominence in Iowa are:

Eastern Iowa Breeders, Inc.   Organized in 1943 by J. W. Pirie, DVM, the Iowa State University dairy extension staff, and local dairymen, Eastern Iowa Breeders was an outgrowth of the Linn County Cooperative ABA. The new organization was first known as Eastern Iowa Artificial Breeding Association, but the name was changed to Eastern Iowa Breeders, Inc., about 1955.

Pirie owned and operated the bull stud and provided semen for cooperating county units under contract. Private capital financed the bull stud, laboratory, and equipment. Operational expenses of the program were taken care of by the counties organized on a cooperative membership basis.

Eastern Iowa Breeders proved to be a progressive organization. It was one of the first to convert to frozen semen. Outstanding bulls were bought and many young sires proved. The organization was strongly supported by the Iowa dairymen. Gradually business was expanded into other states and other countries.

In 1964, Eastern Iowa Breeders bred 139,082 cows, which included 26,670 beef to dairy and 9,662 beef to beef matings. Fifty-four bulls, nineteen beef, were maintained. The decline in dairy cattle population, however, reduced volume thereafter.

In 1965, the assets of Eastern Iowa Breeders were sold to Carnation Farms Breeding Service and the bulls moved to Watertown, Wisconsin.

Pirie then became affiliated with Carnation Farms Breeding Service as a consultant and a southwestern representative.

Northwest Iowa Federated Breeders Cooperative (NIFBC). Located at Sheldon, Northwest Iowa Federated Breeders Cooperative was started in 1946. It was the result of a merger of four small studs in Kanawka, Humboldt, and Sheldon counties.

Men active in promotion were Philo C. Tabor, Humboldt; Lynn Thompson, Clarion; Walter Kruse, Sheldon; and Norman Weilander, of Pocahontas. Robert C. Fincham, extension dairyman, and Sam Thompson, agricultural economist, both of the Iowa State University in Ames, helped the board of directors prepare articles of incorporation and bylaws. The entrance of frozen semen into the field and competition, however, upset this program. Some locals started purchasing semen from other bull studs. The "federated local" was doomed to produce problems and most AI organizations throughout the country developed a direct membership program.

NIFBC, however, was the first cooperative bull stud in Iowa. Tabor, formerly president of the board of directors, became manager. In July 1956, NIFBC converted to frozen semen and claimed to be the first in the United States to make a complete conversion. During 1956, the volume amounted to twenty-six thousand cows, down from earlier years. On 10 October 1957, NIFBC merged with Iowa Breeders Cooperative, which was located in Des Moines.

Dairy Genetics. In 1946, Dairy Genetics, Inc., a privately financed business, was organized by Herman Hayes, Roger Leinbach, Edward Snydergaard, Harlan Geiger, and Lael Moon, all former Iowa State University agricultural extension workers. A group of bulls was assembled at Cornell and Hoffman Road, Des Moines. This corporation absorbed Western Iowa Breeders at Dennison and provided service in southern and southwestern Iowa. The articles of the organization provided for its change to a direct member cooperative when it was strong enough to move that way.

Dairy Genetics was a semen-producing business, serving several local county AI cooperatives. In 1950, the stud contained forty-one bulls of all dairy breeds, including Milking Shorthorns. During 1951, about fifty-eight thousand cows were serviced. In 1955, the assets of Dairy Genetics, Inc., were acquired by the newly formed Iowa Breeders Cooperative.

Iowa Breeders Cooperative. Iowa Breeders Cooperative was started in 1954, with headquarters located at the Dairy Genetics plant. In 1955, Iowa Breeders Cooperative took over the bulls, equipment, and other assets of Dairy Genetics. It was a direct membership organization servicing southern Iowa.

Leaders in setting up the cooperative were Robert Irwin, John Kitt, Everet Nash, William Rankin, Art Duessing, Dewey Grim, and Ralph Mason. Cooperative loans furnished initial capital. Greg Raps, a staff veterinarian at the Iowa State University, became the manager. During 1955, over 75,000 cows were serviced. Northwest Iowa Federated Breeders merged with Iowa Breeders Cooperative in 1957. The organization grew and exceeded 100,000 first service cows the next few years. However, increasing costs of operation and dwindling dairy cow numbers brought a consolidation in 1961 of Iowa Breeders Cooperative; Nebraska Dairy Breeders, in Fremont; MFA Dairy Breeders, in Springfield, Missouri; and Consolidated Breeders Cooperative, in Anoka, Minnesota. The new organization was Midwest Genes. It represented a consolidation of bulls and became the semen-production unit for the four cooperatives. The bulls from all units were moved to Anoka in the spring of 1962.

In 1965, Iowa Breeders Cooperative and Consolidated Breeders Cooperative merged their assets. Thus, Consolidated Breeders became the largest member of Midwest Genes.

Pioneer Beef Cattle Company.   In the late 1960s, the Pioneer Seed Company of Des Moines established a beef bull stud and performance-testing unit at Johnston, Iowa. Emphasis was on performance testing. Bulls that were performance tested, which represented several breeds, were sold to the trade. Semen was also made available. As business grew, from 1971 to 1975, a second unit was established at Tipton, Indiana.

The leader of the project was John Airy. Garold Parks served as manager of the Iowa division, and Ron Long was manager of the Indiana division. With the 1975–1976 slump in beef prices, the operations of Pioneer Beef Cattle Company diminished.

Kansas

Artificial insemination of dairy herds in Kansas began prior to 1950 by out-of-state organizations operating in the Wichita, Topeka, and Hiawatha areas. There was also some early AI work in the dairy herd at Kansas State University in Manhattan.

Kansas State Artificial Breeding Service Unit (KABSU) was organized in 1950. It is a project of the Department of Dairy Husbandry at the Kansas State University, cooperating dairymen, and of beef cattle producers in the state. The bull stud and physical plant are located on university-owned land at Manhattan. Early financing came from a state appropriation of $62,000. The stud was started with bulls representing the six dairy breeds. As beef AI developed, a strong battery of beef bulls was added.

The university led the way in establishing KABSU. The late F. W. Atkeson, then chairman of the Department of Dairy Husbandry, set the pace. He was assisted by Fred Foreman (in the northern counties) and Ralph Bonewitz (in the southern counties). Both men acted as organizers. J. W. Linn, dairy extension, aided in details of the organization at Manhattan.

While there is no formal board of directors, KABSU is controlled by an advisory committee of livestock men and an artificial breeding committee composed of representatives from the Dairy Department, Animal Husbandry Department, and College of Veterinary Medicine at Kansas State University.

Earl L. Farmer was the first manager. The first year of operation, 24,512 cows were serviced, and four years later the number had doubled. After a few years as manager, Farmer became a member of the teaching staff at the university's dairy department.

He was succeeded by Ancel Armstrong. During his tenure, Armstrong expanded the volume of dairy cows bred artificially and also developed one of the best beef AI programs in the country. He resigned in July 1970 to form his own organization, the *New Breeds Industries, Inc.*, and to engage in the import and sale of semen from exotic beef breeds.

Charles L. Michaels became the manager of KABSU in July 1970 and continues in that role.

During 1971, KABSU accounted for 102,590 first services: 43,860 dairy and 58,730 beef cows.

New Breeds Industries, Inc.   Upon his resignation from KABSU in July 1970, Armstrong established his own organization, the New Breeds Industries, Inc. During the period from 1970 to 1974, the trade in exotic breeds was brisk. Armstrong had a working arrangement with Canadian AI bull studs as well as overseas AI organizations and cattlemen. He assisted in importing large quantities of exotic breed semen into the United States. With cattle of the "new breeds" now more plentiful in the United States, less semen is imported, but New Breeds Industries continues to play a leading role, particularly in the development of beef breed programs. Armstrong was elected president of the American Simmental Association in 1977.

Kentucky

Kentucky dairymen and agricultural extension workers organized the Kentucky Artificial Breeding Association (KABA), on 18 September 1946. A cooperative, it was organized on the federated plan, with headquarters at Louisville. There were forty-six local associations, each with a director and an executive committee.

John Moser, a dairyman at Louisville, devoted much time in starting the organization. He served as president during the first three years of operation. The Fall Cities Milk Producers Association donated the service of their fieldman, Robert Connor, to promote the AI work. He served as the first secretary-treasurer.

Marshall C. Carpenter, then a state dairy extension worker, was hired as the first manager. He still serves in this capacity.

In the first ten months of operation, KABA had 11,486 first services. Beginning with five locals and 550 members, in nine years there were forty-nine local associations and 18,000 members. In 1964, a total of 94,024 dairy and beef cows were bred artificially. Due to mounting costs of operation, Kentucky Artificial Breeding Association merged with the then-forming Select Sires, Inc. The bulls were moved on 15 March 1968.

The Kentucky Association received strong support from the College of Agriculture at the University of Kentucky. Timely and applicable research on AI and reproduction, beneficial to the entire industry, has been carried out by Durward Olds of the university's staff.

Louisiana

Artificial insemination of dairy cattle in Louisiana was started by Louisiana State University Dairy Department in 1943. In 1946, the first organized artificial breeding association was formed at Ruston and was known as the North Louisiana Jersey Breeders Association. It later became a part of the statewide program.

On 16 July 1947, the Louisiana Breeding Cooperative, now known as Louisiana Animal Breeders Cooperative, Inc. (LABC), was organized at Baton Rouge. The change in name was made effective September 1965. This statewide organization has direct membership. The first semen was shipped on 16 October 1947. The statewide program continued to grow. On 1 January 1955, there were fifty local technicians, forty-five local breeding circuits, and service was available in sixty-two parishes.

The artificial breeding program is jointly administered by the Louisiana Animal Breeding Cooperative and the Louisiana State University College of Agriculture, with money made available by the Louisiana state legislature.

In 1947, T. E. Patrick became the manager of LABC, a position he held until late 1977. Under his tenure, the association withstood many technological changes—including the conversion to frozen semen—and personnel. Patrick was succeeded by Arnold Baham as manager.

In order to increase the number of bulls available, LABC, Noba, Inc., and Atlantic Breeders Cooperative formed the Affiliated Breeders. This move made 275 high-quality bulls of twenty-five different breeds available to the three organizations.

Several out-of-state organizations also provide service in Louisiana and, at present, about 45 percent of the dairy and 2.0 percent of the eligible beef cattle are bred artificially.

Maine

The principal AI organization based in Maine was the Maine Breeding Cooperative at Augusta. It resulted from the consolidation, in 1946, of two small artificial breeding units: Central Maine, organized in the fall of 1939, and the Androscoggin Valley Artificial Breeders Association, organized in July 1940. Leading in planning and in setting up artificial breeding in Maine was Stetson Smith, former county agent in Penobscot County. The early program was assisted by the Maine Development Commission. Other financing was done through membership fees and loans obtained from local banks.

In 1950, Paul Browne, a former vocational agriculture teacher in Maine and an agricultural extension worker for ten years in Massachusetts, was employed as manager.

The program was greatly supported by dairymen, including Fred Nutter, of Corinna, and Donald Corbett, of Winslow. During 1964, there were 55,810 dairy cows, 59.4 percent of the total in Maine, and 535 beef cows artificially bred. Because of the ever-increasing cost of operation and the declining number of milk cows, in October 1965, the cooperative voted to merge with all of the New England studs to form the Eastern AI Cooperative, Inc., Ithaca, New York.

Maryland

Maryland Artificial Breeding Cooperative, Inc., the first and only bull stud then in Maryland, was organized in July 1945. Headquarters were at College Park, Maryland.

Floyd Arnold, extension dairyman, served as the cooperative's first secretary. E. J. Weatherby was the first manager. There were 736 cows bred the first year, but by 1954 the volume was 43,000 head. Weatherby, who received his advanced training at Rutgers University, had increased the cooperative's business to 51,000 first service cows in 1956. Following his death on 5 November 1956, the Maryland Artificial Breeding Cooperative and the West Virginia Artificial Breeding Cooperative, in Clarksburg, West Virginia, merged to form the Maryland-West Virginia Bull Stud. Final arrangements were completed in September 1957.

W. L. Campbell, who had been manager of the West Virginia Cooperative since 1948, became manager of the new association. Shortly thereafter, a farm was purchased near Frederick and the headquarters was transferred there. Both the Maryland and West Virginia cooperatives retained their names and membership structure. During 1968, the Maryland–West Virginia Bull Stud reported 89,467 first services. In 1969, the organization became part of Sire Power, Inc.

Sire Power, Inc. With headquarters at Frederick, Sire Power, Inc., a cooperative resulting from the consolidation of the bull studs of Maryland–West Virginia and Northeastern Breeders Cooperative (NEBA), of Tunkhannock, Pennsylvania, was formed the fall of 1969. The new arrangement provided an enlarged battery of proven bulls and made possible the practical conduct of an intensive young-sire-sampling program. Sire Power, Inc., represents the bull stud for its three-member organization. Each member retains its identity and continues to serve its own patron area as previously. Sire Power, Inc. is controlled by a seven-man board of directors. Fred Downey, Maryland, was the first president of the board. The NEBA facilities at Tunkhannock are utilized as the semen-producing unit of Sire Power.

W. L. Campbell became manager of Sire Power, Inc., bringing a quarter century of experience as an AI organization administrator.

Wye Plantation. One of the early leaders in beef cattle improvement and the use of AI in America is Wye Plantation, located in Queenstown. The plantation, consisting of thirteen hundred acres on the Maryland Eastern Shore, was purchased by Arthur A. Houghton, Jr., an executive of the Corning Glass Company, in 1937. Shortly thereafter, he hired James B. Lingle as manager. A purebred Angus herd, with the foundation animals purchased in Maryland and Oklahoma, was started. In 1957, several bulls were imported from the British Isles.

Lingle proved to be a master breeder of Angus.[3] His goal was production of beef. He frowned on the small compact type of Angus as dictated, at that time, by show-ring standards. He determinedly selected hardy, big-bodied cattle, insisting "that you have to have a big frame if you are going to hang much meat on it." Years before, when the breed associations recognized performance programs, Lingle began to record the weight of calves at birth, at weaning, and after feed tests. A most complete set of records, including body measurements, was kept. The University of Maryland, with Willard Green in charge, cooperated in the gains and growth tests. The herd was enrolled on the PRI program and performance pedigrees at Wye were in early use.

In 1961, Wye Plantation started selling frozen semen from performance-tested bulls. There was a working arrangement established with Noba, Inc., to handle semen. But, as a rule, Wye handles direct semen sales. Lingle's philosophy was to sell a breeding program, not just a unit of semen. By 1975, Wye had over one thousand semen customers in forty-five states and eleven foreign countries. Many bulls were sold to breeders and some to AI studs. The top of all sales came in 1968 when a two-thirds interest in the bull Lodge of Wye was sold to Black Watch Farms, New York, for $250,000.

When Lingle retired in 1972, management of the herd was turned over to John R. (Dick) Whaley III. Whaley carried on the Wye herd philosophy and the Lingle-established program. In 1978, Houghton gave the entire Wye Plantation Angus herd to the University of Maryland. The breeding program and the research work that were earlier instituted are still continued.

Massachusetts

The Massachusetts Selective Breeding Association, Inc. (MSBA), a federated cooperative with some twenty-five county units, was organized 8 January 1946. Headquarters was established at Shrewsbury, Massachusetts.

V. A. Rice and Stanley Gaunt aided in promoting the program. They were assisted by James Dayton.

Chester Putney was the first manager. He served from 1946 to 1952, leaving to become secretary of the Ayrshire Breeders Association, in Brandon, Vermont. Putney was succeeded by Fredrick Buschner. During the first year, 18,634 cows were inseminated and the bull stud had sires of all six dairy breeds as well as Angus and Hereford. The MSBA worked closely with the New England Angus Association and was the first to offer AI beef service in New England.

On 1 July 1957, Buschner resigned to become the eastern field representative for the expanding ABS. The Massachusetts AI program serviced 59,411 cows and had a stud of 41 bulls (5 beef) in 1957.

With the availability of frozen semen, the association, as was true of most small bull studs, felt the need for more volume and greater use of bulls. Negotiations began for a merger with other New England organizations. Franklin H. Perkins, Jr., a former fieldman for MSBA, was appointed acting manager during the interim. In the fall of 1958, a merger was brought about consolidating MSBA, the Connecticut Artificial Breeding Association, the Central Vermont Artificial Breeding Association, and the Rhode Island Artificial Breeding Association. The new organization, New England Selective Breeding Association (NESBA), had quarters at Woodbridge, Connecticut, utilizing the facilities of Connecticut ABA. Paul Heller became manager. He was succeeded in 1960 by Henry Soborg, who died on 9 August 1963. Michael Johnstone served as manager following Soborg's death.

The bulls from all four studs were appraised, and the best were moved to Woodbridge. The NESBA purchased the Connecticut facilities. A board of directors representing the four-merged organizations was established.

During 1965, NESBA accounted for 106,384 first service cows. In January 1966, NESBA merged with the Eastern Artificial Insemination Cooperative, Inc.

Soligenics. Roger Ives left the Massachusetts Selective Breeding Service in 1955 to establish Soligenics, one of the earliest custom-semen-freezing and storage businesses in the United States. Through the years, Soligenics has carried out a worthwhile service for cattle breeders in the northeast and is well recognized for its high professional standards.

Michigan

Michigan Artificial Breeders Cooperative, Inc. (MABC), organized in May 1944, led the way for a strong AI program in that state. A year before, a small stud—Wolverine Artificial Breeders Association—had been started near Jenison, along with one or two other local associations. All joined forces when MABC got underway at East Lansing.

The leader in forming the cooperative was A. C. (Ace) Baltzer, an extension dairyman at Michigan State University.

Baltzer was the association's secretary-treasurer. Fred Dombroske, still with MABC, was the production manager. The cooperative was financed by membership enrollment and by a signup fee for cows that were to be serviced. As capital was amassed, a building program was begun. The first year, some of the bulls were stabled in a bank barn with a straw roof. The bull barns, laboratory, and office buildings were erected on land controlled by Michigan State University.

The first year, 4,156 cows were inseminated, but ten years later, the volume was 155,000 head. The program continued to grow with the conversion to frozen semen in the late 1950s.

On 1 February 1955, "Ace" Baltzer died of a heart attack. Therefore, quick adjustments were necessary. President Frank Heim and the board hired E. A. (Pip) Wenner to replace Baltzer beginning 15 July 1955. He had the title of "field service-public relations manager." Wenner had served twenty-five years as a 4-H club agent, county agent, and district extension supervisor. He carried on the program launched by Baltzer until his retirement in the fall of 1963.

Kenneth Baushke was then hired as general manager. Under his guidance, MABC expanded its service program, adding beef bulls and also reorganizing districts. MABC met competition with superior bulls, efficient service, and loyal dairymen. During 1964, a total of 280,150 cows, including about 39,000 beef cows, were serviced.

In 1966, following the course of many AI cooperatives, MABC became a part of Select Sires, Inc. The organization continues to serve its patrons as MABC–Select Sires. The MABC bull barns house "in waiting" bulls for the Select Sires bull battery. Baushke continues as manager of the Michigan Association.

General Genetics. Jacob Van Hoven and his son operate this small AI organization at Jenison. It is a privately owned business that began operations in 1967. Some semen is processed from bulls in which an interest is owned. Semen is distributed for several out-of-state AI organizations and herd service is provided by technicians affiliated with General Genetics. In

1971, a total of 29,640 cows—200 beef—were serviced. A custom-semen-freezing business is also operated.

Minnesota

As was true in a number of states, many of the small AI associations that were formed between 1939 and 1942 failed because of lack of experience, poor organization, poor promotion, and mismanagement.

According to University of Minnesota dairy extension workers, nine AI organizations were formed in the state between 1939 and 1942. Among those organized during this period, only two succeeded—Minnesota Valley Breeding Association and Southern Minnesota Breeding Federation. While the intentions of various farm groups and agencies that promoted AI in the early days were well meant, all soon learned "it isn't nearly so simple as it appears." For example, the Minnesota Artificial Breeding Association, with forty bulls, was sponsored in part by the Farm Security Administration (Federal). After two-and-one-half years of operation, the association had an operating loss of $3,481.33 and owed $20,375 borrowed money. It went out of business in 1944.

The University of Minnesota strongly supported the development of a strong AI program. In June 1937, C. L. (Stub) Cole set up a breeding ring, using AI, among dairymen in the area of the Minnesota Experiment Station at Grand Rapids. This trial proved that artificial breeding would work under field conditions. Considerable pressure was put on the university to establish a bull stud to service the state. The authorities took the position that the role of the university was to educate and to advise and that the AI service program should be owned and operated by dairy farmers. This philosophy prevailed.

Minnesota Valley Breeders Association (MVBA). This farmer-owned cooperative started business in January 1941 at New Prague. Previously, Wallace Miller, the county agent, had attempted to get dairymen to participate in bull associations. A young dairyman, Martin Beckman, suggested artificial breeding instead of bull associations. After meetings with interested dairymen, a visit was made to the University of Wisconsin bull stud, which greatly impressed the group, and, as a result, MVBA was established. Beckman served many years as president.

The business began with limited capital: a gift of $300 from local businessmen and a loan of $2,500 from the local bank. Memberships were sold at $5 each to about one hundred local dairymen as the project got underway.

Miller was hired as the manager. He served from 1941 until his retirement in 1972. He must be credited, along with the board of directors and faithful employees, in building MVBA into a business that has withstood the economic changes and developments of the past thirty-five years. The first year, only 6,650 cows were serviced, but by 1951 there were 167,523, and in 1961, the peak year, 330,554 head.

Due to declining dairy cow numbers and competition, the number of cows began to drop, totaling 302,987 with 27,633 beef services in 1971. During the years, MVBA had converted to frozen semen and had expanded its operations to a larger area of Minnesota and the Dakotas. An eight-hundred-acre farm was acquired, and a dairy herd of several hundred milk cows was established. The dairy herd is used to prove young sires as well as to produce income. Bull calves were fed out and marketed as steers. In 1970, total farm income was $151,000. A "Harvestore Division," selling silos in seventeen counties, was set up and grossed $2 million a year. A liquid nitrogen plant was installed and provided not only for MVBA, but outside sales made a profit.

Miller was strong for diversification. Dissatisfied with the liquid nitrogen field units available, he and his sons-in-law, Joe Schuster and Curt Swenson, established Minnesota Valley Engineering Company and made their own containers. This company has expanded to a worldwide operation.

During the years, the MVBA physical plant was enlarged and the bull stud was increased in number and production potential. Beef AI became an important part of the service.

Miller retired in 1972, after thirty-one years as manager. He was succeeded by Sterling Gillingham, the present manager, and MVBA is carrying on with its diversified program.

Southern Minnesota Breeding Federation. Organized in 1942 with headquarters at Owatona, Southern Minnesota Breeding Federation (SMBF) grew from a consolidation of the local associations in the area. The cooperative had strong leadership from local dairymen. It began operations with a $4,000 bank loan and with $3,400 that was furnished by members.

R. W. Seath, former secretary of the Minnesota Holstein Association, became the manager. William Jones, a dairyman, was the first president, and he was succeeded by Roy Backhouse.

The business grew from 7,885 cows serviced in 1942 to 60,192 in 1951.

In November 1954, SMBF and Land O' Lakes Creameries, Inc., a bull stud in Anoka, Minnesota, consolidated. SMBF serviced 69,287 cows and Land O' Lakes serviced 85,144 that year. The consolidated organization was named Consolidated Breeders Cooperative. Headquarters was established at Anoka. Seath continued as manager.

Land O' Lakes Creameries, Inc. This organization was started in 1946 as a development of the Land O' Lakes Herd Improvement Division. A number of small AI associations throughout the producers' area of Land O' Lakes were bought and merged into one large organization. John Brandt, president of Land O' Lakes Creameries in St. Paul, was largely responsible for the organization. John W. Dysart, manager of the Herd Improvement Division, was in charge of the AI program. After eight years, the merger with SMBF occurred.

Consolidated Breeders Cooperative. This cooperative was governed by an eighteen-man board chosen from representatives of Land O' Lakes and SMBF. Under the guidance of Seath, an outstanding battery of bulls was attained, the physical plant was enlarged, and the field services were expanded. During 1955, the bull stud had 64 bulls and 171,677 cows were serviced.

In 1962, Consolidated Breeders Cooperative became a part of the bull stud pooling brought about by the formation of Midwest Sires. In 1965, IBC in Des Moines merged with Consolidated Breeders Cooperative.

Gregg Raps, former manager of IBC, became manager of Consolidated Breeders with Seath carrying on as operations manager. The four members of Midwest Genes, of which Consolidated Breeders was a major shareholder, serviced 415,738 cows in 1966.

Due to the economic situation, on 1 January 1967, the Consolidated Breeders Cooperative and Badger Breeders Cooperative combined their assets, and a new organization, Midwest Breeders Cooperative, Shawano, Wisconsin, emerged. A one-hundred-sire-production center was established at Stewartsville, Minnesota, to augment the plant at Shawano and became operative the fall of 1968.

Northwest Breeders Association. In 1945, six cooperative creameries in Roseau County organized a dairy herd improvement program. This led to the formation of Northwest Breeders in 1945, a cooperative with headquarters at Roseau, Minnesota. Financing was first done by the creameries and later by membership and service fees. Services were extended to eighteen local associations. The first year, about 5,000 cows were serviced; by 1958, the volume was 20,231 head.

The manager from the beginning was Irving W. Meade, a former county agent and farmer. Due to increasing costs and competition, Northwest Breeders Association merged with Consolidated Breeders in July 1959.

Northwestern Artificial Breeding Association. In February 1949, Wisconsin Scientific Breeding Institute (a part of ABS) established a bull stud and formed the Northwestern Artificial Breeding Association at Duluth, Minnesota. This establishment grew, breeding 143,415 cows in 1953. With improved air service and with frozen-semen developments underway, ABS closed the Duluth bull stud in 1954.

International Cyro-Biological Services. During the summer of 1972, A. H. J. Rajamannan, a prominent research worker in reproductive physiology and the AI field, announced the founding of International Cyro-Biological Services with headquarters at St. Paul. Rajamannan had a working relationship with several British cattle breeders and the Milk Marketing Board of England. A bull stud featuring performance-tested beef bulls of the Devon, Lincoln Red, Sussex, Welsh Black, and South Devon was established. Except for the South Devon all of the bulls were from studs affiliated with the Milk Marketing Board.

The advantages of the above bulls were promoted on the basis of a smaller calf at birth and of good gains from crossbred progeny.

Cyro-Biological Services was also involved in the embryo-transplant program. As the demand for exotic semen declined the embryo-transplant business became of major importance and was expanded.

At present, about 50 percent of the dairy cows and heifers and 4 percent of the beef cows in Minnesota are artificially serviced.

Mississippi

The first organized effort for the AI of dairy cows in Mississippi began in the Tupelo area about 1941.

Tupelo Area Artificial Insemination Association. Through the efforts of Gale Carr, field representative of the North Mississippi Livestock Association, and George McLean, editor of the *Tupelo Daily Journal*, a cooperative association was organized in October 1941. This was the Tupelo Area Artificial Insemination Association. Headquarters was at Tupelo. Financing was provided by fifty businessmen and farmers who each signed notes for $100 to purchase three Jersey bulls and to provide working capital. Six county units were established. Only Jerseys were involved until 1951 when Holstein and Guernsey semen were obtained from Mississippi Artificial Breeders Cooperative, in State College, Mississippi.

Cecil Trawick was employed as manager. Progress was slow, but in 1945, there were 1,350 cows bred. Of these, 850 were registered Jerseys. In 1959, serviced cows totaled 12,955 head. In 1960, the association sold its bulls to the Mississippi Artificial Breeders Cooperative and obtained its semen supply from that organization. The name of the organization was changed to Tupelo Area Breeding Service.

Mississippi Artificial Breeders Cooperative. The Mississippi Artificial Breeders Cooperative, at Mississippi State, was organized in 1946. Extension dairymen, L. A. Higgins and Arlis Anderson, both of Mississippi State University, led the efforts. A board of directors was established with Dee Mercier, of Corinth, elected president. Financial support was provided by the state and by public subscription. Earle Thomas, Jr., was the first manager. He was succeeded by Ernest Griggs, the present manager.

The cooperative grew slowly but serviced 27,823 cows in 1965. In 1966, the bull stud was discontinued and Mississippi Artificial (Animal) Breeders became federated with Select Sires, Inc. The name was changed to Midsouth Animal Breeders Cooperative.

During 1946, another small organization, Pike Artificial Breeders Cooperative, was formed at McComb, Mississippi. Four Jersey bulls were maintained, and in 1950 six thousand cows were bred. This cooperative merged with Mississippi Artificial Breeders Cooperative.

Missouri

Modern-day artificial insemination in Missouri began in the Dairy Department of the University of Missouri–Columbia, in July, 1937. A highly proved Holstein bull, U-Mo Campus Aagie Segis Sultan, had become stifled. His semen was collected using the ampullae massage method by Harry A. Herman and Victor Berliner. A number of calves were obtained. The first to be born was U-Mo Aagie Sultan Valiant, on 18 April 1938.

Fred F. McKenzie, of the Department of Animal Husbandry, returned from a visit to Europe in late 1937 with a Cambridge model artificial vagina. It was put into use and modified artificial vaginas made up and used for further semen collections. Considerable research was done on semen characteristics and artificial insemination at the Missouri Agricultural Experiment Station. The first bulletin, entitled "Artificial Insemination of Dairy Cows," written by Harry A. Herman and A. C. Ragsdale, appeared in June 1939.[4]

Several artificial-breeding organizations were formed in Missouri in the ensuing years.

Hughesville–Farm Security Administration. The first organization in Missouri, and the second organized AI project in the U.S., was formed and began operations in June 1938 at Hughesville. It was a cooperative effort between the Department of Dairy Husbandry and the Farm Security Administration. Service was provided for cows in two large "colony farms" of about 125 cows each and thirty individual farms with small herds. About 800 cows were involved. James Muster, Sidney B. Johnson and E. M. Copeland supervised the project for the Farm Security Administration. Semen from University of Missouri dairy bulls was collected, and deliveries made every other day. Liquid semen, undiluted, was used. Conception rate, the first service, was about 50 percent. Later, the project bought its own bulls. Breeding efficiency improved, and the program was considered a success, though there was much skepticism the first year. The project was closed out in 1944 when the government sold its farm holdings in that area.

Midwest Breeding Farms. The Midwest Breeding Farms in Trenton began operations in 1943. It was started by the milk plant at Trenton, a private enterprise, to improve producer herds. The beginning was the purchase of Herndale Farms Breeding Service in Fayette. Herndale was started by McKenzie and associates the spring of 1939 to supply AI service for horses, beef, and dairy cattle. It was not a financial success.

Midwest Breeding Farms served the dairy area in the northwest part of the state. About fourteen thousand cows were serviced annually and fifteen bulls were kept. Walter K. Dinwiddie was the manager. This enterprise was closed in 1952 when it sold its assets to ABS.

MFA Breeding Service. The largest AI organization based in Missouri was the Missouri Farmers' Association Artificial Breeding Association of Springfield. It began business on 1 January 1946. Formed by the Missouri Farmers' Association, one of the largest farmer cooperatives in the nation, it was designed to improve milk production in southwestern Missouri. The name of the association was changed to MFA Dairy Breeders and eventually to MFA Breeding Service.

John Fawcett, a veteran Jersey breeder and manager of the herd at the School of the Ozarks in Point Lookout, Missouri, was the first manager of MFA Artificial Breeding Association. He served for two years and was succeeded by J. Warren Nordyke.

In September 1953, F. G. Stevenson, who had been the successful manager of Southern Illinois Breeding Association, was hired as manager. Under his guidance, a modern bull barn, laboratories, and offices were built, and the service area was expanded. By 1960, over eighty-one thousand cows were serviced annually. In 1961–1962, MFA Dairy Breeders became part of Midwest Genes, and Stevenson was elected president. On 1 January 1967, MFA Breeding Service affiliated with the newly formed Midwest Breeders Cooperative in Wisconsin.

East Missouri Artificial Breeding Association. East Missouri Artificial Breeding Association, supported by Sanitary Milk Producers (SMP) in St.

Louis, began operations in October 1946. Headquarters were at O'Fallon. R. D. Pennewell, president, and A. D. Lynch, secretary-manager of SMP, worked with M. J. (Pat) Regan and E. T. (Scratch) Itschner in organizing the association. August F. Dieckman became president, and A. D. Lynch, secretary. Raymond G. McCarty, a county agent, was hired as manager.

Five years later, the association was breeding about fifteen thousand cows. The volume did not increase greatly over the years. Lowell Donnell succeeded McCarty as manager in 1954, converted to frozen semen, added several proved sires, and promoted many breeding programs. Since the cow population was declining, none of the promotion programs worked. East Missouri ABA filed bankruptcy. The bulls and equipment were sold in 1961 to Noba, Inc., and that organization took over the service area.

Missouri Dairy Breeding Association. Harry Rollins, a newspaper man, with assistance from the Carthage Chamber of Commerce, established the Missouri Dairy Breeding Association at Carthage in 1946. Seven to ten bulls were in the stud. The volume had reached 9,239 cows in 1950. Due to financial problems, the bull stud was discontinued in 1951, and Rollins became a distributor for ABS.

Montana

Montana is predominantly a beef state. There are dairy herds only around the larger cities. However, one AI organization was formed in Montana.

Western Montana Dairy Breeders Association. Through the efforts of the creamery at Missoula, J. O. Tretsven, an extension dairyman at Montana State University, and local county agents, Western Montana Dairy Breeders Association was organized in 1947. The first president of the cooperative was Paul Spannish and the first manager was Don Wold.

The bull stud and headquarters were located at Missoula. The first shipment of semen was made to seven county units in September 1947.

S. Ray Greenlaugh succeeded Wold as manager about 1950. Cows bred totaled 10,012 in 1955. Shortly thereafter, with frozen semen available, the bull stud was closed and out-of-state organizations supplied the semen.

Nebraska

Cows in the University of Nebraska's dairy herd were being inseminated long before organizations were established in the state. The first stud organized in Nebraska was the Douglas County Breeders Association in April 1941. Nine bulls were kept at the Douglas County fairgrounds at Elkhorn. The same year, Burt County organized and purchased semen from the Douglas County stud.

In 1943, the Washington and Lancaster Breeders associations were organized. Washington County secured semen from the Douglas County Breeders Association, and the Lancaster Coop made arrangements to get semen from the bulls owned by the Dairy Husbandry Department of the

University of Nebraska. The Lancaster group was assisted by members and directors of the Lincoln Non-Stock Co-op Milk Association.

To service herds in northern Nebraska, the Norfolk Breeders Association was organized in 1944. This was the last stud established except for the Grand Island stud that maintained a few Holstein bulls. From 1945 until 1950, many county associations were formed and purchased semen from the established studs in the state. The number of cows increased slowly from 2,280 in 1941 to 35,200 in 1950.

Nebraska Dairy Breeders Association. In June 1950, all of the bull studs in Nebraska merged to form a federated cooperative—the Nebraska Dairy Breeders Association (NDBA). During the first three years, offices were at Lincoln, and semen was purchased from the University of Nebraska Dairy Department. For the purpose of establishing their own bull stud, the cooperative placed twenty-five cents of each service fee in a reserve fund.

Reuben Peterson was hired as field manager to work with the local associations.

In 1953, the directors purchased a farm and buildings at Fremont, Nebraska, and Nebraska Dairy Breeders Association moved to a new home. Peterson continued as manager, a position he held until 1966 when the association merged with Midwest Breeders.

In 1960, the association's name was changed to Nebraska Artificial Breeding Association (NABA). During 1961, the cooperative serviced 37,282 cows. Beef AI was beginning to grow, and there were 16 beef bulls, 5 Milking Shorthorn, and 19 dairy bulls in the stud.

In 1962, NABA became part of Midwest Genes, Inc., which, in turn, merged with Midwest Breeders Cooperative in 1966.

Nevada

No artificial-breeding associations were organized in Nevada. In 1960, there were 3,427 dairy cows reported as artificially bred. In 1971, the total was 8,087 dairy and 2,031 beef cows. All service is from out-of-state organizations.

New Hampshire

New Hampshire–Vermont Breeding Association. Formed in 1944, the New Hampshire–Vermont Breeding Association, with headquarters at Concord, started operations in 1945. The cooperative was aided by dairy extension personnel and county agricultural agents. William T. Jordan, Sr., was the first president. George Bassett, of Fremont, served as secretary for nine years. W. R. (Win) Amidon was general manager.

In 1945, 1,559 cows were artificially inseminated. By 1954, the total had reached 47,000 and was 52,616 head in 1965. Bulls of all dairy breeds, Angus and Hereford, were maintained.

While New Hampshire dairymen started the association, it was soon joined by dairymen in northern Vermont. There were about as many cows serviced in Vermont as in New Hamshire by the association. The board of directors was made up of representatives from each state. In 1966, New Hampshire–Vermont Breeding Association merged with other New England cooperatives to form the Eastern AI Cooperative.

New Jersey

The first *cooperative* cattle-breeding association to use AI in the United States began operations in New Jersey in May 1938.

New Jersey Cooperative Artificial Breeding Association No. 1, Inc. When Enos J. Perry returned from a three-month stay in Denmark, steps were taken to organize the first cooperative in the autumn of 1937. While abroad, Perry was privileged to see the first Danish artificial-breeding cooperative ever organized (see Chapter 6). The techniques and methods of operation were studied while spending several days with K. A. F. Larsen, a technician, and some of the officers.

In December 1937, the New Jersey Holstein Association passed a resolution favoring the organization of a cooperative artificial-breeding association. The result was the founding of Cooperative Artificial Breeding Association No. 1. Clifford E. Snyder, of Pittstown, was elected president, and Perry was elected secretary. The association had 102 members with 1,052 cows enrolled. Many of the first members signed up only a part of their herd, awaiting to see results.

An important incentive for many dairymen to join the cooperative was the desire to use the Holstein bull, N.J.E.S. Sir Mutual Ormsby Jewel Alice #654594, owned in partnership by the New Jersey Agricultural Experiment Station and P. P. Van Nuys, of Belle Mead. The first bulls used were located at the Van Nuys's farm.

The organization was fortunate in obtaining James A. Henderson, a Canadian DVM, who had just finished his master's degree at Cornell Veterinary College as the first technician and manager. To further assist the association, K. A. F. Larsen was engaged to act as adviser and to work with Henderson for two months. He used the recto-vaginal technique for insemination, and the method was gradually adopted throughout the United States.

Several other cooperative breeding associations were organized in New Jersey following Cooperative Unit 1. They were Sussex County Cooperative Breeding Association, 1939; Burlington County Association, 1939; South Jersey Association, 1939; and Tri-County Association, 1946.

At the famous Walker-Gordon Farm, the Plainsboro Bull Association (organized in 1939) operated a bull stud for fifteen years but discontinued in 1954 and purchased semen from the New Jersey Cooperative Breeders Association, Inc. Charles E. Hunt, of Allentown, who began working at Walker-Gordon Farm in 1928, was the AI technician. According to him,[5] the first live calf from AI arrived in November 1938. The Walker-Gordon Farm then was milking 1,610 cows, rented to the farm by individuals. About 50 to 65 bulls were in use. When AI was put into use, it reduced the feed bill in an amount over $5,000 per year by reducing the number of bulls kept.

In 1954, the local cooperatives in New Jersey merged to form the New

Jersey Cooperative Breeders Association. The headquarters and the bull stud were located at Annandale. D. C. Reid, a veterinarian, was the manager. The first year (1938), 1,703 cows were bred artificially in New Jersey. In 1954, the New Jersey Cooperative Breeders Association reported 60,885 head.

In 1964, a merger of New Jersey Artificial Breeders Cooperative, First Pennsylvania, Lehigh Valley, and Northeast Pennsylvania Breeders Association took place. The new organization—Northeastern Breeders Association (NEBA)—merged later with Maryland–West Virginia Bull Stud to form Sire Power, Inc.

New Mexico

New Mexico has large beef and dairy cattle herds. AI is used heavily in the dairy herds with about 61 percent of the state's 42,000 milk cows serviced. About 1 percent of the 900,000 beef cows were bred by AI in 1971.

Roosevelt County AI Association. This county association was started in 1939. Difficulties in communications with herd owners, failure of cows to settle and the costs involved forced an early cessation of the stud. Semen is now provided by out-of-state organizations, and most herds have their own inseminators.

New York

New York with about one million head ranks second among all U.S. states in the number of milk cows. Cash from the sale of milk accounts for 53 percent of the state's farm income. In a state historically dependent upon the production of milk, it is not surprising that New York became one of the early leaders in adopting the AI program.

Like most states, New York had several, early organized, small AI cooperatives that eventually were consolidated to form a strong statewide cooperative. The Pioneer Dairy Cattle Breeding Cooperative, in Dryden, was the first in the United States to be organized and completely financed by the members, exclusively for the purpose of artificially breeding dairy cattle.

The promotion and organization of the AI program in New York had aggressive support by Cornell University extension dairymen, Stanley J. Brownell and Raymond Albrectsen. Brownell had vigorously promoted the Purebred Bull Association movement, and he quickly saw the genetic impact that AI, if the best of proved sires were used, would have on a dairyman's income. During 1938, the Pioneer Association bred 714 cows. In 1939, the three breeding associations bred 3,841 head.

New York Artificial Breeders Cooperative. On 11 April 1940, a statewide artificial-breeding organization—New York Artificial Breeders Cooperative (NYABC)—was formed by consolidating the local association's (Pioneer) bull studs. The new organization was a federated type of cooperative with thirteen directors, who represented eight districts and five dairy cattle breeds.

Harold L. Creal, of Homer, was the first NYABC president, and J. L. Sears, of Baldwinsville, was the first secretary-treasurer. Maurice Johnson was hired as manager, a position he capably filled until his retirement in 1960.

Syracuse, New York, 1940–1945. Headquarters were established on a farm at Syracuse and operations were carried on there until 1944, when headquarters were moved to Ithaca. The new location is on land owned by Cornell University. A longtime lease agreement prevails. This arrangement permits close cooperation with Cornell research workers and the extension staff. Some of the most significant research in the field of animal reproduction and AI has resulted.

A new physical plant including bull housing, laboratories, and offices were constructed at Ithaca resulting in a large and modern facility. Additions have been made as the program grew and now enable the housing of about 350 bulls. A farm and facilities for housing "bulls in waiting" are a part of the present physical plant.

During 1945, NYABC serviced 46,507 cows. Brome Tioga serviced 8,569 and Seneca 1,400. These two cooperatives gradually phased out their bull studs and utilized semen from NYABC.

Creal was succeeded by J. Stanley Earl as president. Earl, a dairy farmer at Unadilla, served for many years in this capacity. Charles J. Krumm succeeded Johnson as general manager. Under Krumm's guidance—assisted by a strong board of directors, with Norman Allen, of Schaghticoke, as president—NYABC forged ahead in a time of rapid technological developments, increased competition and declining dairy cow numbers. In 1964, a consolidation was made of NYABC local associations thus providing for central control. NYABC serviced 557,641 cows in 1964. The conversion had been made to frozen semen, beginning in 1953, and in 1962 liquid nitrogen refrigerant was used in the field. During the years, NYABC has been a leader in sire evaluation and sire proving through progeny testing.

In 1966, one of the largest mergers of AI organizations in the country occurred. NYABC and all of the artificial-breeding cooperatives in New England merged to form the Eastern Artificial Insemination Cooperative, Inc. (EAIC), with headquarters at Ithaca. This move pooled bull power and assured in addition to New York the beleaguered associations in Maine, New Hampshire, Massachusetts, Rhode Island, and Vermont that their patrons would have good AI service. A true cooperative with directors from each of the states involved on the board of directors, EAIC has continued a path of progress. Unlike many of its counterparts, herd service has been a primary goal. Over three hundred affiliated technicians serviced some 30,000 herds in 1971 with 692,000 cows involved.

Charles Krumm retired as general manager of EAIC in 1979. He was succeeded by James Mellinger, who had served as field operations manager since 1971.

North Carolina

The AI program in North Carolina began with three local breeding associations in 1944.

Forsyth County Artificial Breeding Association. This association started in 1944–1945 with two Guernsey and two Holstein bulls. The association was fostered by the Forsyth County Commissioners. The construction of a barn, purchase of bulls, and equipment were financed by the county.

The Farmers Federation. This organization was formed by dairymen with the aid of county agents and the North Carolina State University extension personnel. Headquarters were at Asheville.

Another small association was organized at Burnside, Yaney County, starting with one Guernsey bull in 1944–1945.

John Arey, extension dairyman, and Dean Colvard, North Carolina State University, did much to promote the early associations.

Southeastern Artificial Breeding Association. On 28 May 1947, Prentice came to North Carolina State University for a conference with agricultural extension officials. He proposed to set up a bull stud, with twenty-two bulls, at Asheville, without cost to farmers or others. This offer was made with the stipulation that the extension service organize local cooperative-breeding associations throughout the dairy areas of the state and that semen be purchased from the bull stud for $1.50 per first service. The North Carolina State University Extension Service accepted Prentice's proposal. The bull stud was in operation on 1 January 1948, with twenty-two proved bulls— eight Guernsey, eight Jersey, and six Holstein.

Meanwhile, organization of the local units had been underway and in December 1947 there were twenty-two county associations. In 1948, John F. Brown was hired by the agricultural extension service to help organize local associations and to promote the AI program. Brown later became the ABS representative. F. I. Elliott was employed by North Carolina State University at Raleigh, to direct the educational and technical phases of the AI program. He assisted with the many technician-training programs that followed.

During 1950, nearly 32,000 cows were artificially bred and service had been extended to fifty-six counties. The program grew with 45,505 cows serviced in 1954. That year, with improved semen-preservation methods and air transportation, ABS closed the Asheville stud but continued to supply semen to the local associations.

North Carolina Institutional Breeding Association. An example of using genetically superior bulls, by means of AI, to improve production is dramatically portrayed by the results of the North Carolina Institutional Breeding Association.[6]

In 1949, a group of North Carolinians met at Raleigh to decide what steps could be taken to improve the Holstein herds at the state's training schools and mental institutions. Included in the group were institutional managers, herdsmen, college teachers, and research workers. A survey indicated 10 herds, with 956 animals of breeding age and 32 herd sires. It was agreed that AI, using high-production-proved sires, was the most logical approach.

A critical analysis showed that only three of the bulls merited use in a bull stud designed to improve genetic production potential. Twenty-nine bulls were sent to the butcher. One additional proved bull was bought. Proceeds from the sale of the bulls, plus a legislative appropriation, provided a bull barn and laboratory at Raleigh. Elliott and R. M. Myers with the assistance

Table App.2. Production by Years: N.C. Institution Herds

Year of Freshening	No. of Records	305-day, 2X, ME		
		Milk	%	Fat
1949	343	10,572	3.46	365
1950	570	10,842	3.49	378
1951	551	11,114	3.49	388
1952	570	11,534	3.44	396
1953	553	12,130	3.47	421
1954	615	12,957	3.51	455
1955	608	13,094	3.53	462
1956	626	13,624	3.50	477
1957	603	14,178	3.53	501
1958	641	14,575	3.63	530
1959	702	14,513	3.69	536
1960	687	14,935	3.73	558
1961	623	15,927	3.81	608
1962	639	16,390	3.74	614
1963	699	16,965	3.68	623
1964	629	17,018	3.66	623
1965	637	17,030	3.71	632
1966	611	17,380	3.64	632
1967	637	17,087	3.66	625
1968	580	17,276	3.59	621
1969	596	17,720	3.65	646
1970	571	17,443	3.67	639

Source: R. M. Myers, *The N.C. Institutional Breeding Association: Its History and Its Philosophy;* program of the "Eighty-seventh Annual Convention of the Holstein-Friesian Association of America" (26–29 June 1972), Brattleboro, Vermont.

of J. E. Legates guided the program. With the bull stud located at Raleigh, a supply of semen was furnished to every institutional herd.

The genetic aspects of the breeding program were carried out by representatives of the institutional herds and by Legates and his associates at North Carolina State University. The feature of the breeding program has been a planned progeny-testing system to provide plus-proven sires for use in the herds. Many young sires were purchased and proved out. Only the best were retained for extensive use in the herds. The success of this program speaks for itself. In 1949, the average production per cow was 10,558 pounds of milk, 366 pounds of butterfat—2X—305 ME. Twenty years later, production had increased to an average of 17,720 pounds of milk and 646 pounds of butterfat. The production by years is indicated in Table App.2.

This program is being continued, but by utilizing frozen semen from high-plus-proven bulls throughout the county.

The AI program in North Carolina has grown steadily with over 40 percent of the dairy cows artificially bred.

North Dakota

Artificial breeding of dairy herds in North Dakota began in 1948. Semen was purchased from out-of-state bull studs including Land O' Lakes and Minnesota Valley Breeders.

Clarence C. Olson, extension dairyman, North Dakota State University, helped to organize the local units, and in 1953 there were forty-two such units and 26,231 cows were bred artificially. Beef AI has expanded the last few years. In 1971, nearly 35,000 dairy cows and 32,000 beef cows were serviced. All semen is purchased from out of state—or results from custom freezing of semen from bulls owned within the state.

Ohio

In February 1939, a group of dairymen in Medina County organized an active AI association under the guidance of Virgil Burris, county agent; Max Drake, assistant county agent; with assistance from C. L. Blackman, extension dairyman, Ohio State University, and Dr. V. W. Zuercher, the first inseminator. The organization operated nearly two years giving way to the formation of larger units.

Three major AI cooperatives based in Ohio have played a prominent role in cattle improvement throughout the world. Two of these, Coba, Inc., in Columbus, and Noba, Inc., in Tiffin, resulted from mergers of small associations that were formed between 1939 and 1942. The third, Select Sires, Inc., in Plain City, is a federation of eleven state or regional cooperatives.

Noba, Inc. In 1939 and 1940, a group of dairymen in Fremont organized a breeding association to service cows in Sandusky and Ottawa counties. A second group, Seneca County Breeding Association, was organized in July 1940. B. W. Kagy, DVM, inseminated the first cow, a Guernsey owned by Walter Hansen, on 14 July 1940. Howard J. Ziegler, of Clyde, was the first president of the Seneca Association. The stud was located just west of Tiffin.

By April 1941, E. J. Hoffert, who was working with the Sandusky-Ottawa Association, arranged to obtain semen and share expenses with the Seneca County group. In February 1942, the two associations combined to form Northern Ohio Breeders Association (Noba), located seven miles east of Tiffin. A farm was purchased for the headquarters. Sam B. Rose, of Findlay, became president—a position he ably filled until March 1960. Other early officers were Ziegler, vice-president; O. J. Smith, of Fremont, secretary; and Lester Hall, of Tiffin, treasurer.

Max Drake, former assistant county agent in Medina and managing partner of Louis Bromfield's Malabar Farm, was hired March 1943 as the first full-time manager. Noba began using frozen semen in 1954; was the first to develop plastic vials for frozen semen; the first to set up a variable pricing system for semen; and one of the first cooperatives to expand operations by establishing divisions in other states.

The first year, Northern Ohio Breeders Association inseminated 6,774 cows. Growth was steady with 121,434 in 1954, and over 200,000 (76,000 out of state) in 1959.

The name *Noba, Inc.*, was adopted in 1958. In 1963, Noba, Eastern Iowa Breeders, and Coba, Inc., formed United Semen Exchange to utilize the best bulls more fully and to progeny-test young sires. To further strengthen exchange of semen, joint ownership of sires, advertising, and promotion, NAL

Affiliated Breeders, consisting of Noba, Inc., Atlantic Breeders Cooperative, and Louisiana Animal Breeders Cooperative, was established in 1969.

The physical plant has been expanded, and in 1979 Noba was operating in thirty northern Ohio counties and fifteen states. South Carolina Cooperatives became a part of Noba, Inc., in 1957 and East Missouri ABA in 1959. Frozen semen is exported to several overseas countries. A distribution and export center is maintained in Florida, a beef bull AI unit is located in Oklahoma, and a modern facility for young sires proving, Noba-Georgia, is maintained near Elko, Georgia. Semen for over 300,000 first services, beef and dairy, are produced by Noba's combined facilities.

Noba, Inc., as is true of several other long-established AI organizations, has benefited by a number of employees of long and dedicated tenure.

Coba, Inc. Central Ohio Breeding Association, Inc. (COBA), was incorporated on 29 May 1946. It resulted from a merger of Northeast Ohio Breeding Association and Western Ohio Breeders Association.

Richard Kellogg, who had been managing Northeast Ohio Breeding Association, was hired as COBA's first manager. He retired in 1977. Under his guidance, COBA expanded its operations, built a new headquarters plant with the necessary bull barns, laboratories, and storage facilities. Kellogg was succeeded by Wallace Erickson as general manager.

The first year of operation, COBA had twenty-five local units and bred 22,044 cows. Ten years later, there were fifty-two bulls in the stud and 168,138 cows were serviced. In 1954, COBA changed to frozen semen, and that same year expanded business by selling semen to Terry Graham in Tyler, Texas, who was developing an AI business. In October 1957, COBA bought out Graham's business and established its own office in Tyler, Texas, serving Texas, Oklahoma, New Mexico, Arizona, and Old Mexico.

In February 1955, COBA made one of the early exportations of frozen semen when forty ampules of the Jersey bull, Signal Commander, packed with dry ice, were shipped by air express to Erik Tillisch, in Copenhagen, Denmark.

Due to continued industrialization of rural areas in Ohio and a decline in dairy cattle, in October 1965, COBA joined with the Kentucky Artificial Breeding Association of Louisville; Illinois Breeding Cooperative of Hampshire; and Southern Illinois Breeding Association of Breese to form Select Sires, Inc. On 1 January 1966, all of the bulls owned by the four cooperatives were transferred to Select Sires, which became the semen-producing business for its four members. For several years, Select Sires, Inc., headquarters were located at the COBA, Inc., facilities.

Select Sires, Inc. Select Sires, Inc., of Plain City, Ohio, is the giant of the AI cooperatives. It had its beginning in 1965–1966 when four cooperative AI studs, located in three states, began a semen-exchange program. They soon realized that with a number of studs joining forces a stronger sire program would be a reality and would provide greater efficiency in semen production and utilization. A national advertising and promotion programs were added benefits.

On 23 June 1969, Select Sires grew by the addition of six new members: Michigan Animal Breeders Cooperative of East Lansing; Virginia Animal

Breeders Association of Rocky Mount; Cache Valley Breeding Association of Logan, Utah; Mississippi Animal Breeders Cooperative, Mississippi State; Tennessee Artificial Breeding Association of Brentwood; and East Tennessee Artificial Breeding Association of Knoxville.

In 1973, some members of the Illinois Breeding Cooperative of Hampshire withdrew as members of Select Sires, Inc., and assumed the wholesaling of semen from Curtiss Breeding Service. The remainder continued with Select Sires and later became known as Prairie States-Select Sires.

On 1 July 1975, All West Breeders in Burlington, Washington, became affiliated with Select Sires, Inc. In early 1977, East Central Breeders Association Cooperative, in Waupun, Wisconsin, became the eleventh member of the federation.

Select Sires is a federated AI cooperative, owned by eleven member cooperatives. Each member organization continues service in its respective area and maintains its own office and field staff. The members own no bulls directly but purchase semen from Select Sires. This makes possible a larger supply of top-ranking beef and dairy sires, along with a young-sire-proving program, that no single member could afford.

Select Sires, Inc., is governed by a twenty-member board of directors elected from the eleven member organizations. The 1976–1977 officers were Jack Dendel (Michigan), as president; Brainard Palmer-Ball (Kentucky), as first vice-president; Fred Friday (Ohio), second vice-president; Ronald M. Hawkes (Idaho), as secretary; and James Clayton (Tennessee), as treasurer.

The first manager of Select Sires, Inc., was James Mellinger. He resigned in 1971 to become field operations manager for Eastern AI Cooperative. Mellinger was succeeded by James R. Nichols, associate dean of agriculture at the Virginia Polytechnic Institute in Blacksburg, Virginia. He took over as general manager the fall of 1971.

Under Nichols's leadership, the federation completed its corporate structure, obtained financial backing, acquired a 287-acre farm at Plain City, Ohio, and erected its modern headquarters facility. An experienced staff, many from the federated member organizations, was assembled. An open house dedicating the new facilities was held 1 August 1973.

In 1973, Nichols was succeeded by Richard H. L. Chichester, who was named executive vice-president and general manager. Chichester had joined Select Sires in 1971 as director of public relations.

Select Sires, Inc., in 1976 owned about 500 bulls of all ages, including 80 proved dairy bulls and 123 beef bulls representing 24 breeds. During 1975, sales of 1,459,208 units of semen were reported. It is estimated well over 800,000 cows are inseminated annually.

Approximately 55 percent of the dairy cattle and 10 percent of the beef cattle in Ohio are artificially bred.

Oklahoma

Organized AI programs for dairy cattle began in Oklahoma in December 1945. The program was aided by a state appropriation.

Lloyd H. Stinnett, extension dairyman, was in charge of the project and was assisted by an advisory committee of dairymen.

Eastern Oklahoma Dairy Breeders Association. Located at Muskogee, Eastern Oklahoma Dairy Breeders began operations in December 1945. Leo H. Elwell, a veterinarian, was the first manager. During 1946, the association had eighteen bulls representing the Guernsey, Holstein, and Jersey breeds. Three thousand two hundred cows were artificially bred in five Oklahoma counties and in one Texas county.

Elwell was succeeded after a few years by Earl Farmer, an agriculture graduate of the University of Missouri–Columbia, and he was in charge until the association ceased business in 1958.

Northwestern Oklahoma Dairy Breeders Association. This association, with the bull stud located at Enid, began operations in February 1946. There were two veterinary part-time managers until George W. Norwood took charge. He continued until the stud was closed in 1958.

In 1946, there were 19 bulls in the stud and 3,419 cows in eight county units bred.

With herds scattered, it became difficult to maintain technicians. Oklahoma was one of the first states where the "do-it-yourself" system of AI was used—this in the days of liquid semen. With the availability of frozen semen, out-of-state organizations supplied the market.

Many beef breeders in Oklahoma began using AI, employing frozen semen, often custom collected, about 1960. With the introduction of the exotic breeds, beef AI expanded greatly.

Oregon

The first organized effort to use AI for dairy cattle in Oregon began in 1946. Prior to that time, AI had been used on an experimental basis in the herds at Oregon State University at Corvallis.

Oregon Dairy Breeders Association. The Oregon Dairy Breeders Association, a cooperative of the federated type, was organized on 1 February 1946. R. W. Morse and H. P. Ewalt, extension dairymen at Oregon State Agricultural College in Corvallis, were leaders, assisted by Henry Hagg, dairyman, of Beaverton. Hagg was the first president, serving until 1955.

The name of the association was changed to Oregon Breeders Association in 1961.

The central stud was established on a purchased farm near Corvallis, and the buildings were remodeled.

Service began with Ben Simonson as manager and Fred Roberts as laboratory technician and assistant manager. The membership was on a cooperative basis, with a $10 membership fee and a breeding fee of $7.

The first year of business, 12,000 cows were bred with gradual growth, until in 1954, 36,436 cows were serviced.

In 1955, Ben Simonson resigned and joined the Roger Jessup Farms Breeding Service.

Roberts became manager in December 1955 and served until early 1964 when Oregon Breeders Association merged with what was then Evergreen Northwest Breeders, later All West Breeders. Prior to the merger, a semen-exchange program had been arranged with the Cache Valley Breeding Association and the Evergreen Northwest Breeders. As last reported, with All West Select Sires and several other organizations providing service, about 55 percent of Oregon's dairy cows and 2 percent of the beef cows are bred artificially.

Pennsylvania

The organized artificial-breeding program in Pennsylvania had its beginning in August 1942. Through the efforts of L. R. Bennett, a county agent, and a committee of dairymen, the Union County Artificial Breeding Cooperative was formed. According to R. H. Olmstead, a veteran Pennsylvania State University dairy extension specialist, some AI had been attempted earlier in Union and Susquehanna counties, but it did not prove practical.

The dairymen in Union County agreed they would not begin operations again until they had at least twelve hundred cows enrolled. Soon, thirteen hundred cows were signed up. The Union County Association operated successfully until 1944 when the members voted to allow their headquarters to become the First Pennsylvania Artificial Breeding Cooperative with headquarters at Lewisburg.

From this early beginning, dairymen in all sections of Pennsylvania had been utilizing AI service. Between 1942 and 1945, five central cooperatives were organized. At the same time, county cooperatives that would be affiliated with one of the Central (Bull Stud) Cooperatives were being formed. Each county group was expected to enroll at least fifteen hundred cows before starting AI service. By 1948, most of the county units had been organized. Profiting from the experience of some states that had already found that the AI program was not without problems, Pennsylvania built solidly and carefully. There was good teamwork between the dairy leaders, the Dairy Department, and the extension staff at Pennsylvania State University.

In 1945, a state association of all AI cooperatives, "The Pennsylvania Association of Artificial Breeding Cooperatives," was formed. It has had a strong supporting influence on the program from the standpoint of close cooperation between organizations, research support, and public relations.

J. O. Almquist came to the university in January 1944 as a member of the Dairy Science Department, and a research laboratory was established. In 1947, a committee from the State Association of Artificial Breeding Cooperatives obtained a grant of $95,000 from the state's general assembly, and a new Dairy Breeding Research Center was built. The five cooperatives contributed $34,500 for laboratory equipment and later $49,200 for construction of a sterility research barn that was completed in 1951. Through the years, the AI organizations in Pennsylvania have strongly supported research at the Dairy Breeding Research Center and the work of Almquist, T. Y. Tanabe, R. J. Flipse, and graduate assistants.

First Pennsylvania Artificial Breeding Cooperative. The First Pennsylvania Artificial Breeding Cooperative was formed in 1942. It was an outgrowth of

Union County ABC. Individuals' notes and a loan from the Production Credit Association financed the business.

A. Glynn Esch became manager in October 1945. Previously, James Muffly, one of the organizers, was the first manager, and Cecil Resseguie, of Lewisburg, was manager one year. Esch served for ten years. He was succeeded in July 1955 by F. Pressly Shannon, who continued until the cooperative merged in October 1964. Under his leadership, new facilities were constructed and the number of cows that were bred increased.

In 1942, 500 cows were bred. By 1954, the total was 43,220 and reached 58,463 in 1963.

On 1 October 1964, First Pennsylvania Artificial Breeding Cooperative, Lehigh Valley Cooperative, Farmers and Northeast Pennsylvania Artificial Breeding Cooperative consolidated to form Cooperative Northeastern Breeders Association (NEBA) at Tunkhannock. F. Pressly Shannon (First Pennsylvania) became general manager.

Lehigh Valley Cooperative Farmers. In July 1943, Lehigh Valley Dairy in Allentown established a Dairy Herd Improvement Center and started offering AI service to its producers. A forty-three-acre farm was bought and facilities were provided for the housing of thirty-eight bulls. Offices and a modern laboratory were built. The board of directors of Lehigh Valley Cooperative Dairy had direct control of the breeding business. Glen A. Boger, the president, was a strong booster for the AI program.

John Schwenk, formerly county agent in Berks County, was employed in 1943 as farm manager and superintendent of breeding service. Ten years later, 28,480 cows were serviced on over two thousand farms. A two-way radio setup facilitated the work of the AI technicians in the field.

In 1963, the last full year of operation, 31,306 cows were artificially bred. Lehigh Valley Cooperative Farmer AI Service was consolidated with two other cooperatives in October 1964 to form NEBA. Schwenk became a member of the NEBA board of directors.

Southeastern Pennsylvania Artificial Breeding Cooperative. Southeastern Pennsylvania ABC, commonly called SEPA, and forerunner of Atlantic Breeders Cooperative, was organized in September 1943 and began operations on 1 January 1944. Headquarters were at Landisville. Service was provided for Lancaster and York counties. In 1945, the first local cooperative units were established, purchasing semen from the SEPA bull stud. The first full year of operation, 5,069 cows were bred.

David Yoder became manager after the first year and has capably filled the post since. One of the few managers with over thirty years' tenure, he has, with the assistance of an able board of directors, guided the organization through many changes and continued progress.

A few years after its organization, SEPA headquarters were moved to a farm near Lancaster, and modern facilities were created. Additions to the barns and laboratories have since been made. In 1963, about 142,000 cows were serviced and a battery of 50 select bulls were in use.

On 1 July 1964, Southeastern Pennsylvania ABC and Western Pennsylvania ABC, of Clarion, merged to form Atlantic Breeders Cooperative, Inc.

Northeast Pennsylvania Artificial Breeding Cooperative (NEPA). NEPA was organized in 1944 by the dairymen of northeastern Pennsylvania with the help of the Agricultural Extension Service and began operations on 1 February 1944.

During 1945, NEPA technicians inseminated about 10,000 cows in 1,000 member herds. The volume grew steadily and ten years later exceeded 80,000. In 1963, the last year of operation, under its original name NEPA reported that 107,817 cows were bred artificially.

The first manager of NEPA was William F. Schaefer, Jr. He had managed Southeastern Pennsylvania ABC for a year before taking over the Tunkhannock operation. Schaefer continued as manager until 1964 when NEBA was formed. During his tenure, the facilities were improved and modernized. Schaefer served as executive secretary and treasurer of an eight-state group, the Eastern AI Cooperatives, which was organized to sample and prove young bulls.

NEPA led the drive to establish the Dairy Breeding Research Center at Pennsylvania State University. The bill was sponsored by Albert Madigan, the state assemblyman, who became president of NEPA in 1950 and also served as president of the State Association of Artificial Breeding Cooperatives.

Western Breeders Artificial Breeding Cooperative. Organized in 1944 as the Clarion Cooperative, Western Breeders Artificial Breeding Cooperative was operating in 1945. The bull stud and headquarters were at Clarion. The first year, 14,327 cows were bred.

Local dairymen, particularly Clyde L. Moore, who was the first president, and J. P. Gruber, who became president later, with the aid of the area county agents and Pennsylvania State University extension dairymen, organized the cooperative. Twenty-three counties, most with a director on the board, formed Western's operational territory.

George Thompson was employed as manager in 1945. He was succeeded by Harry Roth in 1960 who served until the merger in 1964 with Southeast Pennsylvania ABC to form Atlantic Breeders Cooperative. The last full year of operation, Western serviced 134,281 cows and had 73 bulls making it the second largest AI cooperative in the state.

Atlantic Breeders and Sire Power. The five early formed AI cooperatives in Pennsylvania were quite successful as related above. However, changing technology—particularly the universal use of frozen semen—improved sire evaluation methods and economic pressure resulted in all five undergoing some form of consolidation in 1964.

Atlantic Breeders Cooperative. Atlantic Breeders Cooperative in Lancaster is the result of a consolidation of Southeastern Pennsylvania ABC and Western Pennsylvania ABC, which occurred on 1 July 1964.

The headquarters of Southeastern Pennsylvania ABC are utilized, and there are about 75 dairy and beef bulls in the stud. An extensive young-sire-proving program is conducted.

During 1972, Atlantic serviced 281,874 cows and in 1973 produced over 600,000 units of dairy and beef bull semen. Sires available are enhanced by

Atlantic being a member of NAL Affiliated Breeders. The trade territory of the cooperative includes Pennsylvania and semen is shipped to other states. An ever-growing export trade is being developed with considerable quantities of semen shipped to South America.

David Yoder is general manager and Harry Roth is assistant manager.

Sire Power, Inc. Sire Power, Inc., was organized 1 October 1969. It is a federation of three already established AI cooperatives: the previously consolidated Northeastern Breeders Association, Maryland Artificial Breeding Cooperative, and West Virginia Artificial Breeding Cooperative.

Thus, several semen-producing businesses with 175 bulls, including 105 young sires being progeny-tested, were merged into one stud. The facilities at NEBA were leased for the semen production and distribution. The facilities at Frederick, Maryland, are used as administrative headquarters and to house bulls that have been sampled.

W. Lewis Campbell is general manager of Sire Power, Inc., and each of its three affiliate members.

Sire Power, Inc., is governed by a board of seven directors; four from NEBA; two from Maryland ABC; one from West Virginia ABC.

During 1971, Sire Power, Inc., reported 247,333 first services. The sales program has been extended to other states and an export market.

Other Pennsylvania AI Organizations. Several privately owned bull studs were promoted in Pennsylvania as follows:

Golden Genes, Inc. Organized by J. B. Schaefer and J. N. Earnshaw in 1954, Golden Genes, Inc., of Dimock, Pennsylvania, specialized in the sale of semen from several outstanding Holstein bulls. Semen for special matings and some export business was done. The firm was discontinued after a few years.

Zimmerman Dairy Farm. Zimmerman Dairy Farm, of Leighton, Pennsylvania, entered the AI field in 1956, offering semen from its herd sires. The farm was known for its high-producing cows. Due to the death of Joe Zimmerman, the AI business was discontinued in 1968–1969.

Taurus, Inc. In 1974, Richard Witmer and associates organized a small bull stud—Holsteins only— with headquarters at Mehoopany, Pennsylvania. It is a privately owned corporation and provides services for several thousand cows annually.

At the present time, about 65 percent of the dairy cows in Pennsvlvania are bred artificially.

Rhode Island

This smallest state in the nation had one AI organization, the Rhode Island Breeders Cooperative, Inc., which was formed in 1948 with headquarters at Kingston. A six-man board of directors from Farm Bureau districts governed its operations. Dean, Mason Campbell, and John Atwood, extension dairyman at the University of Rhode Island, assisted in the organization. Orville D. Curtis was the first and only manager.

The first year, 815 cows were bred and 6,408 in 1959. The stud was discontinued in 1959 and semen was purchased from the New England Selective Breeding Association. The organization became a part of Eastern AI Cooperative in the 1966 merger.

South Carolina

The AI program started in 1938 in the Clemson Agricultural College herd in an experimental way. The Clemson Bull Stud was organized in 1945 under the direction of J. P. LaMaster, head of the Dairy Department at Clemson University. He was largely responsible for getting AI started in South Carolina. C. G. Cushman and C. H. Lomas, extension dairymen, organized the county units.

The bull stud was located at Clemson University. LaMaster selected most of the bulls. The University employed the manager of the stud. The county units were organized into the South Carolina Federation of Artificial Breeding Cooperatives. Each county association paid Clemson Bull Stud for semen. The county court appropriated some money for operating expenses and furnished the AI technicians office space. R. D. Steer, of Greenwood, a marketing specialist at Clemson University, served as secretary of the federation for many years.

In May 1945, the first shipment of semen was made to Spartanburg County Cooperative Breeders Association. Sidney P. Marshall, now at the University of Florida, managed the bull stud for several years. He was succeeded by Victor Hurst, who managed the stud most of its existence and was responsible for training inseminators and farmers how to use AI.

In November 1957, with commercial frozen semen available, Clemson University closed the bull stud and the South Carolina Federation of Artificial Breeding Cooperatives affiliated with Noba, Inc. The federation was allocated a Noba, Inc., board representation, and Noba placed a representative in South Carolina to direct the program. In 1971, about 42,000 beef and dairy cows (45 percent of the state's dairy cows) were bred artificially.

South Dakota

South Dakota Cooperative Breeders Association, located at Brookings, was organized in May 1946. P. L. Kelley, Dairy Department head at South Dakota State University, worked with dairymen to perfect the organization.

Finances were obtained by a membership assessment and a $20,000 state appropriation. In 1946, Marvin Taylor was employed as manager and served throughout the existence of the cooperative.

During 1946, only 377 cows were bred, but by 1954 the number reached 9,188. The headquarters were on college property. Friction developed between college people and the cooperative board and management. In 1952, the association purchased land and erected its own buildings. Competition had become strong. A working relationship with Minnesota Valley Breeders was developed, and in 1962 the stud was phased out. Thereafter, AI service has been by out-of-state organizations.

Tennessee

The first cooperative AI association in the South was started at Spring Hill, Tennessee, also home of the first Jerseys imported into Tennessee. The association, Maury-Williamson Artificial Breeding Association, was organized in September 1938, and operations started 1 February 1939. There was only one bull, a Jersey, in the stud the first year. Fred Cowart, the manager-herdsman-technician, collected and processed the semen and inseminated 555 Jersey cows in 1939.

In 1940, two additional Jersey bulls were added and a second technician was hired. In addition to Maury and Williamson counties, where the stud was located, two additional counties were added. In 1944, there were seven Jersey and two Holstein bulls in the stud and 1,679 cows were serviced. In May 1946, the Tennessee Artificial Breeding Association was formed; Maury-Williamson ABC provided the nucleus.

From 1939 to 1947, three AI cooperatives were established in Tennessee. Fred Cowart, present manager of Tennessee ABA (now part of Select Sires), has worked with the program since 1938 and his contributions to cattle improvement in Tennessee are many.

Tennessee Artificial Breeding Association. Organized in May 1946, this cooperative had its beginning in 1938–1939 as the Maury-Williamson Artificial Breeding Association. Headquarters were established at Brentwood on the old fairgrounds. The State of Tennessee owns the land and buildings and the cooperative operates the AI program. Cowart continued as manager when the new organization took form.

Service was extended to twenty-one counties in middle Tennessee. In 1954, there were 62 bulls, including 5 beef in the stud, and 50,807 cows were bred.

The conversion to frozen semen was made in 1957. During the next ten years, as a result of declining dairy population and competition from out-of-state organizations, the yearly volume decreased. On 1 September 1969, Tennessee ABA affiliated with Select Sires, Inc., and now receives its semen from that bull stud. Service continues in the twenty-one original counties. A custom-semen-freezing business is operated.

West Tennessee Artificial Breeding Association. The West Tennessee Artificial Breeding Association was organized in 1945 with headquarters at Yorkville. It was earlier known as the Yorkville Cooperative Artificial Insemination Association, Inc.

P. R. Farlow and Edward Jones, agricultural agents for the Illinois Central Railroad, were among the active promoters. The railroad furnished some of the bulls; limited state funds were secured; and the local bank association helped with the financing. Clyde Chappell, dairy extension specialist at the University of Tennessee, helped with the organization. In 1946, the cooperative had six Jersey bulls and bred one thousand cows.

Fred J. Hatler became the first manager and served for several years. R. S. Freeman succeeded Hatler as manager and served until about 1970.

The number of cows bred averaged from ten thousand to twelve thousand yearly. As dairy cow numbers declined in the area, fewer bulls were kept, and frozen semen was purchased from other studs. During 1969–1970, the bull stud was phased out and West Tennessee Artificial Breeding Association became a distributor for Affiliated Breeders, Inc.

East Tennessee Artificial Breeding Association. The East Tennessee Artificial Breeding Cooperative was organized 29 April 1946 at Knoxville. Dairymen who promoted the cooperative included J. G. Lowe (president), A. B. Harris, and Leonard Rogers. A board of nineteen directors, one from each of the federated local associations, governed affairs of the cooperative.

Leland C. Stanfill was employed as manager, and he served until 1969. During his tenure, the number of cows that were bred increased from 3,226 in 1946 to 40,066 in 1961. In 1959, the headquarters' location at Knoxville was sold and the bull stud moved south of Knoxville to a larger farm. New bull barns, laboratories, and offices were constructed.

On 1 September 1969, the East Tennessee Artificial Breeding Association affiliated with Select Sires, Inc. The same territory is serviced. Donald Ardrey became manager.

At the present time, while Tennessee has no bull studs, about eighty-seven thousand dairy cows—25 percent of the total—and forty-two thousand beef cows—4.0 percent—are bred artificially.

Texas

The great state of Texas, with its millions of beef cattle, had only one bull stud based in the state in the early days. This was the North Texas Artificial Breeding Farm of Dallas. It began operations in November 1946. L. H. Smokler, a Guernsey dairyman, financed and promoted the bull stud.

Counties and service areas were organized on a federated basis by R. E. Burleson, an extension dairyman at Texas A&M University, and by local county agents. George W. Norwood was the first manager of the stud. The first year, four thousand cows were bred and reached fourteen thousand per year by 1954. In April 1949, David Smokler assumed management of the AI business. With commercial frozen semen becoming available in the 1950s, the stud was discontinued.

Texas A&M University. Beginning in 1947, the Dairy Science Department of Texas A&M University established a local breeding service, the Texas A&M Breeding Service. In 1957, about 1,150 cows were bred artificially and 777 in 1964. The project was phased out in 1965.

In the early 1940s, the workers at Texas Technological College, in Lubbock, used AI in the college dairy herd and for a few nearby herds. This program was discontinued after a few years.

Pan American Breeders, Inc. In October 1971, J. W. Pirie started Pan American Breeders, Inc., with headquarters at Terrell. Until October 1974, it was chiefly a custom-collection, semen-storage, and distribution center. A privately owned corporation, emphasis is now on development of a strong lineup of dairy and beef bulls. A custom-freezing business is also conducted. During the first year, about two thousand cows were bred. The volume is growing.

Pirie, a veteran in the AI industry, formerly operated Eastern Iowa Breeders.

At the present time, about 25 percent of the dairy cows and 1.3 percent of the beef cows in Texas are bred artificially.

Utah

The first AI of dairy cattle in Utah was attempted by W. E. Rasmussen in Weber County during 1942–1943. Renewed interest was created at Brigham Young University, Provo, in 1947 by Grant Richards. This operation ceased about 1953.

Several milk plants in Cache Valley that were interested in improving milk production supported the AI program. One of these, Sego Milk Co., in Richmond, started a small breeding association in 1947 but joined in plans for a large cooperatively owned central organization.

Cache Valley Breeding Association (CVBA). Cache Valley Breeding Association (CVBA), with its first headquarters at Hyde Park, Utah, began service on 4 December 1948. A $10 membership fee and a loan of $11,000 from four of the milk companies furnished the initial capital. Five outstanding sires were leased without charge from the State University, and the Sears-Roebuck Foundation contributed $9,000 for the purchase of additional dairy sires.

The organization was set up as a farmer-owned cooperative. The original board of directors consisted of two dairymen, appointed by each of the milk-processing plants, along with Rulon Osmond of Star Valley, Wyoming, and LaVor Doney of southern Idaho. L. B. Caine, Logan, was elected president. He served until his death in 1964.

Elected president of the board in 1964 was Ronald M. Hawkes, of Paul, Idaho. Hawkes, a dairyman, built a high-producing herd of over two hundred cows, and he and his family operate a five-hundred-acre farm. At present, he is the secretary of Select Sires.

The first manager of CVBA was Glen Downs, an employee of Sego Milk Co. He was succeeded in 1949 by J. Elmo Packer, who retired in 1972. He was replaced by Rulon Osmond, who has been associated with the organization since its beginning.

CVBA inseminated about 6,900 cows in 1949, its first year. By 1956, the volume had reached 73,000. Service was offered throughout Utah, Idaho, Wyoming, and New Mexico. It was the principal bull stud in the intermountain area. In 1957, new facilities were erected on a seventeen-acre tract of land, on Highway 91, near Logan, and the headquarters was moved to its present location.

CVBA pioneered in beef AI beginning in 1960 and in 1966 had 27 beef bulls in the stud. In 1966, there were 89,980 first services, 21,552 were to beef bulls, and in addition 8,808 units of beef semen were sold to ranchers or to other studs.

For years, CVBA has cooperated with other AI cooperatives in progeny-testing young sires and exchange of semen. In September 1969, the association became a part of the federation of Select Sires, Inc. The facilities at

Logan are used as a semen-distribution center and some Select Sires–owned beef bulls are housed. A custom-semen-freezing service and storage business are also conducted.

Vermont

Vermont is not a large state, but it has nearly 200,000 milk cows, and 78 percent of the farm income is derived from the sale of milk.

The first efforts with AI of dairy cows started in 1939–1941 when E. H. Loveland, an extension dairyman at the University of Vermont, helped establish a small association in Addison and Washington counties. Conception rate was poor, and the association was dissolved. Interest continued, however, and a cooperative comprising three counties was formed under the leadership of Lucien Paquette, an Addison County agent. This group later affiliated with the New York Artificial Breeding Cooperative.

The eastern and northern counties organized local breeding associations in 1944–1945 and joined with New Hampshire dairymen in forming the New Hampshire–Vermont Breeding Association, a cooperative, with headquarters at Concord, New Hampshire.

Central Vermont Breeding Association, Inc. In Washington County, with the assistance of county agent R. O. Sinclair, dairymen in 1944 formed the Central Vermont Breeding Association, Inc. Its headquarters were at Montpelier. Andrew Christens, of east Montpelier, served as president for many years. He had the support of many dairymen including Elmer Towne, commissioner of agriculture in Vermont and president of the Artificial Breeding Council of New England.

The Central Vermont Association had 4 Jersey bulls and serviced 801 cows in 1946. The business grew steadily with 31,797 cows bred in 1958. The bull stud was maintained at Burlington, where W. D. Bolton supervised semen production. He was assisted part of the time by Robert Lacasse.

The association remained in strong financial shape through its history. By 1958, over $130,000 in service rebates had been paid to the patrons. Membership grew to twenty-five hundred.

In 1959, the association became a part of New England Selective Breeding Service, and all bulls were transferred to the stud at Woodbridge, Connecticut. In 1961, Paul Heller became manager and served until 1965. In 1966, Central Vermont Breeding Association, Inc., became a part of the consolidation involving all New England AI organizations to form the Eastern Artificial Insemination Cooperative.

At the present time, about 56 percent of Vermont's milk cows and 10 percent of the beef cows are bred artificially.

Virginia

The AI program in Virginia started in 1946 with the formation of five small bull studs. They were designated as Culpeper, Franklin, Laudon, Shenandoah Valley, and Southeast associations. These associations had a total of 31 bulls and bred 2,525 cows in 1946. W. N. Angle, of Rocky Mount, and J. C. Eller, of Culpeper, were among the early leaders. Through the efforts of dairymen and members of the Dairy Department at Virginia Polytechnic Institute, a statewide cooperative was formed.

Virginia Artificial Breeding Association. A farmer-owned cooperative, the Virginia Artificial Breeding Association, Inc. (VABA) was formed in February 1950 and began operations in May of that year. Organized as a non-profit, non-stock cooperative, it was set up on a federated basis with thirty-six districts and a director from each.

R. G. Connelly and Arden Foster, extension dairymen, Virginia Polytechnic Institute (VPI), were instrumental in organizing the districts.

Headquarters were established on a farm near Rocky Mount, and W. H. Armstrong, an extension dairy specialist at VPI in charge of the DHIA program, was hired as manager. He served as manager for fifteen years. Under his guidance and an active board of directors, the headquarters facilities were enlarged and modernized. The association's financing came from memberships and cooperative loans.

During 1950–1951, the stud comprised of Guernsey, Holstein, Jersey, and Angus bulls. Growth was steady with over 70,876 cows serviced in 1964. Armstrong changed employment in 1965, and George Miller became manager. In September 1969, Virginia Animal Breeders Association became a part of the Select Sires, Inc., federation. Mr. Miller was transferred to the office of Select Sires in Plain City, Ohio, and was replaced by Emory Brubaker, as manager.

Among the several presidents of VABA, Rowland F. Hill III, of Orange, aided in the federation with Select Sires and served for many years as president and as a member of its board of directors.

At the present time, about 50 percent of the dairy cows and 2 percent of the beef cows in Virginia are artificially bred.

Washington

The State of Washington has long been known for its outstanding dairy herds, many of which are purebreds.

AI in the state began in 1939 when E. M. Gildow, a veterinarian at Carnation Farms, used it in that herd.[7]

There were a number of small AI organizations started before the larger bull studs were initiated in Washington. In 1941, a small breeding association was formed in Skagit County. Eleven bulls were in use and 565 cows bred the first year. This was the beginning of Northwest Cooperative Breeders. In 1942, another small association, Walla Walla Breeding Service, at Walla Walla, was formed. It was absorbed by Evergreen Breeders about 1950. In 1945, a privately owned breeding association, Far West Breeders, appeared. A few hundred cows were bred annually. It discontinued in 1947.

At Enumclaw, the Mount Rainier Bull Ring was organized in 1950 by a group of dairymen. Thirteen bulls were in use, and by 1954 about fourteen thousand cows were bred annually. This service was taken over by other organizations about 1955.

In 1961, W. E. DeJong, Snohomish, initiated DeJongs Supreme Sire Service. There were fourteen hundred to nineteen hundred cows serviced yearly. This bull stud discontinued in 1965.

Northwest Cooperative Breeders. This farmer-owned cooperative evolved from the Skagit County organization that was started in February 1941 at Mt. Vernon. Along with shortage of money, there was some skepticism about launching what was then considered an untried program. The Dairy Department and extension specialists at Washington State University were reluctant to give support at the beginning. This viewpoint changed as experience was gained.

H. L. Wills, assisted by other dairymen including Iver Youngquist in the Mt. Vernon area, persisted and, as a result, plans were made. Wills served as manager by default the first year as there was no money to hire a full-time man. There were six bulls, two each of the Guernsey, Holstein, and Jersey breed to start. Money was borrowed from the Seattle First National Bank and also from the Mount Vernon Bank. In 1943, a consolidation of the local breeding units occurred and Northwest Cooperative Breeders was organized. A farm near Burlington was purchased and the buildings were remodeled.

D. M. Waldo was the first manager. He served until 1952 and was succeeded by Earl B. Hope who soon left for other employment. On 10 October 1953, Jay C. Harris, assistant manager of Evergreen Breeders in Chehallis, became manager, a position he held until his retirement in February 1971. In 1942, only 3,000 cows were serviced, but by 1955 the volume passed the 100,000 mark.

In 1955, Northwest Cooperative Breeders merged with Evergreen Breeders Cooperative. The name of the combined organizations became Evergreen Northwest Breeders, with headquarters at Burlington. Harris continued as manager.

In 1964, the Oregon Dairy Breeders, in Corvallis, and Evergreen Northwest Breeders merged to form All West Breeders (AWB). That same year, Western Montana Breeders, in Missoula, merged with AWB.

In 1965, California Dairy Breeders merged with AWB. The result was the largest AI cooperative in the northwest. In 1966, there were 85 bulls in the stud, and 188,232 cows were bred artificially. Through the years, the physical plant was steadily improved, making it one of the best in the nation. In 1972, All West reported 228,549 first service cows, of which 33,190 were beef.

In 1971, following Harris's retirement, the present manager, Archie Nelson, who was serving along with Victor Gray, as assistant manager, took over. Clifford Bailey, of Snohomish, longtime president of All West, headed the board of directors, which is made up of representatives from the merged organization.

In July 1975, in order to strengthen its sire and sales potential, All West became a federated member of Select Sires, Inc.

Evergreen Breeders Association. Organized in 1946, with headquarters at Chehallis, Evergreen Breeders Association went through the same early stages as Northwest Cooperative Breeders. The first year, there were 13 bulls in use, and 663 cows were bred. Growth came as local breeding associations were organized, and in 1953 a total of 33,619 cows were bred.

Ross Coie was the first manager. Following the death of Coie on 12 February 1954, Robert Lintott, former field supervisor at Northwest Cooperative Breeders, became acting manager.

As is true nationwide, the dairy cow population in Washington has declined, but less than in the East and Midwest. About 64 percent of the dairy and 5 percent of the beef cows in the state are artificially bred.

West Virginia

The West Virginia Artificial Breeders Cooperative, Inc., owned and controlled by dairymen, was organized in August 1945 and began operations on 1 January 1946. The bull stud is located on a 110-acre farm, near Clarksburg.

Financial assistance, $18,000 to start, was provided by farm organizations, the Clarksburg Chamber of Commerce, milk plants, and the sale of memberships. The project later received some state support. West Virginia University supported the cooperative strongly in its organizational program.

D. C. Master, a veterinarian, and one of the founders of NAAB, was the first manager. The first year, with 403 members in five counties, 4,543 cows were bred. As the program grew, Master resigned, and W. L. Campbell became the full-time manager. He served until September 1957. Berlin Galford took over as manager at West Virginia ABC in 1957 and served until 1961, when he was succeeded by Richard H. L. Chichester (now manager of Select Sires, Inc.). Following Chichester as manager at West Virginia ABC was Clark H. (Red) Taylor, a former dairy herdsman at West Virginia University.

In July 1961, headquarters of the Maryland-West Virginia Bull Stud were moved to a new location—the "Old Stevens Airport Farm"—six miles north of Frederick, Maryland, which had been purchased by the cooperatives.

The AI program in West Virginia has grown. In 1971, about 20,000 milk cows (30 percent of the total) and 3,967 beef cows were bred artificially. The West Virginia Artificial Breeding Cooperative, Inc., is now a federated member of Sire Power, Inc., in Frederick, Maryland, and Tunkhannock, Pennsylvania. W. L. Campbell is manager and in addition serves as manager of both Maryland ABC and West Virginia ABC.

Wisconsin

Wisconsin is the nation's leading dairy state. It has the most milk cows— nearly two million head—and the most dairy farmers. In 1977, the cash income to farmers for milk sold was about $1.9 billion and accounted for 60.4 percent of the state's total farm income.

The organized AI program began in 1939.

In 1934, Howard Clapp, manager of Pabst Farm in Oconomowoc, Wisconsin, reported[8] that cow-to-cow inseminations had been in use in that herd for some years and that semen was being collected by ampullae massage. L. E. Casida and George M. Werner were experimenting with semen collection and AI in the University of Wisconsin dairy herd about the same time.

Wisconsin, as did most states during 1939–1941, had a number of small AI organizations. Experience soon demonstrated this was not a practical approach. The small organizations were merged to form large central bull studs serving a large area. Some of the smaller units, because of low conception rate, lack of capital, and bad management, failed before consolidations took place.

Three such organizations formed in 1939 were **Rock County Breeders Cooperative**, first in the state, and encouraged by Janesville businessmen who owned farms in the area. The directors and businessmen provided the capital. The first cow was bred on 5 March 1939. One thousand cows were bred the first year, and 1,880 in 1942. Due to operational difficulties, the bulls were sold in 1943 and semen was obtained from the American Scientific Breeding Institute in Madison.

Langlade County Breeders Cooperative. Started in March 1939 and inspired by farmers who wanted herd improvement, operations of Langlade County Breeders Cooperative began with three Holstein bulls (one proved) and three Guernsey bulls (two proved). Although a successful organization, it merged with the Badger Breeders Cooperative in 1945.

Barron County Breeders Cooperative. Organized the spring of 1939, the organization started with eight bulls. Service was discontinued in November 1942, due to labor shortages brought on by World War II.

Beginning in 1940 and continuing through 1942, at least ten associations maintaining a bull stud were organized. Many county breeding associations followed from 1942 on. They purchased semen from already established bull studs, including the University of Wisconsin. Early growth of the AI program in Wisconsin is summarized in Table App.3.

Until the spring of 1943, the inseminator's work in Wisconsin was done by

Table App.3. Growth of the AI Program in Wisconsin, 1939–1951[1]

Year	No. Assns.	Cows Bred	No. Bull Studs
1939	3	2,000	3
1941	9	10,000	10
1943	13	40,000	8
1945	21	86,000	8
1947	30	256,988	7
1949	33	408,375	7
1951	46	691,208	7

Source: Personal communication on 7 June 1973 with Gene Starkey, an extension dairyman at the University of Wisconsin–Madison.

[1]Compiled by Fritz Hendrickson, of the Dairy Science Department at the University of Wisconsin–Madison.

veterinarians. The State Veterinary Association had heretofore insisted that AI belonged in the province of the veterinarian. At a meeting of the State Board of Examiners for Veterinarians and University of Wisconsin representatives, it was agreed to train laymen technicians.

The program of AI in Wisconsin proceeded to move ahead with six to eight major bull studs providing service. There have been some consolidations, but the organizations existing in 1951, or their successors, are fairly well intact.

Midwest Breeders Cooperative (Badger Breeders Cooperative). Located at Shawano, Wisconsin, Badger Breeders Cooperative (since 1967, Midwest Breeders Cooperative) was organized in April 1940. Consolidated Badger Cooperative (CBC), a large milk plant, led the way for an AI program in the area. The beginning was Tri-County Breeders, organized under the private corporation act, financed and controlled by CBC. While local dairy farm leaders were involved, George Rupple, general manager of CBC, promoted the program. The University of Wisconsin helped. George Baumeister, a dairy extension specialist, was employed for one year to help get the association underway. It was time well spent as Badger Breeders Cooperative, which was to follow, proved to be a splendid example of farmers working together to improve their herds, and their income.

The first year of operation, 1,524 cows were bred. W. H. Dreher, DVM, for many years production manager for Badger, was the first manager. By 1945, the association was serving ten counties, and about forty thousand cows were serviced annually. In 1945–1946, realizing the need to enlarge the scope of operations, a reorganization occurred. The name *Badger Breeders Cooperative* was adopted. An increasing number of county breeding associations were purchasing semen from the bull stud and the time was right for consolidation. Kenneth E. Wallin was hired 1 January 1946 as general manager.

Under Wallin's guidance and with the aid of a progressive board of directors, the ten-county service area, over the next thirty-five years, was expanded into a multistate organization with markets not only in the Midwest, but in many states and several foreign countries. The physical plant was enlarged as more bulls, laboratory, and office space became necessary. For the year of 1959, Badger had 62 bulls in use and reported 373,576 first services. It had become one of the larger AI cooperatives.

Effective 1 January 1967, Badger Breeders Cooperative and Consolidated Breeders in Anoka, Minnesota, consolidated to form Midwest Breeders Cooperative (MBC), the present organization, with administrative offices located at Shawano, Wisconsin. On October 1, 1967, the assets of M.F.A. Breeding Service, a division of Missouri Farmers Association in Springfield, Missouri, were purchased by MBC. Within the next year, bull stud facilities were built at Stewartville, Minnesota. The result was additional offices at Stewartville, Minnesota, and Springfield, Missouri. In 1971, MBC had 737,797 first services—including 46,307 beef—to rank second only to the federated cooperative, Select Sires, in total services for an AI cooperative.

During Wallin's tenure, MBC, under the guidance of W. H. Dreher, developed the technique of freezing semen in plastic insemination tubes.

Called the "Magic Wand," this innovation resulted in an improved conception rate. Wallin's successor is Thomas Lyon, the present manager.

University of Wisconsin Bull Stud (Experimental). In 1940, in order to gain firsthand knowledge, as well as carry on experimental work, the Dairy Department of the University of Wisconsin established the University of Wisconsin Experimental Ring. E. E. Heizer and his colleagues in the Dairy Department planned on making service available to a few herds in Dane County. Since demand was so great, service was extended to include five or six additional counties. Some of the local breeding associations, unable to finance a good bull stud, purchased semen from the university.

The last year of operation (1951), there were 35 bulls in the stud, and 34,402 first services were reported. The university conducted some excellent research while the bull stud was operating. Studies included: proper stage of heat to inseminate, comparison of semen extenders, volume of semen per insemination dose, insemination techniques, effect of ageing on semen, and relationship between nonreturn rates and diagnosed pregnancy.

The University of Wisconsin bull stud was discontinued in 1950–1951, when the Southern Wisconsin Breeders Cooperative was formed in 1950 and when semen became available from the Wisconsin Scientific Breeding Institute in Madison.

East Central Breeders Association Cooperative. Organized in January 1941, East Central Breeders Association Cooperative (ECBAC), in Waupun, is the first direct-member–artificial-breeding association established in Wisconsin. Operating chiefly in ten counties in east-central Wisconsin, it is one of the few AI organizations that still retains its same service area thirty-five years after its establishment. There has been no acquisition of the assets of other AI businesses through the years. In May 1977, however, ECBAC became a federated member of Select Sires, Inc. Located in the heart of one of Wisconsin's best dairy areas, East Central Breeders has staunch support from its members.

Operations began in July 1941. At the start, six Holstein, five Guernsey, two Brown Swiss, and two Angus bulls were in the stud. George J. Landaal, of Waupun, led the promotion. He was the first president, and V. S. Kutchin the first secretary. Cooperative membership provided the finances. D. H. Doudna was hired as the first manager. The first year, 5,900 cows were bred. By 1954, first services reached 59,262.

Doudna resigned in August 1945 and was replaced by E. R. Carlson, who had been with the organization since 1942 (until the spring of 1943, the Wisconsin laws required insemination work to be done by veterinarians).

Carlson served as manager until 1965, the year when he retired. He was replaced by Alton Dale Block. He served until 1970 when he took over the management of Tri-State Breeders Cooperative at Westby.

Wallace E. Erickson succeeded Block as manager in 1970 and continued the progressive program that was underway. In June 1977, Erickson, following the affiliation of ECBAC with Select Sires, Inc., became the general manager of COBA, Inc., in Columbus, Ohio.

ECBAC conducted a swine AI breeding project for many years and was one of the leaders in this area. While successful in its application, the swine

AI program did not generate sufficient volume to be profitable and was sold in November 1975 to International Boar Semen, of Eldora, Iowa. That firm is carrying on a successful domestic and international program.

American Breeders Service, Inc. The history of American Breeders Service (ABS), the giant of the AI industry, involves developments in several states. Since the ABS headquarters are now in Wisconsin, we shall attempt to list the chronological events leading up to its establishment. The story properly begins with J. Rockefeller Prentice, and the events that motivated him to devote his life and substantial funds to cattle improvement.

J. Rockefeller Prentice, 1902–1972. "Rock" Prentice, as he was commonly known in cattle-breeding circles, was one of the most dynamic and controversial figures involved in the AI program in America. His goal was to improve cattle by the best, scientifically proved, methods available. He had no patience with half-truths and weak approaches.

Stimulated by his father's research at Mt. Hope Farm to breed more productive farm animals, Prentice became interested in improving the production of dairy cattle. He frequently said that he hoped, in his lifetime, to be able to vindicate the views on cattle improvement set forth by his illustrious father. According to him, "We ought to be able to turn out dairy cattle with the same uniformity for production one finds in the uniformity for flavor and aroma in a good brand of coffee—can after can—it's dependable."

J. Rockefeller Prentice was a man of action. He founded the American Dairy Cattle Club in 1936; the American Dairy Guernsey Associates of Northern Illinois in 1941, which became American Breeders Service; and in 1947, the American Foundation for Biological Research, which is well known for ultra low temperature research of sperm, other living organisms, and tissues. Prentice was the co-founder of the Chicago Farmers Club and served as a board member of Passavant Memorial Hospital, Chicago.

The American Dairy Cattle Club (ADCC) took a radical departure from the usual program of registering any animal whose sire and dam are recorded. The ADCC program required continuous records of production on cows and that each sire be measured by the performance of his daughters. Cows and bulls that did not measure up were dropped. This program was highly controversial among purebred dairy cattle breeders of that era.

Recognizing the broad advantages it offered in breed improvement, Prentice set up his first AI organization, which led to ABS in 1941. Dedicated to progeny testing, he never ceased to search for more accurate evaluations of genetic merit in sires. Finding production-record processing of several million dairy cows in USDA files held up for lack of funds, he contributed $150,000 to USDA to keep the program moving.

He pioneered with Linde Air Products in New York in researching (a costly venture) liquid nitrogen refrigeration. However, the entire AI industry and the cattle industry benefited.

Prentice's business acumen and strong belief in private enterprise, as well as tough competition, did not endear him to many university people and AI cooperative organizations. While his contributions to the betterment of livestock were many, he purposely avoided any credit. He believed that where there is no competition, there is neither the will nor the desire to take risks or to pioneer better ways to be of service. J. Rockefeller Prentice must be given much credit for daring to set the pace of the AI program and associations.

On 5 January 1950, the American Breeders Service was formed. During its first two years, it kept records and coordinated activities. In March 1952, the Indiana AB, Southeastern AB, and Northwestern AB associations were dissolved and merged into ABS.

However, during 1952, new ABS studs were established at Kansas City, Missouri, and Palo Alto, California. In 1954, with improved air transportation, the Asheville, Duluth, and Kansas City studs were merged with the Carmel, Indiana, stud.

Beginning in 1952, much research was done at the Wisconsin Scientific Breeding Institute to develop frozen semen. ABS was the leader in this field. After four years of experimentation, ABS made the conversion to a frozen-semen program in 1956. With over 1,100 technicians using ABS semen that year, a total of 980,761 first services were made.

Early in 1963, ABS was incorporated. Offices, during its earlier years, were established in Chicago, Illinois. In August 1965, the Chicago office was closed and facilities were moved to a 1,100-acre farm at DeForest, Wisconsin. There a new office building, bull barns, laboratories—and, in short, the most modern AI plant anywhere—had been constructed. The Madison facility is now the main headquarters for the widely spread operations of ABS.

To ensure the continuity of the organization he developed, in 1967, Prentice sold ABS through a stock exchange to W. R. Grace and Company. The Grace Company, known for its vast complex to serve agriculture worldwide, has continued to operate ABS in a progressive manner. Markets for semen have been expanded worldwide and growth has been made in the number of dairy cows serviced. In the beef field, ABS has been a leader with beef studs until 1976 at Calgary, Canada, and now near Ft. Collins, Colorado, in addition to the DeForest, Wisconsin, facilities. During 1972, a total of 2,152,573 first services were made—1,569,074 dairy and 583,499 beef.

Many individuals have contributed to the success of ABS, such as Philip I. Higley (manager of the various AI organizations founded by Prentice and president of ABS from 1963 until his retirement in 1968); Leland W. Lamb (president of American Dairy Cattle Club and for many years bull procurement leader); David E. Bartlett (vice-president in charge of sire health programs and active on the NAAB Sire Health and Management Committee); F. I. Elliott (research, and in charge of semen production and processing and chairman of NAAB's research and technical committee); and Robert E. Walton (president of ABS since 1968, and president of NAAB from 1972 to 1974. An able geneticist, he has led the way for a modern and expanding program).

Pabst Farms, Inc. Pabst Farms, in Oconomowoc, Wisconsin, owned by the Pabst Brewing Co., reported using AI in its famous Holstein herd as early as 1934. Semen was collected by ampullae massage. With the use of the artificial vagina, the AI program in the herd was expanded.

Howard Clapp, herd manager, started selling semen from Wisconsin Admiral Burke Lad, one of the great bulls of his day, in the late 1940s. This no doubt accounted for so many sons of this bull being in bull studs throughout the country. From a small beginning, semen sales totaled $26,000 in 1954. It is rumored that semen from Wisconsin Admiral Burke Lad, in those days, sold for $100 per inseminating unit.

After the death of Clapp in 1950, Sylvester Weiler took over as herd manager and directed semen sales. Frozen semen was put into use. The herd was dispersed in the early 1960s, but AI did much to spread the "Burke" breeding to many herds.

Tri-State Breeders Cooperative. Tri-State Breeders Cooperative had its beginning when the Vernon County Breeders Cooperative started operations on 29 March 1941 and when Trempealeau County Breeders Cooperative started on 10 January the same year. These two county cooperatives merged in April 1947 to form Tri-State Breeders, with offices located now at Baraboo and Westby, Wisconsin.

Oren Johnson, county agent at Vernon County, did much to develop the AI program in his county. In 1941, Vernon County dairymen bred only 205 cows, but by 1945 the number exceeded 10,000. The merger in 1947 resulted in a rapid expansion. The service area, as additional county associations were added, soon included large sections of southwestern Wisconsin, northeastern Iowa, and southern Minnesota. This accounts for the adoption of the name *Tri–State Breeders*. Located in one of Wisconsin's best dairy areas—populated with many progressive breeders—the cooperative grew soundly. Johnson became an employee of Tri-State in 1947 and was appointed general manager in 1951. During 1947, a total of 37,505 cows were serviced. Ten years later, the yearly volume was 228,366 head. The Vernon County group in 1941 had erected a bull barn and headquarters facilities near Westby, and this site became headquarters for Tri-State Breeders. As business grew, the facilities were expanded and modernized.

Tri-State Breeders was joined by Southern Wisconsin Breeders Cooperative, of Madison, in 1960. This consolidation, worked out by the directors of the two cooperatives, enhanced the bull battery and the number of cows serviced annually.

In March 1961, Neilus Larson succeeded Johnson as general manager. He served as general manager until 31 December 1969. On 1 January 1970, Alton Dale Block became general manager and served until his resignation in September 1973. The present manager, Gerald L. Larson, joined the Tri-State staff in November 1960 as field services manager.

Taxed for space and needing to enlarge its computer equipment, Tri-State in 1968 joined with two farmer cooperatives in Baraboo to erect and equip a modern facility to take care of the record work for all three. The administrative offices and staff are located at Baraboo. The bulls, laboratories, and semen distribution remained in Westby.

In the 1960s, Tri-State pioneered an extensive swine AI breeding project from Platteville, Wisconsin. Services were rendered throughout Wisconsin, Iowa, Minnesota, and Nebraska. Due to costs involved, this project was eventually discontinued. For a short while, horse AI services were offered, but this too was dropped due to the failure of frozen semen to result in a satisfactory conception rate.

Tri-State Breeders is governed by a thirty-one-man board of directors, each representing a county.

Entirely farmer financed, Tri-State has established an outstanding record of service, paying dividends to patrons most years, and building members' equities to well over two million dollars. Semen is sold now to many states, and a strong technician service program is maintained in Wisconsin. Approximately 400,000 cows are bred annually.

Southern Wisconsin Breeders Cooperative. Southern Wisconsin Breeders Cooperative in Madison was organized on 1 December 1950. This development resulted from the group effort of dairymen in the Dane and Sauk counties breeding association that had been obtaining semen from the University of Wisconsin bull stud. Direct cooperative membership provided capital. A board of nine directors governed the cooperative. Charles J. Krumm was the manager.

The first year of operation, Southern Wisconsin Breeders Cooperative inseminated 34,302 cows and semen was being purchased from the University of Wisconsin. Five years later, the cooperative had its own facilities, a bull stud of 37 sires, and 78,333 cows were serviced.

The cooperative grew in area serviced. In 1958, business was done in fourteen counties in Wisconsin and in an area of northern Illinois. The decision was made to relocate the bull stud and headquarters that consisted of a farm with a renovated barn, and the residence was used as an office and laboratory. A new building program was considered, and an eighty-acre farm, about two miles northwest of the original location, was purchased. These plans were not to culminate. The fall of 1959, the boards of directors for Southern Wisconsin Breeders Cooperative and Tri-State Breeders Cooperative decided to merge, to be effective 1 January 1960.

The better sires of the two organizations made up the bull stud. Most of the southern Wisconsin key employees were added to the Tri-State staff. Krumm became operations manager for the merged organization; Sterling Gillingham, Marlowe Nelson, Walter Weihrouch, and others continued similar or newly assigned duties. The board of directors of Tri-State was made up of a balanced representation from the two organizations.

Piper Brothers Farm, Inc. Piper Brothers Farm, Inc., in Watertown, Wisconsin, a privately owned business, started with AI being used to improve their own herd. The Piper brothers—Carl, Milton, Alvin, and Arden—established the Piper Brothers Farm, Inc., Artificial Breeding Service in 1949. The county agent, Forest Fellows, became associated with the business and helped to organize breeding units in Wisconsin and Illinois.

The Piper Farm—owned and operated for over one hundred years by several generations of Pipers—found a demand for AI from many herds. By 1959, over fifty technicians were working and 92,060 cows were serviced. Brown Swiss, Guernsey, and Jersey bulls had been, in the meantime, added to the Holstein stud, which had provided the start. Arden Piper did considerable early experimental work with frozen semen and placed it in limited field use in 1953–1954.

In 1966, Carnation Farms Breeding Service purchased Piper Brothers Farm, Inc., Artificial Breeding Service and established headquarters on the farm.

Carnation Farms Breeding Service (Wisconsin Division). Carnation Farms Breeding Service, with principal headquarters at Hughson, California, uses the Piper Brothers Farm as a midwestern operational center. As related earlier, in 1965, Carnation acquired the assets of Eastern Iowa Breeders. The bulls and equipment from Eastern Iowa Breeders were moved to Watertown; additional facilities were built; the staff was enlarged; and AI service was offered on a national and international basis. While Carnation Milk Farms had sold frozen semen from its herd sires on a select-mating basis earlier, it now entered the commercial field offering service to purebred and grade dairymen. Clarke A. Nelson, vice-president of the Carnation Company in Los Angeles, headed up the artificial breeding division. Overall supervision was provided by Leness Hall, general manager for many years, and Reuben Peterson, assistant manager, who for many years managed the Nebraska Dairy Breeders Association.

In addition to the Wisconsin stud, Carnation enlarged its operation by merging Carnation Farms Breeding Service and Genetics, Inc., in Hughson, California, on 1 November 1972. The new company is Carnation Company—Genetics Division. The Wisconsin unit continued to be an important part of the enlarged operations. During 1972, Carnation—Genetics accounted for the artificial breeding of 632,262 dairy and 168,394 beef cows. It is the third largest conglomerate in the AI field in America.

In addition to the AI organizations we have mentioned, there are no doubt a few additional, privately sponsored, small businesses that started in the early 1950s but eventually discontinued. Among these are Associated Breeding Service, Inc., in Lena, Wisconsin. That organization started about 1950, with 12 bulls, and 9,567 cows were serviced in 1951. Volume grew to 23,560 cows in 1958 but declined thereafter, and the business was discontinued in 1966. Another such business was Clinton Stock Farm Breeding Service, Barron County, Wisconsin. Starting in 1953 with 10 bulls, 629 cows were serviced the first year. The business discontinued in 1960.

During 1971, about 65 percent of the slightly over 2 million dairy cows were artificially bred. This participation in the nation's leading dairy state demonstrates the confidence dairymen have placed in the AI program.

Wyoming

So far as we can determine, no bull studs have been established in Wyoming. However, there has been considerable AI service provided by out-of-state organizations. In 1949, Cache Valley Breeding Association in Utah reported that 959 cows were artificially bred by its services in Lincoln County.

Known better for its beef production, though there are some good dairy herds around the principal cities, Wyoming had a population of 20,000 dairy cows and 800,000 beef cows in 1971. There were 5,647 milk cows (28 percent of the total) and 22,413 beef cows (3.0 percent of the total) artificially bred that year.

Practically all AI service is the "within herd" type of service with frozen semen supplied by out-of-state organizations. Inseminator-training schools and frozen semen have made AI an important adjunct to cattle improvement in Wyoming.

Puerto Rico

The organized plan for AI got underway in Puerto Rico in July 1952, when the AI program of Puerto Rico was initiated.

In 1946, the agricultural extension service sent Gaztambie Arrillage to Pennsylvania State College to study artificial insemination. Following his return, conferences were held by Arrillage giving instructions to agricultural extension leaders, department of agriculture officials, teachers, and students. Demonstrations on AI techniques were held at the farm of Fredrico Lopez del Valle at Toa Alta with over fifteen hundred dairy farmers and others attending.

The association is a cooperative and is financed by state aid and cooperative membership. Membership is represented by directors. The executive committee directs the local program, while the Puerto Rico Department of Agriculture employs the manager, who directs operations at the headquarters.

In 1950, the Puerto Rico Department of Agriculture started building the AI center at Barrio Juan Sanchez of Bayamon. Four technicians were trained and a provisional stud set up at the Puerto Rico Agricultural College Farm. Sires were provided, and $40,000 was allocated for the construction of the center at Bayamon. The inseminations at first were given free of charge, mainly for the purpose of promoting the program.

In December 1951, J. O. Almquist of Pennsylvania State University was brought to Puerto Rico to assist. He recommended that a sound program be set up in a pilot area and then extended to the remainder of the island. In April 1952, the Department of Agriculture appointed John Naugle, of Pennsylvania, to organize and administer the insemination center.

On 1 July 1952, the center was ready to start giving service. Twenty-two dairymen joined the program and a sponsoring committee of nine farmers was set up, with Mario Torrech of Bayamon as president. Month by month, always preceded by an educational program, the service was extended to new areas. By the end of the year, 132 dairies had joined. Due to lack of telephones, the technicians covered their routes daily, the farmer raising a red flag when a cow was to be serviced. In 1952, about 800 cows were bred, and 7,711 in 1954.

The bull stud located at Bayamon had as its first manager Cesar Estrada Gonzalez. A bull committee consisting of the organization manager, representatives of the department of agriculture, the extension dairymen, and farmers select the bulls that are to be used. Hato Rey, chief of animal husbandry, and the Agricultural Services Administration have general supervision of the AI program.

About 1954, semen was imported from the mainland to augment the bulls in use. Since that time, frozen semen is shipped regularly from the United States by several AI organizations that have helped direct the program.

Dairying is the second largest industry on the island, but it does not supply the demand. In 1975,[9] there were about 575,000 head of cattle and water buffalo in Puerto Rico. Not all are used for milk purposes. In 1971, about 69,000 cows were bred artificially.

Canada

Artificial insemination of cattle, in an organized way, began in Canada in 1941. The early developments closely parallel those in the United States.

There are ten Canadian provinces and an organized AI program is in operation in each except Newfoundland. Bull studs are maintained in the larger provinces, and they supply the bulk of the semen used throughout the country. A small quantity of semen is imported from the United States.

It is estimated that about 55 percent of the dairy and 10.4 percent of the beef cows in the Dominion are serviced artificially.[10]

Considerable frozen semen is exported. During 1973, a total of 1,466,762 units (one insemination dose) were exported to thirty-four different countries. Of this total, 1,075,241 units, nearly all from exotic breeds, were exported to the United States. Due to the fact that Canada built two quarantine stations in the late 1960s, cattle breeders were enabled to import bulls of the exotic breeds from Europe, where foot-and-mouth disease prevails. There are, however, areas free of the disease from which European cattle and semen can be exported to the United States and other countries. During 1970–1975, when the demand for the exotic breeds for crossbreeding and developing new strains in the United States was at an all-time high, the United States did not have adequate quarantine facilities. The chief avenue for obtaining exotic semen for use by U.S. cattle breeders was via Canada. The traffic was heavy. With a slump in beef prices from 1974–1978, imports of exotic beef semen to the United States have greatly declined.

Artificial insemination of swine beginning in 1952 has been encouraged by the provincial governments, and considerable research and field experimentation have been done at the University of Guelph by J. W. Macpherson, C. J. King, C. A. V. Barker, and associates. The Waterloo Cattle Breeding Association inseminated 11,641 sows from 1961 to 1967, but discontinued the swine program. In 1965, the Oxford County Pork Producers provided loans to the Oxford and District Cattle Breeding Association to implement swine AI. About 400 sows were inseminated as a pilot project to determine fertility, costs, and feasibility. By 1973, there were swine-semen-producing units operating in Ontario and Quebec.

On 1 October 1974, swine AI in Canada was expanded on a commercial basis with formation of the Ontario Swine AI Association. This farmer-owned cooperative was basically organized along the lines of the cattle AI organizations. James N. Wilkins, who had experience in nearly all of the early efforts with swine AI in Ontario, was employed as manager. The association expanded its memebership to all provinces with over four hundred members in 1977. During 1976, a total of 5,223 services from boars of five breeds were sold in Canada, and swine semen is being exported. The Ontario Swine AI Association utilizes both frozen and liquid semen.

The Canadian AI program is regulated by the federal and provincial governments working closely with cattle owners, the AI organizations, and the breed registry associations. Requirements governing the AI of pure-bred dairy cattle are established by the Joint Dairy Breeds Committee (JDBC). This committee is composed of representatives from each of the dairy breed registry associations in Canada. In principle, the JDBC is similar to the Purebred Dairy Cattle Association in the United States.

The regulations with respect to sire identification, blood typing, identification of semen, and other provisions are similar to those in effect in the United States. Canada has somewhat stricter regulations, however, by requiring permits to store semen on private premises, permits for inseminating businesses, and all managers and technicians must have been approved by JDBC and, in many cases, by the provincial governments.

The semen-producing business must maintain an active program of sire proving with respect to production, type, and control of undesirable inherited characteristics. Further, the program must be acceptable to JDBC. It is required that each semen-producing business keep its customers informed of all known facts, good and bad, about the breeding qualities of sires whose services are offered. Also, the business is not to engage in undesirable competitive practices nor in excessive promotion of specific sires.

Custom collection of semen from privately owned bulls is permitted only at licensed AI centers or at facilities approved by the Veterinary Director General for Canada.

Since the Canadian government and cattle breeders have made heavy expenditures in developing an AI system, the philosophy prevails that the goal must be genetic improvement and not for the making of profits. Encouragement is given to invest funds above operating costs in sires, research, and development. Privately owned, profit-motivated AI businesses are not encouraged.

Through the years, the AI program in Canada has progressed at a pleasing rate. Modern technology is used. The first large AI organization in the world to "go 100 percent frozen semen" was the Waterloo Cattle Breeding Association in 1954. Working relations between the AI organizations, cattle breeders, breed associations, and governmental personnel are congenial with the vigor and enthusiasm of a young and growing country.

Cooperation between the AI organizations in the United States, as represented through the NAAB members, and Canada has been excellent. In 1957 and again in 1974, the Ontario Association of Animal Breeders hosted the NAAB annual convention. In 1976, the British Columbia Artificial Insemination Centre, in Milner, joined with All West Breeders in hosting the annual meeting and convention for that year.

Between 1941 and 1949, at least twelve AI organizations were established in Canada. Some of these, as is true in the United States, were begun as a service for a specified group of dairy cattle owners. Most were farmer-owned associations and were financed by memberships and government funds. The provincial governments encouraged the AI program from the start. Funds were supplied, in addition to those furnished by cattlemen, to purchase land and buildings, buy bulls, and to hire personnel. In each area, the Provincial Livestock Branch and the agricultural representatives of the various counties assisted in the promotion and organization.

With the development of the frozen-semen program, better transportation, and communications, some of the smaller organizations, in Ontario particularly where eight once functioned, have consolidated, forming larger units. Since 1960, with an upward surge in beef AI, several new organizations have become prominent in Western Canada and in the Province of Quebec.

Canadian Association of Animal Breeders. As a result of cooperative efforts among the Canadian AI organizations in 1955, a national organization, the Canadian Association of Animal Breeders (CAAB), was formed in the early 1960s. Following his retirement from the Ontario Association of Animal Breeders and Semex in 1977, Roy Snyder became CAAB's secretary-treasurer.

Ontario

The Province of Ontario now has three principal AI organizations. These result from the consolidation of eight, early formed, smaller organizations. They are Eastern Breeders, Inc., Kemptville; United Breeders, Inc., Guelph; and Western Ontario Breeders, Inc., Woodstock. They are all farmer-owned associations. Each operates within its defined service area, but all combine efforts on common goals.

Ontario Association of Animal Breeders. The three aforementioned organizations now make up the membership of the Ontario Association of Animal Breeders (OAAB), which was formed in 1951 by the then eight original bull studs in the province. William P. Watson, livestock commissioner for Ontario, and a strong supporter of the AI program, suggested the need for such an organization. R. J. McDonald, DVM, was elected as the first president; Roy Snyder, vice-president; and D. C. McKay, DVM, secretary-treasurer. The association acts as the official voice of the AI industry in Ontario. OAAB also operates Semex-Canada, which coordinates semen-exporting activities for AI organizations throughout Canada. Since 1955, OAAB has marketed semen in Alberta, Manitoba, Saskatchewan, and other provinces.

Since 1978, with the Canadian Association of Animal Breeders growing, many of the activities of the OAAB are now handled by that association.

As the export of frozen semen grew in volume, Snyder, in 1957, discontinued as manager of the Waterloo Cattle Breeding Association and became secretary-manager of the OAAB and Semex. Snyder was succeeded by M. G. (Moe) Freeman, who had been serving as assistant secretary-manager.

Eastern Breeders, Inc. Eastern Breeders, Inc., in Kemptville, Ontario, is the result of amalgamating Quinte District Cattle Breeders Association, in Belleville, and Eastern Ontario Cattle Breeding Association, in Kemptville, on 6 July 1962. The early development of the merging organizations follows.

Eastern Ontario Cattle Breeders Association had its beginning in 1944 when a small AI unit was set up in Leeds County. One of the key organizers was J. R. Ostler, agricultural representative in Leeds County, from 1920 to 1962. According to Ostler, "Nearly all the milk went to cheese factories; nearly all of the cattle were grade Holsteins; production was average or below; and most were spring fresheners." The sponsors were Ontario cheese producers, and the provincial livestock branch assisted largely through the efforts of Watson.

A small barn was leased and three Holstein bulls were purchased. The organization was known as the Leeds Cattle Breeders Association. The first year, 474 cows were bred in seventy-three herds. In 1947, the Eastern Ontario Cattle Breeding Association was formed. M. C. McPhail, principal of the Kemptville Agricultural School, led the way. Finances came from membership fees, loans from breeders, and the provincial Department of Agriculture.

R. G. Smiley, DVM, was hired as manager. The service area and membership grew with 3,886 members, and 41,526 cows were bred in 1955. In 1956, an enlarged headquarters building was put into use.

Quinte District Cattle Breeding Association was organized in June 1948. About one-third of the cost of buildings and bulls was provided by the Ontario Department of Agriculture. Headquarters were established at Belleville on two acres of land, consisting of a barn and house. The house was used as the laboratory, office, and dwelling for the manager. Eight bulls were bought, but only two of them proved satisfactory for AI use. Conception rate was low. Fred H. Hubbs, of Bloomfield, was the first president.

The barns and buildings were remodeled and additional bulls were purchased as the business grew. There were 4,902 cows bred in 1948, increasing to nearly 32,000 by 1954.

With the formation of Eastern Breeders, Inc., in 1962, J. M. McLean, DVM, became general manager. He served until 11 March 1967. The post was then filled by Smiley.

The number of first service cows bred by Eastern Breeders increased to a peak of 159,491 in 1970.

United Breeders, Inc.   United Breeders originated on 1 December 1968 with the merger of three smaller associations: Central Ontario Cattle Breeding Association in Maple; Lambton Cattle Breeding Association in Wyoming; and Waterloo Cattle Breeding Association in Waterloo. Presently, United Breeders owns 235 acres of land and has housing for 185 bulls. New barns, an office, and a laboratory were built in 1969. Office facilities are on the same site and are occupied by the Ontario Association of Animal Breeders and Semex, Canada (the semen-sales organization). C. R. (Clem) Reeds, DVM, is the manager.

The history of the three organizations just mentioned provides some interesting highlights.

Central Ontario Cattle Breeding Association.   This association resulted from the merging in 1957 of the Maple Cattle Breeders Association in Ontario and Tor Dist Cattle Breeders Association, Ltd.

The Maple Association began operations in 1945, under the management of C. R. Reeds, as a cooperative financed by memberships and cooperative loans. It grew in size until nine counties were served. In 1949, Reeds resigned from the Maple Association and was instrumental in forming the Toronto District Holstein Breeders Club, which was financed by private capital. In 1953, the name was changed to Tor Dist Cattle Breeders Association, Ltd., as other breeds in addition to Holsteins were being serviced. Reeds managed the association until the merger in 1957.

After Reeds left the Maple Association, management was handled by a team consisting of Wilfred Kerffer, the president, and D. C. McKay, a veterinarian; both men continued until the merger. In 1945, Maple serviced 887 cows and reached 47,800 head by 1954. Both Maple and Tor District operated in the same area.

Central Ontario Cattle Breeding Association was owned by its members and was managed by Reeds.

Lambton Cattle Breeding Association was organized in Wyoming, Ontario, in 1949. Memberships and cooperative loans furnished the capital. Fred A. W. Marsh was the manager. The first year, 750 cows were bred, and in 1960 a total of 10,661. Frozen semen was put in use in 1954. The majority of service was to Holstein and to Hereford bulls.

Waterloo Cattle Breeding Association was established in 1941. It is credited with being the first Ontario insemination unit established. It began when William Henderson, a Jersey breeder of Galt in Waterloo County, became interested in AI for Jersey cattle. Henderson and M. E. I. McLoughry, an agricultural representative, organized and set up a unit for breeding Jerseys. Albert Hess, DVM, of Kitchener, did the inseminating, and McLoughry kept the records and handled finances. Only one bull was in use. Since the conception rate was low, the unit stopped its operation.

In 1942, the Waterloo County Holstein Breeders Association organized an AI unit to service Holstein herds in the area. At first, the bulls were housed by E. W. M. Snyder, Roy's father, and Henry B. Bowman, of Petersburg. In 1946, a site was purchased, and a barn to hold eight bulls was built. In 1946, Snyder was employed as assistant and worked with Johnson. In 1950, he became secretary-treasurer–manager of the association.

The volume of business done by the association expanded steadily, and by 1960 it reported 78,962 first services.

The Waterloo Association pioneered—working closely with the Ontario Veterinary College—in semen extenders, frozen semen, semen storage, and shipping methods. It was the first AI business in the world to convert entirely to frozen semen for field use (1954).

United Breeders, Inc., managed by Reeds, has a strong battery of dairy and beef bulls. There are many European imports among the beef bulls. During 1973, a total of 183,443 first services were made. Most of the semen is packaged in French straws.

Western Ontario Breeders, Inc. The service area of Western Ontario Breeders (WOBI) extends from Niagra Falls on the east to Windsor on the west. WOBI was organized in 1973 when the merger of Essex Cattle Breeding Association, Hamilton District Cattle Breeding Association, and Oxford and District Cattle Breeding Association occurred.

WOBI has a modern physical plant with barn space for over 150 bulls. In the stud's modern laboratory, semen is frozen in ampules and plastic straws. The twelve-county area, serviced in Ontario, involves sixty-five technicians and about as many additional employees in the office, laboratories, barn, and sales service. Semen is exported through Semex to twenty-five foreign countries and many parts of Canada.

The organizations that merged to form WOBI have played a leading role in developing the AI program in Ontario. They are discussed in alphabetical order.

Essex Cattle Breeding Association. Organized in 1947, and financed by the provincial government and cooperative loans, Essex Cattle Breeding Association serviced 800 cows its first year of operation. M. C. Dalton was one of the leaders in promoting the organization. Norman Shain served as manager. Headquarters were established on a site near Essex. For 1960, a total of 8,235 cows were serviced. As frozen semen became available, it was supplied from other Ontario units.

Hamilton District Cattle Breeding Association. Located at Hannon, in Ontario, this association was organized in May 1948. Assistance was given by the Livestock Branch of the Ontario Department of Agriculture in acquiring buildings and bulls. The association was organized as a corporation with twenty-six directors, each representing one hundred members.

G. R. Rourke, DVM, was one of the early managers. The first year of operation there were 2,779 first services and 42,533 by 1960.

Oxford and District Cattle Breeding Association (ODCBA). Located at Woodstock, ODCBA was organized in late 1945. The first cow was bred 15 May 1946. At that time, the organization was known as Oxford Holstein Breeders Association and was originally founded to furnish service for the purebred Holstein herds in Oxford County. Sixteen-hundred memberships at $100 each provided the working capital. A barn was leased from the Woodstock Agricultural Society and four bulls were procured. Later, the association bought the property and erected new facilities. Service from all major breeds of cattle was made available.

The first year, 3,200 cows were bred, and the total grew to 117,709 in 1960. About half the services were to purebred cows.

R. J. McDonald, DVM, was manager of the association. J. Bruce Amos, now the director of sire procurement, was the original field supervisor. Two resident veterinarians were employed to treat problem breeding cows, to diagnose pregnancy, and to assist patrons.

With the formation of Western Ontario Breeders, Inc., McDonald was promoted to general manager, and continues in this capacity. Emphasis is on high-quality service from the best sires available. A young-sire-proving program has been underway for several years and is being expanded.

Ontario receives much support from purebred breeders. At the present time, over three-fourths of all cattle registered are a result of AI.

In addition to the three Ontario organizations discussed, the Canadian Department of Agriculture in 1973 issued licenses to two additional semen-producing businesses: St. Jacobs Artificial Breeding Co-op (Ontario), Earl S. Feick, manager; and Agro Brothers (Hamilton, Ontario), Angelo Agro, manager. The St. Jacobs organization provides service in parts of three counties. Semen is also exported through Semex.

During 1973, in Ontario, a total of 369,747 dairy and 64,390 beef cows were artificially bred. This accounted for about 51 percent of the dairy and 8.8 percent of the beef cattle of breeding age. The Canadian beef registry associations modified their rules in 1970–1971 to permit registration of offspring resulting from AI. This has been a boon to AI programs in registered herds.

Quebec

The Province of Quebec has an area of 594,860 square miles and about six million people. It is second only to Ontario in human population. Dairying is an important industry. There are about 910,000 dairy and 195,000 beef cows in the province.[11] As a means of improving milk production and livestock production, the Quebec Department of Agriculture, through its livestock branch, in 1947 authorized support for an AI program.

Centre d' Insemination Artificielle du Quebec. The Centre, financed by government funds and membership fees, was established at St.-Hyacinthe, Quebec, in 1947. The Province of Quebec Department of Agriculture in launching support stated, "On account of great technical difficulties for private enterprise to organize, for the time being, an artificial breeding center in the province, the Department of Agriculture, through its Livestock Branch, is willing to maintain such an organization." The necessary bull barns, laboratories, and offices were built. Ayrshire, Holstein, and Canadian bulls were purchased, and the business slowly got underway.

Growth came largely by "Dairy Cattle Breeders' Clubs" becoming members of the Centre. These clubs were required to have at least five hundred cows signed up to start and to have a potential of at least fifteen hundred to two thousand; have an inseminator hired; and agree to abide by the rules of the Centre. Members paid $1 for a lifetime membership and $1 per service for semen from bulls at the Centre. The staff at the Centre trained inseminators free of charge.

Pierre Labrecque, Ernest Mercier, and Louis-N. St-Pierre were the people most responsible for the organization. St-Pierre became the first manager. The Centre became a member of NAAB in 1957.

The program grew steadily as experience was gained, and farmers found that the service provided better producers. In 1966, the final conversion to frozen semen was made. Claude Hayes, who worked closely with St-Pierre, became manager some years ago and has guided the program on a steady course. In 1973, a total of 434,164 dairy cows were bred. Beef AI has been gaining and accounted for 13,515 services the same year.

Two additional licensed semen-producing businesses were formed in the early 1970s: Centre d' Insemination Artificielle de Rimouski Enrg. (Rimouski, Quebec), Yves Caron, manager; and Centre D' Insemination Artificielle, Quebec Charolais Ltd., (Farnham, Quebec), Jacques Baar-Nason, owner. These two have since ceased operations in Quebec.

British Columbia

Artificial insemination was first introduced into British Columbia in 1939, on a provincial-owned farm—Colony Farm—because a famous old bull called "Vrouka" refused to work naturally. J. S. (Spence) Stroyan took a carload of Holstein cows to Madison, Wisconsin, learned the technique of AI at the University of Wisconsin–Madison, returned to British Columbia, and began to practice it at Colony Farm.

British Columbia Artificial Insemination Centre (BCAI). In April 1944, the dairymen in the Fraser Valley of British Columbia became interested in the possibilities of AI. It was decided to set up a central station to service the valley. This eventually led to the formation of BCAI.

On 20 September 1944, the bylaws of the new association were approved, and its first board of directors was elected. The organization, first known as the Lower Fraser Valley Artificial Insemination Centre, had headquarters on a purchased farm near Surrey. Stroyan was hired as manager, and two technicians were employed. The unit was located in an old renovated barn and the milk house served as a laboratory. The Production Service of the Dominion Department of Agriculture loaned the unit some bulls to start operations. The first cows were artificially bred in early 1945.

Operations were relatively successful for a time, but low conception rate and financial difficulties plagued the unit. The provincial Department of Agriculture gave some financial and professional help, by loaning the unit the services of J. C. Bankier, the provincial pathologist. As a result, conditions improved, and in September 1948 a new barn to house twenty-four bulls and laboratory were constructed on the present site at Milner. The name *British Columbia AI Centre* was adopted in 1949. Once again, the unit was in high gear. But, due to a very dry 1951 and low conception rates, the business was again in a precarious position.

In March 1953, R. J. Irwin was hired as manager and veterinarian for the unit. Many changes were implemented in order to bring operations up to the standard of other Canadian AI centers. Conception rates rose from the low forties to over 70 percent, and finances gained accordingly. The unit had at this time ten technicians; a fieldman, Dean Shantz; and an office staff of three, headed by Archie Todrick.

The Centre has continued to grow and services expanded. New barns have been built and over one hundred bulls can be housed; a modern laboratory and office are part of the facilities.

In the late 1950s, Dean O. Shantz became manager, succeeding Irwin who had given the Centre effective management for several years. Shantz served as manager from 1956 until 1977, and under his guidance the program expanded. He retired in July 1977 and was succeeded as manager by Gordon Souter, assistant manager, who started with the organization in 1959 as a technician. Souter guided the sire selection and development for BCAI many years.

Beef sires have been in the stud since 1950, and in 1954 AI service for ranch herds was started.

With the advent of frozen semen, the Centre established sales and service units in Alberta, Manitoba, and Saskatchewan. To serve western Canada more effectively, a branch office and a bull stud were established at Calgary, Alberta, in 1970.

In 1966, the first exotic beef breed bull entered the stud. The Centre began to purchase or lease bulls of the new breeds, and its semen sales expanded to many Canadian provinces, the United States, and many foreign countries. A custom collection service was initiated, and many exotic beef bulls were housed at the Milner and Calgary facilities.

Artificial insemination training schools are conducted regularly and about half the beef cows bred artificially are by the "do-it-yourself" plan.

In 1975, BCAI supplied semen and services for approximately ninety-six thousand dairy cows and ninety-seven thousand beef cows in several provinces, the United States, and overseas countries. It is estimated that about 85 percent of the dairy cows in the Fraser Valley are now bred artificially.

In 1955, BCAI affiliated with NAAB. The Centre for a long time was the only licensed bull stud in British Columbia, until the formation of Chilliwack Artificial Insemination Centre.

Chilliwack Artificial Insemination Centre. In March 1944, W. H. Hicks, then superintendent of the Dominion Experimental Farm at Agassiz, aroused interest in AI of dairy cows when he addressed a group of dairymen at the East Chilliwack Farm Forum. His talk prompted a group to investigate the advisability of starting an AI center in Chilliwack.

In the spring of 1944, a club was formed with a membership of forty-nine farmers, along with an enrollment of 779 cows. The first board of directors was elected 17 May 1944.

Membership grew until a total of 180 Chilliwack farmers had joined the organization, which was known as "The Chilliwack Artificial Insemination Club." A life membership fee of $10 was paid by each, giving the club $1,800 with which to erect buildings on one and one-half acres of land that was rented from J. J. Andrews. Operations began 1 January 1945. Six bulls were supplied by the Dominion Department of Agriculture, and T. G. McLean, a technician, inseminated 897 cows during the year.

In 1949, additional land was purchased and buildings were erected on the present site of the AI club farm. As the business expanded, the name *Chilliwack AI Centre* was adopted. By 1954, about eight thousand cows were serviced annually.

J. E. H. Lawrence was the early manager of the Centre. He, in turn, was succeeded by John A. Peter.

As frozen semen became available and the demand for a wider choice of sires increased, the Chilliwack Centre began buying semen from other Canadian AI studs and gradually disposed of its bulls. It now operates as a successful inseminating business with about eighteen thousand cows being serviced annually.

Nova Scotia

Nova Scotia Animal Breeders Cooperative, Ltd. (NSABC), located at Truro, Nova Scotia, is the only licensed AI center in the province.

It had its beginning in June 1941 by the formation of a small breeding unit at Yarmouth. Several other small units were started, but by 1945 all were under one management, the NSABC. In 1947, the Truro Cooperative became the semen-producing business for the entire province, and in 1949 it became a provincial unit and took on its present name.

As a farmer-owned cooperative, NSABC is governed by a ten-man board of directors. Until about 1970, the unit was in charge of a board of management. W. D. Murray, who guided the operations from 1945 until his retirement in 1975, was both manager and president of the unit.

According to Murray, "The first year of operation only Guernsey and Jersey bulls were kept. A total of 659 cows—408 Guernseys and 251 Jerseys—were bred. One herd owner had 21 Jerseys bred and 19 settled on the first service. This gave us a good boost."

By 1963, about thirty thousand dairy and beef cows were being bred annually. There were twenty local inseminating areas set up with a technician serving each. In 1970, the entire operation was made a central federated type of organization. In some of the thinly populated cow areas, farmers have been trained to inseminate their own cows.

Bulls of all the leading dairy and beef breeds are available. A young-sire-proving program is underway. There is some exchange and purchase of semen from New Brunswick and other provincial studs. The conversion to frozen semen was made in 1960–1962.

In 1970, under the organization plan, Donald Newcombe became president and Murray, manager. Donald Cameron, laboratory manager, succeeded Murray as manager in 1975.

In 1973, Nova Scotia had about 39,000 dairy cows with 18,395 bred to dairy bulls and 6,347 bred to beef bulls. The beef cow population is about 24,000 head and 12.5 percent are bred artificially as compared to 63 percent of the dairy cows. In a land of severe weather and ocean storms, the Nova Scotia AI program has grown in spite of handicaps.

New Brunswick

Artificial insemination organizations were formed in New Brunswick between 1942–1950 along the same general scheme as those cooperatives in other provinces.

There is only one licensed commercial semen-producing business in the province—the New Brunswick Central Artificial Breeding Cooperative, in Fredericton. D. G. Moore, DVM, is manager.

In 1963, New Brunswick had about fifty-seven thousand head of dairy and thirteen thousand beef cows, and 32 percent of the total was artificially bred. Forty-two percent of the dairy and 13 percent of the beef cows were artificially bred in 1973. Some semen is purchased from other provincial studs.

Prince Edward Island

Prince Edward Island does not maintain a licensed commercial semen-producing AI center. There is a cow population of twenty-three thousand dairy—52 percent bred by AI to dairy bulls and 24.3 percent to beef bulls—and twelve thousand beef cows. Of the total, 24 percent was bred by AI. Frozen semen is obtained from New Brunswick and from licensed bull studs in other provinces.

Newfoundland

This province, somewhat isolated, has not attempted to operate a bull stud. However, it is indicated that 559 dairy cows were bred to dairy bulls and 27 to beef bulls by AI in 1973. Frozen semen is imported, largely from other bull studs in the Maritime Provinces.

Alberta

With the increased use of AI for beef cattle, utilizing frozen semen, and the introduction of the exotic breeds, there came a great increase in the number of licensed commercial semen-producing centers in the prairie provinces. Alberta leads in the number of these. As of 1973, there were six licensed businesses and one private operation.

Western Breeders, Ltd. Located at Balzac, Alberta, Western Breeders, Ltd., was founded in 1968. Douglas G. Blair, a former fieldman for British Columbia AI Centre, Milner, is the founder and general manager. T. Gordon DeLair is assistant general manager. Earl F. James, Balzac, and Rodney James, of Lacombe, are directors.

Western Breeders specializes in semen production and sales from imported exotic beef bulls. In addition, the standard beef breeds such as Angus, Hereford, and Shorthorn, as well as the dairy breeds, through Brown Swiss and Red and White Holstein, are represented. Some of the bulls are owned by private breeders and are housed at Western Breeders' headquarters for custom collection and semen sales. Emphasis is on performance-tested and progeny-tested sires.

A large semen-export business has been built, particularly to the United States, where the semen from exotic breeds is in demand for crossbreeding and for the establishment of new strains of cattle. Western Breeders, Ltd., has a close working relationship with New Breeds Industries in Manhattan, Kansas. As a result, large quantities of semen from exotic breeds have been made available to AI organizations and breeders throughout the United States.

In 1969, Western Breeders, Ltd., joined with the Ontario Association of Animal Breeders to form Western Breeders Service for semen sales in western Canada. A bull stud was established at White City, Saskatchewan, to service that province and the Province of Manitoba.

During 1972, service was provided for 110,824 beef and 82,450 head of dairy cows in these provinces.

Simmental Breeders Cardston, Ltd. Simmental Breeders Cardston, in Alberta, began operations in 1966. It all began when rancher Travers Smith, dissatisfied with the weaning weights he was getting on his Hereford and Shorthorn calves, decided to investigate crossbreeding. He read the Swiss government literature on the Simmental breed and discussed importing cattle with Wayne Malmberg, who had been instrumental in bringing Charolais into Canada. Hobart Peters, then of the Manyberries Range Station, had studied Simmentals in Europe, and he told Smith they had a lot to offer. The Brown Swiss beef breed crosses at the U.S. Range Station, in Miles City,

Montana, also influenced Smith to look for milk as well as meat if he wanted heavy calves at weaning time.

In July 1966, Smith went to Switzerland and then to France, looking for a place to use his import permit. He encountered difficulties in finding cattle under nine months of age that had not been vaccinated for foot-and-mouth disease. Eventually, he located a bull—Parisien—just five miles from the Pie Rouge Herd Book Office at Dijon. Soon the bull was on the way to Canada to go through quarantine.

This Simmental bull was the first to be imported into North America.

Simmental Breeders Cardston, Ltd., commonly called SBL, is a privately owned corporation. Among the stockholders are Travers Smith; his brother, Frank Smith; and Hans Ulrich, of Claresholm. Headquarters are at Cardston.

In 1968, three additional bulls (Fern, Sultan, and Capitane) and four heifers were imported by SBL. By this time, interest in Simmental was growing and additional cattle were being imported into Canada.

Parisien, born in 1966, furnished frozen semen for a clamoring market in Canada and the United States. Arrangements were made for Curtiss Breeding Service to act as the exclusive U.S. distributor, and for Western Breeders, Ltd., to act as the Canadian distributor. Parisien turned out to be a highly influential bull. He is "No. 1" in the Canadian and American Simmental Association Herd books. He became the first Simmental Golden Certified Meat Sire by Performance Registry International standards. His progeny number in the thousands. He has nearly three hundred sons and over seven hundred daughters gain tested. (Figure App.2). His son, Amor, was the first Simmental imported into the United States. He was purchased by an Oklahoma-Texas syndicate for $154,000.

SBL grew in cattle numbers and semen sales. Simmental associations were formed in both Canada and the United States. The American Simmental Association has one of the most complete performance-testing programs of any breed. All cattle recorded must be weight-gain tested and sire summaries are published regularly. Travers Smith died in 1974, but he lived to see some of his goals for cattle improvement come true.

Universal Semen Service. Following the death of Travers Smith, SBL underwent some reorganization. A new organization, Universal Semen Service (USS), a privately owned corporation, was formed, largely by SBL members, in 1974. There are twenty-six shareholders, mostly ranchers and farmers located in southern Alberta. SBL continues as a separate organization with semen from its bulls processed by USS.

USS is strictly a custom semen-collection, processing, and storage business. A semen-marketing service is operated for bull owners. USS does not own or lease bulls. It has an associate organization, Universal Semen Sales, Inc., which is located at Great Falls, Montana, to market semen in the United States. Foreign marketing is handled by Semaltex, an Alberta-owned company.

USS operates throughout Canada and the United States. During 1976, about 130 bulls were collected and approximately 145,000 units of semen were processed. In 1975, modern barns, laboratories, and other facilities were completed. The manager is R. M. Gebauer, a Colorado native, with many years experience in the AI field.

Figure App. 2. Parisien—first Simmental bull imported into North America. (Photograph courtesy of Curtiss Breeding Service.)

Canadian Stock Breeders Service International, Ltd. Canadian Stock Breeders Service (CSBS) located at Winterburn, Alberta, was founded in 1971 by F. G. Day, an orthopedic surgeon. It was established as a corporation with nine directors, and Day as chairman of the board. Leroy Rollins was managing director; Helen Day, assistant director of production; and R. L. (Ron) Holland, assistant director of livestock.

CSBS became the sales outlet for a number of Canadian companies and for individuals involved in the international sale of frozen semen from top quality beef and dairy bulls. In addition to its own bull stud, CSBS had a working arrangement with several other Canadian AI organizations for exchange of and distribution of semen. It discontinued business in 1976.

Southern Breeders, Ltd. Southern Breeders, Ltd., a privately owned corporation, started in January 1971 at Lethbridge, Alberta. Mark R. Stringham served as general manager for several years and Brien Thompson as production manager.

In May 1971, a disastrous fire destroyed all facilities. A rebuilding program followed. A new corporation, Southern Sires, Ltd., was formed with Stringham as managing director. In May 1974, Southern Sires, Ltd., transferred its bull stud to Southern Breeders, Ltd., and it continued operations with T. L. Church, DVM, general manager. The business discontinued in 1976.

American Breeders Service of Canada, Ltd. In 1970, American Breeders Service, Inc., in DeForest, Wisconsin, established ABS of Canada, Ltd., at Calgary, Alberta, to take care of its needs for exotic beef bull semen as well as to operate in Canada. This proved to be a growing business for several years, but declined as beef prices slumped in 1974–1975. The facility was closed in 1975.

BCAI Centre. The British Columbia Artificial Insemination Centre established a second semen-producing and service unit at Calgary in 1970. This unit, as mentioned under British Columbia, continues to serve the western provinces.

Prairie Breeders. Also located near Calgary is Prairie Breeders, a private corporation. It was established about 1967 to provide beef services, including the exotic breeds, to the western provinces. William F. Hart is manager.

Bov Imports, Inc. Bov Imports, Inc., a Canadian corporation, was formed in 1971. As the name suggests, its function is to import and market semen in foreign countries. The semen was produced at Rimouski, Quebec, and sold through the Bov Imports Office in Denver, Colorado.

In 1975–1976, Bov Imports, Inc., ceased much of its business in Canada and operations were moved to Denver, Colorado. The firm took over the International Beef Breeders facilities with J. J. (Bud) Prosser serving as manager. During 1975, about forty-four thousand units of semen were merchandised.

Canadian Beef Sires Progeny Test. A very important organization, located at Camrose, Alberta, is Canadian Beef Sires Progeny Test. Organized in 1970–1971, its function is to control and administer a beef-progeny-test program for AI units in Canada.

The organization is managed by a committee comprised of representatives of the participating members. Reeds was chairman, and Snyder served as secretary, with Blair in charge of operations. The program is now carried on through the National Sire Monitoring Program in Canada.

Saskatchewan

Saskatchewan has a dairy cow population of about 81,000 head and a beef population of 1,135,000. About 26 percent of the dairy cows and heifers and about 5 percent of the beef cows are artificially bred. The semen supply comes largely from Western Breeders Service, Ltd., White City (Fred Sumison, manager), and other bull studs operating in the prairie provinces.

Another AI unit started in Saskatchewan is S and M Breeders, Ltd., located at Yorktown.

Manitoba

The first AI unit organized in western Canada was started in 1940 at Neepawa, Manitoba. It had a struggling existence and discontinued in 1949. A second, The Rat River Breeding Association, maintained a bull stud from 1945 until frozen semen became available about 1954–1955. As soon as frozen semen was available, the Manitoba inseminating units began buying semen from the British Columbia and Ontario units and later from the Alberta organizations. At present, the province has about 100,000 dairy cows with about 30 percent bred artificially. There are about 430,000 beef cows and some 5 percent are bred artificially. Much of the semen is now marketed through the Manitoba Semen Distribution Centre.

Canada has made good progress in developing an extensive AI program. The number of dairy cows and beef cows artificially bred continues to grow.

Appendix B: NAAB Committees, Awards, and Uniform Coding System

NAAB Committees

Since its beginning, NAAB has utilized committees to assist in carrying out its objectives. Through the years, the committees have increased in membership and activity. The committees are composed largely of NAAB representatives but include many persons from the colleges and farm-oriented businesses. The NAAB committees and representatives have done much to unify the AI industry and the NAAB program. As an indication of the importance of the work done by the committees and representatives, the 1978–1979 roster is presented.[1]

Beef Development Committee. Roy Wallace (1975) Select Sires, Inc., chairman; Ancel Armstrong (1969) New Breeds Industries; Robert Fincham (1976) Midwest Breeders Cooperative; Don Hutzel (1972) Noba, Inc.; Bill Jackson (1978) Kansas Artificial Breeding Service Unit; R. D. Long (1978) Curtiss Breeding Industries; Marvin Nichols (1976) Hawkeye Breeders Service; Keith Vander Velde (1978) Tri-State Breeders; and Ken Young (1975) American Breeders Service.

Credentials Committee. Armin Fruechte (1975) Tri-State Breeders, chairman; Wilfred Broussard (1977) Northwestern State University; Jim Clark (1977) American Breeders Service; George Miller (1973) Select Sires, Inc.; and Rulon Osmond (1977) Cache Valley Breeding Association.

Dairy Sire Evaluation Committee. Doug Wilson (1974) Midwest Breeders Cooperative, chairman; Tom Aitchison (1977) Curtiss Breeding Industries; Robert Albrecht (1978) Tri-State Breeders; Larry Beat (1974) Kansas Artificial Breeding Service Unit; Jerry Conley (1977) Noba, Inc.; Lloyd Ebersole (1973) Sire Power, Inc.; Rodger Hoyt (1971) Select Sires, Inc.; Jeffrey Keown (1975) Eastern AI Cooperative; Pete Milinkovich (1975) Minnesota Valley Breeders; Paul Miller (1972) American Breeders Service; and Bob Strickler (1977) Carnation–Genetics.

Ethics and Practices Committee. Wallace Erickson (1976) Central Ohio Breeding Association, chairman; William M. Durfey, NAAB; Tom Lyon (1977) Midwest Breeders Cooperative; and Bryce Weiker (1978) Noba, Inc.

Inseminator Training Committee. Willis Ritchey (1974) Atlantic Breeders Cooperative, chairman; Don Deckert (1976) Tri-State Breeders; Fischer Harned (1974) Kentucky Artificial Breeders Association; Gene Lowe (1977) Central Ohio Breeding Association; Dick Newman (1974) American Breeders Service; Orville Rickard (1977) Carnation–Genetics; Don Solberg (1978) Curtiss Breeding Industries; and Robert Webster (1978) Eastern AI Cooperative.

International Marketing Committee. Robert Boese (1978) Curtiss Breeding Industries, chairman; Joe Aguirre (1977) Carnation–Genetics; Don Ausman

(1972) Midwest Breeders Cooperative; Roger Emig (1973) Altantic Breeders Cooperative; Charles Krumm (1977) Eastern AI Cooperative; George Miller (1973) Select Sires, Inc.; Bob Pelkey (1972) American Breeders Service; and Reuben Peterson (1977) Tri-State Breeders.

Marketing and Communications Committee. Jim Mellinger (1974) Eastern AI Cooperative, chairman; Cliff Allen (1974) Michigan Animal Breeders Cooperative; Herb Behnke (1976) Midwest Breeders Cooperative; Herald Catlin (1978) All West Breeders; Bob Clauson (1974) Minnesota Valley Breeders Association; Mary Crowley (1978) Tri-State Breeders; Dennis Fulkerson (1977) Illinois Sire Service; Phil Jayne (1976) Carnation–Genetics; Karen Lage (1978) Curtiss Breeding Industries; Steve Mawer (1978) American Breeders Service; Larry Neel (1977) Central Ohio Breeding Association; and Merle Starr (1977) Noba, Inc.

Managers Committee. Glenn Pirrong (1974) Curtiss Breeding Industries, chairman; Arnold Baham (1977) Louisiana Animal Breeders Cooperative; R. Emory Brubaker (1976) VA/NC/Select Sires; W. L. Campbell (1977) Sire Power, Inc.; Sterling Gillingham (1975) Minnesota Valley Breeders Association; Gerald Larson (1974) Tri-State Breeders; Charles Michaels (1978) Kansas Artificial Breeding Service Unit; Archie Nelson (1978) All West Breeders; and Robert E. Walton (1978) American Breeders Service.

Member Association Director's Committee. Robert Thompson (1975) Midwest Breeders Cooperative, chairman; Duane Berg (1978) Minnesota Valley Breeders Association; Robert Burford (1978) Louisiana Animal Breeders Cooperative; Fred Friday (1978) Central Ohio Breeding Association; Hubert Goode (1976) Tennessee Artificial Breeders Association; Lowell Grim (1973) Cache Valley Breeding Association; Roger Kohler (1978) Noba, Inc.; Henry Kopp (1977) Tri-State Breeders; Gordon Macklin (1975) Ontario Association of Animal Breeders; John Savage (1976) Eastern AI Cooperative; and Norman Sollenberger (1975) Atlantic Breeders Cooperative.

Nominating Committee. Erton Sipher (1977) Eastern AI Cooperative, chairman; Marshall Carpenter (1973) Kentucky Artificial Breeders Association; Glenn Pirrong (1977) Curtiss Breeding Industries; Robert E. Walton (1974) American Breeders Service; and David Yoder (1976) Atlantic Breeders Cooperative.

Research Committee. H. C. Kellgren (1974) Select Sires, Inc., chairman; Ed Call (1978) Kansas State University; F. I. Elliott (1974) American Breeders Service; Nesbit Fleming (1977) Curtiss Breeding Industries; Ted Knapp (1978) Eastern AI Cooperative; Jeff Keown (1977) Eastern AI Cooperative; and Harry Roth (1976) Atlantic Breeders Cooperative.

Technical Committee. Clif Marshall (1976) Select Sires, Inc., chairman; Johnny Abilay (1977) Carnation–Genetics; Bruce Bean (1970) Eastern AI Cooperative; Michael Cowan (1978) Tri-State Breeders; Glen Gilbert (1977) Midwest Breeders Cooperative; Jere Mitchell (1977) Curtiss Breeding Industries; Bill Nolan (1978) Illinois Sire Service; Hollis Schwartz (1976) Minnesota Valley Breeders Association; John Sullivan (1976) American Breeders Service; Floyd Weidler (1975) Atlantic Breeders Cooperative; and Norm Werkheiser (1976) Sire Power, Inc.

Sire Health and Management Committee. D. E. Bartlett (1958) American Breeders Service, chairman; Robert Hillman (1978) Eastern AI Cooperative; B. W. Kagy (1958) Noba, Inc.; Tom Lyon (1977) Midwest Breeders Cooperative; Don Monke (1977) Select Sires, Inc.; Jack Morse (1977) Carnation–Genetics; R. E. Vollmar (1967) Minnesota Valley Breeders Association; and Maurice Woulfe (1977) Curtiss Breeding Industries.

Special representatives and appointees include: American Society of Animal Science, Robert E. Walton, American Breeders Service; American Society of Bovine Practitioners, D. E. Bartlett, American Breeders Service; American Dairy Science Association, Gordon A. Doak, NAAB; American Veterinary Medical Association, D. E. Bartlett, American Breeders Service; Beef Improvement Federation, William M. Durfey, NAAB; Roy Wallace, Select Sires, Inc.; DHI Coordinating Group, William M. Durfey, NAAB; Memorial Report, Gordon Doak, NAAB; National Cattlemen's Association, William M. Durfey, NAAB; Polish Friesian Project, Paul Miller, American Breeders Service; Jeff Keown, (alternate) Eastern AI Cooperative; U.S. Animal Health Association, D. E. Bartlett, American Breeders Service.

NAAB Technician Award Program

Realizing the importance of the work done by AI technicians (also known as inseminators), the NAAB board of directors in 1956 took steps for the association to recognize outstanding achievements. Effective 1 January 1958, the following Technician Award Program became effective:

1. Insemination of 35,000, or more, first service cows; or,
2. Twenty years of service as an AI technician. The award consisted of a plaque and a special lapel pin.

Technicians could claim credit for tenure with one or more organizations, but the nominating organization was required to be an NAAB member.

In 1958, twenty-three technicians from eleven organizations qualified for the "35,000 Cow Award." During the second year of the program, D. K. Fuller, of New York ABC, qualified for the "Twenty-Year Award." With the passing of time, an increasing number of technicians qualified for the "35,000 Cow Award" as well as the "Twenty-Year Award." In 1966, an award program was established on the basis of:

1. 35,000, or more, first services;
2. 100,000, or more, first services; and,
3. The Quarter Century Award for twenty-five years of service as a technician.

By the end of the NAAB fiscal year of 1978–1979, about thirteen hundred AI technicians had been awarded the NAAB Technician Award. Included were 50 technicians who had received the "100,000 Cow Award" or "Century Award" and 227 persons (67 technicians) who had received the "Quarter Century Award."

In 1975, the "Quarter Century Award" qualifications were revised to include, in addition to technicians, all employees of NAAB member organizations that had twenty-five years of tenure. The result was the addition of 160 new names.

While the number of professional technicians has decreased more than 70 percent with the direct-semen-sales program and fewer dairy herds, the NAAB Technician Award continues to be coveted and prized.

Uniform Coding System for Identifying Semen

This semen-coding system developed by NAAB and by its member organizations is in general use throughout the United States. It is endorsed by dairy-record-processing centers, the USDA, and breed registry associations. The coding system is administered by NAAB, Columbia, Missouri.

How the Coding System Works.   The code is made up of a total maximum number of eight characters as follows:

Stud: The first two characters represent the code number for the stud. The stud code may be either numeric, alpha, or a combination of alpha and numeric.
Leading zeroes and blanks are omitted.
Breed: Single alpha characters for dairy breeds.
Two alpha characters for beef breeds.
Blanks are omitted from the code.
Bull: Numeric characters from 1 to 9999.
Leading zeroes and blanks are omitted.

Example: 10H232

10: John Doe Breeding Service has Stud Code Number 10.
H: Breed Code for Holsteins.
232: Bull Code for Highhill Master Pilot, Registry Number 1910165.

Dairy

Breed	Code	Breed	Code
Ayrshire	A	Jersey	J
Brown Swiss	B	Milking Shorthorn	M
Guernsey	G	Red & White	W
Holstein	H	Crossbred	X

Beef and Dual Purpose

Breed	Code	Breed	Code
Africander	AF	Maine-Anjou	MA
Angus	AN	Marchigiana	MR
Barzona	BA	Meuse-Rhine-Issel (MRI)	MI
Belted Galloway	BG	Murray Grey	MG
Beefmaster	BM	Normande	NM
Beef Friesian	BF	Norwegian Red	NR
Blonde d'Aquitaine	BD	Pinzgauer	PZ

Breed	Code	Breed	Code
Brangus	BN	Polled Hereford	HP
Brahman	BR	Polled Shorthorn	SP
Red Brahman	RB	Red Angus	AR
Charbray	CB	Red Dane	RD
Charolais	CH	Red Poll	RP
Chianina	CA	Romagnola	RN
Devon	DE	Santa Gertrudis	SG
Dexter	DR	Scotch Highland	SH
Galloway	GA	Simmental	SM
Gelbvieh	GV	South Devon	DS
Hereford (horned)	HH	Scotch Shorthorn	SS
Hays Converter	HC	Sussex	SX
Limousin	LM	Tarantaise	TA
Lincoln Red	LR	Welsh Black	WB
Luing	LU		

Appendix C: List of Tables and Figures

List of Tables

List of Figures

Notes

Notes to Chapter 1

1. Victor A. Rice, Frederick N. Andrews, Everett J. Warwick, and James E. Legates, *Breeding and Improvement of Farm Animals* (New York, N.Y.: McGraw-Hill, 1957), p. 160.
2. Ibid.
3. Supplementary artificial insemination is still used by some horse breeders and particularly racehorse-breeding establishments in the U.S.
4. Telesforo Bonadonna, "Le basi scientifiche e le possibilitè techniche della fecóndazione artificiàle," quoted in Enos J. Perry, ed., *The Artificial Insemination of Farm Animals* (New Brunswick, N.J.: Rutgers University Press, 1968), p. 5.
5. L. L. Lewis, "Artificial Insemination," Oklahoma Agricultural Experiment Station Bulletin No. 93, 1911.
6. *A. I. Digest* 16 (April 1968): 11.
7. Personal correspondence in 1977 with J. W. Tyler, Animal Science Department, University of Wisconsin–Madison.
8. H. A. Herman and A. C. Ragsdale, "The Artificial Insemination of Dairy Cows," Missouri Agricultural Experiment Station Bulletin No. 407, June 1939.
9. F. W. Miller and E. I. Evans, *Journal of Agricultural Research* (hereafter referred to as *J. Agr. Res.*) 48 (1934).
10. C. L. Cole, "Artificial Insemination of Dairy Cattle," *Journal of Dairy Science* (hereafter referred to as *J. Dairy Sci.*) 21 (1938): 131–32; 22 (1939): 107–10.
11. Missouri Agricultural Experiment Station Research Bulletin No. 376, 1943.
12. Enos J. Perry and Harry A. Herman, "Twenty-five Years Experience with AI in Cattle Improvement," *Proceedings Vth International Congress on Animal Reproduction and Artificial Insemination*, Trento, Italy, September 1964.

Notes to Chapter 2

1. I. M. Lerner and H. P. Donald, *Modern Developments in Animal Breeding* (New York, N.Y.: Academic Press, 1966).
2. Richard C. Foley, Donald L. Bath, Frank N. Dickinson, and H. Allen Tucker, *Dairy Cattle: Principles, Practices, Problems, Profits* (Philadelphia, Pa.: Lea and Febiger, 1972).
3. H. Wilmot Carter, "Effectiveness of Artificial Insemination in Dairy Cattle Improvement," *J. Dairy Sci.* 45 (1962): 276.
4. Raymond Albrectsen, "Selection and Repeatability of Sires Used in Artificial Insemination," *J. Dairy Sci.* 30 (1947): 594.
5. William F. Schaefer, Jr., "Comparison of Original Proof, AB Daughters and Apparent Transmitting Ability of 200 Bulls Used in New York and Pennsylvania Breeding Cooperatives," *Proceedings of the Sixth Annual Conference*, pp. 132–37, NAAB, Columbia, Mo., 1953.
6. Carter, "Effectiveness of Artificial Insemination in Dairy Cattle Improvement," p. 277.
7. Charles R. Henderson, "Selecting and Sampling Young Sires," *Proceedings of the 1954 Annual Convention of the National Association of Artificial Breeders* (1954), NAAB, Columbia, Mo.
8. Charles R. Henderson, H. Wilmot Carter, and J. T. Godfrey, "Use of Contemporary Herd Average in Appraising Progeny Tests of Dairy Bulls," *Journal of Animal Science* (hereafter referred to as *J. Animal Sci.*) 13 (1954): 959; Charles R. Henderson and H. Wilmot Carter, "Improvement of Progeny Tests by Adjusting for Herd, Year, and Season of Freshening," *J. Dairy Sci.* 40 (1957): 638.
9. For a summary of the developments and mathematical applications utilized in the formulation of modern sire-proving methods, see S. R. Searle, "Review of Sire-Proving Methods in New Zealand, Great Britain, and New York State," *J. Dairy Sci.* 47 (1964): 402–13; Victor A. Rice, Frederick N. Andrews, Everett J. Warwick, and James E. Legates, *Breeding and Im-*

provement of Farm Animals, 6th ed. (New York, N.Y.: McGraw-Hill, 1970), pp. 145–70; Ivar Johansson and Jan Rendel, *Genetics and Animal Breeding* (San Francisco, Calif.: W. H. Freeman and Co., 1968); Jay L. Lush, *Animal Breeding Plans,* 3d ed. (Ames: Iowa State University Press, 1945); and Lerner and Donald, *Modern Developments in Animal Breeding.*

10. Frank N. Dickinson, "What Is Going to Happen to Future Bull Proofs?" *Proceedings of the Twenty-seventh Annual Convention,* p. 86, NAAB, Columbia, Mo., 1973; the "USDA-DHIA Modified Contemporary Comparison—Sire Summary and Cow Index Procedures," Production Research Report No. 165 (March 1976) Agricultural Research Service (ARS), USDA, Washington, D.C.

11. For derivation and application of the equations for calculating PD,see R. D. Plowman, "Notes and Concepts Used in USDA Sire Summary Procedures," *Dairy Herd Improvement Letter* (1968): 44–202; Foley, Bath, Dickinson, and Tucker, *Dairy Cattle,* pp. 141, 806; "USDA-DHIA Modified Contemporary Comparison—Sire Summary and Cow Index Procedures," ARS, USDA.

12. R. L. Powell, H. D. Norman, and Frank N. Dickinson, "Trends in Breeding Value and Production," *J. Dairy Sci.* 60:8 (1977): 1316–26.

13. USDA-DHIA Cow Index List, Animal Physiology and Genetics Institute, ARS, USDA, Beltsville, Md., Spring 1975.

14. Ibid.

15. Ibid.

16. Based on work by A. E. Freeman, Iowa State University. Reported by Basil R. Eastwood in "Lactation Summary, Herd Ranking and Preliminary Sire Summary Procedures," Iowa State University Market Mimeograph, Dys–1961, Ames, Iowa, September 1968.

Notes to Chapter 3

1. *Dairy Herd Improvement Letter* 53:2 (September 1977), Science and Education Administration, USDA, Beltsville, Md.

2. "1976 Dairy Producer Highlights," National Milk Producers Federation, Washington, D.C.

3. Richard C. Foley, Donald L. Bath, Frank N. Dickinson, and H. Allen Tucker, *Dairy Cattle: Principles, Practices, Problems, Profits,* 2d ed. (Philadelphia, Pa.: Lea and Febiger, 1972).

4. "What DHI Can Do for Dairying," *Advanced Animal Breeder* 25:5 (July 1977):10.

5. *Dairy Herd Improvement Letter* 48:4 (June–July 1972), USDA.

Notes to Chapter 4

1. First published in the *Breeders Gazette* in the 1930s. *Breeders Gazette* ceased publication in 1964.

2. Larry W. Specht, "Who Sires the Tested Cows?" *Advanced Animal Breeder* 25:5 (1978): 5.

3. Personal communications in July 1975 with the secretary of each dairy breed registry organization in the United States.

4. I. M. Lerner and H. P. Donald, *Modern Developments in Animal Breeding* (New York: Academic Press, 1966), p. 143.

5. Personal communication in 1974 with Richard Nelson of the Holstein-Friesian Association of America in Vermont.

6. *Genetic Considerations in Breeding Two Million Cattle to Two Hundred Sires,* with a foreword by Joseph Edwards [Milk Marketing Board, Thames Ditton, Surrey, England] (Cambridge, Eng.: Cambridge University Press, 1959).

7. R. L. Powell, H. D. Norman, and Frank N. Dickinson, "Trends in Breeding Value and Production," *J. Dairy Sci.* 60:8 (1977): 1316–26.

8. Charles Rogers, "A Cow, a Bull, and a Man Changed the Destiny of Three Breeds of Dairy Cattle," *A.I. Digest* (September 1971).

9. Agricultural statistics (1977–1978), USDA, Washington, D.C.

10. Marion Eugene Ensminger, *Beef Cattle Science,* 5th ed. (Illinois: Interstate Printers and Publishers, 1980), p. 59.

11. "Livestock Breeding at the Crossroads," *USDA Yearbook of Agriculture* (1936), pp. 831–62.

12. "Guidelines for Uniform Beef Improvement Programs," USDA, Extension Service, 1976. Copies are available at Superintendent of Documents, U.S. Government Printing Office, Washington, D.C.

13. "Guidelines for Uniform Beef Improvement Programs," 1976.

14. Ibid.

15. Personal communication in 1976 with Dixon Hubbard, animal scientist, USDA, Extension Service, Washington, D.C.

16. Personal communication in 1975 with J. P. Smith, Agri-World Services, Inc., Amarillo, Tex.

17. Personal communication in July 1977 with Don Vaniman, executive secretary, American Simmental Association, Bozeman, Mont.

18. For a summary of AI regulations, see "A Summary of Artificial Insemination Regulations of Beef Breed Associations," NAAB, January 1977. Copies are available at Box 1033, Columbia, Mo.

19. "Crossbreeding for Iowa Beef Cattle Production," Pamphlet 561, Cooperative Extension Service, Iowa State University–Ames, 1975; L. V. Cundiff, "Crossbreeding with Different Biological Traits," *Proceedings of the Tenth Conference on Artificial Insemination of Beef Cattle,* NAAB. Copies are available at Box 1033, Columbia, Mo.

20. For details, see "Crossbreeding for Iowa Beef Cattle Production," 1975.

21. "Crossbreeding," panel discussion, *Proceedings of the Fourth Conference on Artificial Insemination of Beef Cattle* (1970), pp. 33–44, NAAB, Columbia, Mo.

22. "Guidelines for Uniform Beef Improvement Programs," Extension Service, USDA, Washington, D.C.

23. American Beefalo Breeders, Inc., 4152 Lapeer Road, Lapeer, Mich.

24. *Advanced Animal Breeder* (September 1977): 5.

25. *Proceedings of the Conferences on Artificial Insemination of Beef Cattle* (1967–1978), NAAB, Columbia, Mo.

26. For complete details on conducting a beef AI program, see "Better Beef with AI," NAAB, Columbia, Mo.; H. A. Herman and F. M. Madden, *Artificial Insemination of Cattle* (Columbia, Mo.: Lucas Brothers, 1974).

27. Robert G. Zimbleman, J. W. Lauderdale, and E. L. Moody, "Beef AI and Prostaglandin F_2 Alpha," *Advanced Animal Breeder* (May 1977): 15; John R. Beverly, "The Outlook for Estrus Synchronization," *Proceedings of the Twelfth Conference on Artificial Insemination of Beef Cattle,* NAAB, Columbia, Mo.

28. H. H. Stonaker, "Extension Views on Beef AI," *Proceedings of the First Conference on Artificial Insemination of Beef Cattle* (1967), p. 23, NAAB, Columbia, Mo.

Notes to Chapter 5

1. Enos J. Perry, ed., *The Artificial Insemination of Farm Animals* (New Brunswick, N.J.: Rutgers University Press, 1968), chap. 1.

2. J. P. Maule, *The Semen of Animals and Artificial Insemination* (Bucks, Eng.: Commonwealth Agricultural Bureaux, 1962).

3. P. H. Phillips and H. A. Lardy, "A Yolk-Buffer Pablum for the Preservation of Bull Semen," *J. Dairy Sci.* 23 (1940): 399–404.

4. G. W. Salisbury, H. K. Fuller, and E. L. Willet, "Preservation of Bovine Spermatozoa in Yolk-Citrate Diluent and Field Results from Its Use," *J. Dairy Sci.* 24 (1941): 905.

5. D. L. Thacker and J. O. Almquist, "Diluters for Bovine Semen: Fertility and Motility of Bovine Spermatozoa in Boiled Milk," *J. Dairy Sci.* 36:2 (1953): 173.

6. J. O. Almquist, W. T. S. Thorp, and C. B. Knodt, "The Effect of Penicillin upon the Livability, Fertility and Bacterial Content of Bull Semen," *J. Animal Sci.* 5 (1946): 400; J. O. Almquist, P. J. Glantz, and H. E. Schaefer, "The Effect of a Combination of Penicillin and

Streptomycin upon the Livability and Bacterial Content of Bovine Semen," *J. Dairy Sci.* 32:2 (1949): 183.

7. Perry, ed., *The Artificial Insemination of Farm Animals,* chap. 15.

8. Harry A. Herman, "How Long Can You Keep Frozen Semen?" *Hoard's Dairyman* 122:8 (25 April 1977): 550.

9. H. A. Herman and F. M. Madden, *Artificial Insemination of Dairy and Beef Cattle* (Columbia, Mo.: Lucas Brothers, 1974), pp. 105–8.

10. S. W. J. Seager, "Successful Pregnancies Utilizing Frozen Dog Semen," *A.I. Digest* 17:12 (December 1969).

11. Perry, ed., *The Artificial Insemination of Farm Animals.*

12. J. K. Sherman, "History of Artificial Insemination and the Development of Human Semen Banking"; "The Integrity of Frozen Spermatozoa" (Washington, D.C.: National Academy of Sciences, 1978), pp. 201–7.

13. Ibid.

14. J. R. Beverly, "The Outlook for Estrous Synchronization," *Proceedings of the Twelfth Conference on Artificial Insemination of Beef Cattle* (1978), pp. 28–45, NAAB, Columbia, Mo.

15. J. N. Wiltbank and S. E. Mares, "Breeding at a Pre-Determined Time Following Syncro-Mate-B-Treatment," *Proceedings of the Eleventh Conference on Artificial Insemination of Beef Cattle* (1977), pp. 57–65, NAAB, Columbia, Mo.

16. Various papers, *Proceedings of the Eighth International Congress on Animal Reproduction and Artificial Insemination* 3 (1976): 229–337, Institute of Zootechnics, Balice, Krakow, Poland.

17. "Now Embryo Transplants at Your Farm," *Farm Journal,* Dairy–5 (November 1977).

18. Various papers, *Proceedings of the Eighth International Congress on Animal Reproduction and Artificial Insemination* 3 (1976): 229–337.

19. "Ova Transfer News from Carnation–Genetics," *Advanced Animal Breeder* (November 1977), p. 10.

20. "Sexing and Transfer of Bovine Embryos," *Proceedings of the Eighth International Congress on Animal Reproduction and Artificial Insemination* 3 (1976): 258–61.

21. "Better Beef Business," *Better Beef Business* (February 1976), p. 47.

22. "Sex Ratio at Birth: Prospects for Control, A Symposium," American Society of Animal Science, Champaign, Ill., 1971; B. C. Bhattacharya, "Sex Selection: How Soon?" *Farm Journal* (January 1975), pp. 8–9.

23. Ibid.

24. R. G. Saacke and J. M. White, "Semen Quality Tests and Their Relationship to Fertility," *Proceedings of the Fourth Technical Conference on Artificial Insemination and Reproduction* (1972), NAAB, Columbia, Mo.

25. P. L. Senger, "Head-to-Head Agglutination of Bovine Spermatozoa," *Advanced Animal Breeder* (November 1977), p. 13.

26. J. Henion, "The Use of X-Ray Fluorescence Spectrometry for Monitoring Tagged Inks in Semen Packages," *Advanced Animal Breeder* (January 1978), p. 16.

27. J. J. Reeves and R. W. Wright, "Antibodies against Embryos Used for Pregnancy Diagnosis in Cattle," *Advanced Animal Breeder* (November 1977), p. 12.

28. D. M. Holm, "Agricultural Uses of Electronic Identification," *Mini-Review* (November 1976), pp. 76–82, Los Alamos Scientific Laboratory, Los Alamos, N.M.

Notes to Chapter 6

1. W. Belanski, "An Attempt to Determine the Number of Cows Inseminated in the World," *A.I. Digest* 11:1 (1963).

2. Y. Nishikawa, *History and Development of Artificial Insemination in the World* (Japan: Kyoto University, 1965).

3. Telesforo Bonadonna, *Proceedings of the Eighth Congress on Animal Reproduction and Artificial Insemination* (1976), Institute of Zootechnics, Balice, Krakow, Poland.

4. Agricultural statistics (1975), USDA, Washington, D.C.

5. Data taken from Telesforo Bonadonna, "VI International Enquiry into Artificial Insemination in the World (1971–1973)," *Zootec. Vet.* 30 (1975): 1–108; and personal communication with various specialists.

6. Enos J. Perry, ed., *The Artificial Insemination of Farm Animals*, 4th rev. ed. (New Brunswick, N.J.: Rutgers University Press, 1968).

7. Ibid.

8. Harry A. Herman, "A Return to the Origin of AI," *Hoard's Dairyman* 117:6 (25 March 1972): 394.

9. "National Dairy Herd 1955 Census," Milk Marketing Board, Thames Ditton, Surrey, Eng.

10. "Breeding 10,000,000 Cattle, 1959," Milk Marketing Board, Thames Ditton, Surrey, Eng.

11. A. B. Fowler, *Farming News* 48:3 (1942).

12. J. Anderson, *The Semen of Animals and Its Use for Artificial Insemination* (Edinburgh, Scotland: Imperial Bureau of Animal Breeding and Genetics, 1946), p. 106.

13. Personal communication in June 1980 with Peilieu Cheng, of the Chinese Academy of Agricultural Sciences in Peking.

14. Brida von Koskull, "Nylands och Kymmene seminforening" (Wester Kulla, Oitbacka, 1969).

15. "Twenty Years of Artificial Insemination in Finland" (Central Association of AI Societies, Kuriiritie, Tikkurila, 1968).

16. Personal communication in 1975 with F. Mignon, manager of Union des Cooperatives D' Insemination Artificielle in Paris.

17. Horst Kräußlich, "Twenty Years Experience with AI Programmes in Bavaria," part I, *Beef Cattle Science Handbook*, vol. 13 (Clovis, Calif.: Agriservices Foundation, 1976), p. 16.

18. J. D. Sampath Kumaran and Jay S. Kumaran, *Artificial Insemination and Animal Reproduction*, vol. 1, 2d ed. (Nainital, India: U. P. Agricultural Press in Pantnagar, 1968).

19. J. P. Maule et al., *The Semen of Animals and Artificial Insemination* (Bucks, Eng.: Commonwealth Agricultural Bureaux, Farnham Royal, 1962), p. 379.

20. Personal communication in 1978 with Reuven Bar-Anan, Ministry of Agriculture in Hakirya, Tel Aviv, Israel.

21. M. R. Heiman, "Artificial Insemination and Its Effect on Cattle Breeding in Israel by 'ON' AI Coop." (Sarid, Israel: "ON" AI Cooperative, 1969).

22. Ibid.

23. Y. Nishikawa, *Sixty Years of Artificial Insemination in Japan* (Kyoto, Japan: College of Agriculture, 1973).

24. Y. Nishikawa, *Theory and Practice of AI in Farm Animals*, 1st ed. (Tokyo, Japan: Yokendo, 1944).

25. H. Nagase, "Survey on the Experiments on Frozen Horse Semen, No. 3" (Japan Horse Association, Tokyo, 1970).

26. Nishikawa, *Sixty Years of Artificial Insemination in Japan*.

27. Personal communication in August 1969 with A. Rutgers, livestock officer for AI in Utrecht, Netherlands.

28. "New Zealand Dairy Board Report" (1975–1976); and personal observations.

29. New Zealand Dairy Board, "52nd Farm Production Report" (1975–1976), Wellington, New Zealand.

30. "Artificial Insemination of Livestock in the USSR" (1960), translated from Russian in 1964 by the Israel Program for Scientific Translations, Office of Technical Services, U.S. Department of Commerce, Washington, D.C.

31. Agricultural statistics (1975), USDA, Washington, D.C.

32. Bonadonna, "VI International Enquiry into Artificial Insemination in the World (1971–1973)," pp. 1–108.

33. "Report of United States Livestock Production Team to USSR, 1974," ARS, USDA, Beltsville, Md.

34. L. P. Prahov, "Beef Raising in the USSR," *Beef Cattle Science Handbook*, vol. 13 (Clovis, Calif.: Agriservices Foundation, 1976), p. 66.

35. A. I. Beach, "The Raising of Highly Productive Herds of Black-and-White Cattle at 'Lesnoye' Livestock Enterprise," *Beef Cattle Science Handbook*, vol. 15 (Clovis, Calif.: Agriservices Foundation, 1978), p. 213.

36. L. P. Prahov, "Native Beef Cattle of the USSR," *Beef Cattle Science Handbook*, vol. 13 (Clovis, Calif.: Agriservices Foundation, 1976), p. 77.

Notes to Chapter 7

1. "The Semen of Animals and Artificial Insemination," *Technical Communication No. 15* (Commonwealth Bureaux of Animal Breeding and Genetics, Edinburgh, Scotland, 1962), p. 358.

2. "The World Food Situation and Prospects to 1985," *Foreign Agricultural Economics Report No. 98,* Economic Research Service, USDA.

3. R. Daubney, "The Suitability of the White Highlands of Kenya for Grade Cattle of European Breeds," *East Africa Agricultural Journal* (hereafter referred to as *E. Afr. Agr. J.*) 7 (1942): 127–41; S. Brody, *Bioenergetics and Growth* (New York, N.Y.: Reinhold Publishing Corp., 1945), chap. 11, pp. 295–98.

4. D. H. L. Rollinson and W. R. Nunn, *Development of Artificial Insemination in the Tropics,* chap. 20, pp. 358–95; and *The Semen of Animals and Artificial Insemination* (Edinburgh, Scotland: Commonwealth Agricultural Bureaux, 1962).

5. Ibid.

6. J. Anderson, "The Periodicity and Duration of Oestrus in Zebu and Grade Cattle," *Journal of Agricultural Science* 34 (1944): 57–58.

7. J. Anderson, "Artificial Insemination and Community Breeding of Cattle," *E. Afr. Agr. J.* 8 (1942): 2–8.

8. J. D. Struthers, "Artificial Breeding Enterprise in Central Africa" (Chifundi Farm, Sinoia, South Rhodesia, personal communication).

9. "Inseminacao Artificial" (Ministerio da Agricultura, DNPA-DIFRIA, Brasília, 1976).

10. Ibid.

11. P. Bhattacharya, *"Buffaloes": The Artificial Insemination of Farm Animals* (New Brunswick, N.J.: Rutgers University Press, 1968).

12. Ralph W. Phillips, "Breeding Livestock Adapted to Unfavorable Environments" (1949), FAO Publication, USDA, Washington, D.C.

13. Bhattacharya, *"Buffaloes."*

14. I have worked with AI programs in several tropical countries during 1958 through 1974, including three months in Brazil as a representative for the Volunteer Development Corps, Washington, D.C., in the summer of 1974.

Notes to Chapter 8

1. "First Managers' Meeting of Artificial Breeding Associations in U.S." (December 1946), NAAB files, Columbia, Mo.

2. Some confusion occurred in designating the name of early meetings during the formation of the NAAB. The first meeting, in 1946, was called a "Managers' Meeting"; the second meeting in 1947 was entitled the "First Annual Convention, National Association of Artificial Breeding Managers." The third annual meeting (convention) held 13–14 September 1948 at Ithaca, New York, was the first following adoption of the name *National Association of Artificial Breeders.* It was designated on the program as the *first* annual meeting of the NAAB. The 1949 meeting held 17–18 August at Dyersburg, Tennessee, was promoted as the *second* annual convention of NAAB, and the proceedings were so published. In order to follow the sequence of annual meetings as described in the files of the NAAB and in accord with published proceedings, the 1948 meeting is alluded to as the first following the organization of the association. The sequence of numbers was followed from 1948 until 1970. By action of the 1970 NAAB board, with the annual meeting and convention scheduled for Ohio (the birthplace of the NAAB) in 1971, it was decreed that, that year be designated as the Silver Anniversary. As a result, the 1971 annual meeting and its proceedings are indicated as the twenty-fifth.

3. Minutes of meeting at Cedar Rapids, Iowa, 29–30 September 1947, NAAB files, Columbia, Mo.

4. A complete set of the first bylaws of NAAB can be obtained at Box 1033, Columbia, Mo. As time progressed, these early bylaws were changed to fit changing conditions, but they still express the objectives, sentiment, and forward thinking voiced at these first meetings.

Notes to Chapter 9

1. Agricultural statistics (1975), USDA, Washington, D.C.
2. "1976 Dairy Producer Highlights," National Milk Producers Federation, Washington, D.C.
3. *Proceedings of the Second Annual Convention of the National Association of Artificial Breeders* (1949), NAAB, Columbia, Mo.
4. *Proceedings of the Third Annual Convention of the National Association of Artificial Breeders* (1950), NAAB, Columbia, Mo.
5. *Proceedings of the Fourth Annual Convention of the National Association of Artificial Breeders* (1951), NAAB, Columbia, Mo.
6. *Proceedings of the Fifth Annual Convention of the National Association of Artificial Breeders* (1952), NAAB, Columbia, Mo.
7. The code was originally adopted 24 September 1952, was amended September 1959, and was effective as adopted above 1 January 1960.

Notes to Chapter 10

1. *Proceedings of the Sixth Annual Convention of the National Association of Artificial Breeders* (1953), pp. 71–89, NAAB, Columbia, Mo.
2. In earlier days, the term *AB proof* was used instead of *AI proof*.
3. Henry Soborg was a native of Denmark. He was a graduate of the Royal Veterinary and Agricultural College in Copenhagen. During World War II, he was a member of the Danish Resistance Group, serving under both Danish and U.S. Army military leaders. He managed several large dairy herds in Denmark and Germany and for fourteen years operated his own farm with over three hundred head of dairy cattle. He came to the United States in 1948 and in 1950 became manager of Coop. Artificial Breeding Association No. 1 in Clinton, New Jersey. In 1953, he joined the Southern Illinois Breeding Association as sire analyst and served as manager from 1959 to 1960. He became manager of the New England Selective Breeding Association at Woodbridge, Connecticut, in November 1960 and served in that capacity until his death on 9 August 1963. He was sixty-one years old at the time.
4. Minutes of board meeting, *Proceedings of the Eighth Annual Convention of the National Association of Artificial Breeders* (1954–1955), NAAB, Columbia, Mo.

Notes to Chapter 11

1. *Proceedings of the Eleventh Annual Convention of the National Association of Artificial Breeders* (1958), NAAB, Columbia, Mo.
2. Minutes of board meeting, NAAB, Chicago, Illinois, 8–10 October 1958.
3. *Proceedings of the Twelfth Annual Convention of the National Association of Artificial Breeders* (30 August–2 September 1959), NAAB, Columbia, Mo.
4. *Proceedings of the Twelfth Annual Convention of the National Association of Artificial Breeders* (30 August–2 September 1959), NAAB, Columbia, Mo.

Notes to Chapter 12

1. Minutes of board meeting, 29–30 November 1959, NAAB.
2. *Proceedings of the Thirteenth Annual Convention of the National Association of Artificial Breeders* (1960), NAAB, Columbia, Mo.
3. *Dairy Herd Improvement Letter* 37:3 (March 1961), USDA.
4. *Dairy Herd Improvement Letter* 38:3 (1962), USDA.
5. *A.I. Digest* 9:2 (February 1961): 10.

6. Ibid., 9:7 (July 1961): 4.

7. *Dairy Herd Improvement Letter* 38:3 (1962), USDA.

8. Harry A. Herman, *Artificial Insemination of Dairy Goats* (Columbia, Mo.: American Supply House, 1973).

9. *A.I. Digest* 10:9 (1962): 20.

10. *Proceedings of the Fifteenth Annual Convention of the National Association of Artificial Breeders* (1962), NAAB, Columbia, Mo.

11. *Dairy Herd Improvement Letter* 40:4 (1964), USDA.

12. *Dairy Herd Improvement Letter* 40:1 (1964), USDA.

13. *Proceedings of the Sixteenth Annual Convention of the National Association of Artificial Breeders* (1963), NAAB, Columbia, Mo.

14. *NAAB News* 3:4 (July–August 1955): 19.

15. *Dairy Herd Improvement Letter* 41:3 (1965), USDA.

16. *Proceedings of the Seventeenth Annual Convention of the National Association of Animal Breeders* (1964), NAAB, Columbia, Mo.

Notes to Chapter 13

1. *Dairy Herd Improvement Letter* 42:3 (1966), USDA.

2. *Proceedings of the Nineteenth Annual Convention of the National Association of Animal Breeders* (1966), NAAB, Columbia, Mo.

3. *A.I. Digest* 14:10 (October 1966).

4. *Dairy Herd Improvement Letter* 43:3 (April 1967), USDA.

5. *Proceedings of the Twentieth Annual Convention of the National Association of Animal Breeders* (1967), NAAB, Columbia, Mo.

6. *Dairy Herd Improvement Letter* 44:3 (1968), USDA.

7. *Proceedings of the Second Technical Conference on Animal Reproduction and Artificial Insemination* (February 1968), NAAB, Columbia, Mo.

8. *Proceedings of the Twenty-first Annual Convention of the National Association of Animal Breeders* (1968), NAAB, Columbia, Mo.

9. *Proceedings of the Twenty-first Annual Convention of the National Association of Animal Breeders* (1968), NAAB, Columbia, Mo.

10. *Dairy Herd Improvement Letter* 45:2 (May 1969), USDA.

11. *A.I. Digest* 17:4 (April 1969).

12. Minutes of board meeting, 1–2 April 1969, NAAB, Columbia, Mo.

13. The term *DHIA* (Dairy Herd Improvement Association) was changed to DHI (Dairy Herd Improvement) by the National Dairy Herd Improvement Coordinating Group in 1969; and the Standard DHIA Plan was designated "Official DHI" to indicate it is the only plan other than DHIR (Dairy Herd Improvement Registry-sponsored by the dairy breed registry organizations) from which records are to be used to evaluate dairy bulls and cows. See *Dairy Herd Improvement Letter* 49:2 (July–September 1973), USDA.

14. *Proceedings of the Twenty-second Annual Convention of the National Association of Animal Breeders* (1969), NAAB, Columbia, Mo.

15. *Proceedings of the Twenty-second Annual Convention of the National Association of Animal Breeders* (1969), NAAB, Columbia, Mo.

Notes to Chapter 14

1. *Dairy Herd Improvement Letter* 46:4 (July 1970), USDA.

2. Ibid.

3. Carl Warren, *Animal Sex Control* (New York: Orange Judd Publishing, 1940).

4. "Sex Ratio at Birth: Prospects for Control—A Symposium" (American Society of Animal Science, Champaign, Ill., 1971).

5. *Proceedings of the Twenty-third Annual Convention of the National Association of Animal Breeders* (1970), NAAB, Columbia, Mo.

6. *Dairy Herd Improvement Letter* 47 (June–July 1971), USDA.

7. *Proceedings of the Fifth Conference on Artificial Insemination of Beef Cattle* (1971), NAAB, Columbia, Mo.

8. For a full account of the story covering the luncheon, see *A.I. Digest* 19:10 (October 1971).

Notes to Chapter 15

1. *Dairy Herd Improvement Letter* 48:4 (June–July 1972), USDA; 51:2 (1974), USDA.

2. *Proceedings of the Twenty-sixth Annual Convention of the National Association of Animal Breeders* (1972), NAAB, Columbia, Mo.

3. *Dairy Herd Improvement Letter* 51:2 (1975), USDA.

4. *Proceedings of the Seventh Annual Conference on Artificial Insemination of Beef Cattle* (1973), NAAB, Columbia, Mo.

5. *Proceedings of the Twenty-seventh Annual Convention of the National Association of Animal Breeders* (1973), NAAB, Columbia, Mo.; *A.I. Digest* 12:10 (October 1973).

Notes to Chapter 16

1. *Proceedings of the Twenty-eighth Annual Convention of the National Association of Animal Breeders* (1974), NAAB, Columbia, Mo.

2. Ibid.

3. *Proceedings of the Eighth Conference on Artificial Insemination of Beef Cattle* (1974), NAAB, Columbia, Mo.

4. Ibid.

5. *Proceedings of the Fifth Technical Conference on Artificial Insemination and Reproduction* (1974), NAAB, Columbia, Mo.

6. *Proceedings of the Twenty-ninth Annual Convention of the National Association of Animal Breeders* (1975), NAAB, Columbia, Mo.

7. Personal correspondence in 1977 with Jack Eichel, of United Suppliers, Inc., in Eldora, Iowa.

Notes to Chapter 17

1. *A.I. Digest* 33:11 (November 1975).

2. *Hoard's Dairyman* 121:7 (10 April 1976).

3. "CSS: Serving the Livestock Industry," Certified Semen Services, Columbia, Mo.

4. *Proceedings of the Thirtieth Annual Convention of the National Association of Animal Breeders* (1976), NAAB, Columbia, Mo.

5. *A.I. Digest* 24:8 (August 1976): 7.

6. "CSS: Serving the Livestock Industry."

7. *A.I. Digest* 24:7 (July 1976): 6.

8. *A.I. Digest* 24:2 (February–March 1977): 4.

9. *Advanced Animal Breeder* 25:5 (July 1977): 4.

10. *Proceedings of the Seventh Technical Conference on Artificial Insemination and Reproduction* (1978), NAAB, Columbia, Mo.

11. *Advanced Animal Breeder* 25:3 (April 1978): 6–7.

12. Charles Krumm, general manager of Eastern Artificial Insemination Cooperative in Ithaca, New York, was chosen as the 1979 recipient.

Notes to Appendix A

1. Agricultural statistics (1975), USDA.

2. *Dairy Herd Improvement Letter* 18:3 (March 1942), USDA.

3. J. B. Lingle and Charles R. Koch, *The Breed of Noble Bloods* (Princeton, Mass.: Princeton House, 1976).

4. Missouri Agricultural Experiment Station Bulletin No. 407, 1939.

5. Personal correspondence on 28 October 1972 with Charles E. Hunt.

6. Program of the "Eighty-seventh Annual Convention of the Holstein-Friesian Association of America" (1972), pp. 12–14.

7. There are unconfirmed reports that as early as 1920 AI was used in this herd by transferring semen from a cow that was naturally bred to another cow in heat.

8. Report by Fredolph Hendrickson, of the University of Wisconsin–Madison, 1948.

9. Agricultural statistics (1975), USDA.

10. "Nineteenth Annual Report: Artificial Insemination in Canada" (1973), Livestock Division, Canada Department of Agriculture, Ottawa; personal correspondence on 27 May 1977 with Roy Snyder, of Semex, in Guelph, Ontario.

11. Canadian statistics of 1 January 1974. The statistics are based on cows and heifers kept mainly for milk, and twenty-four months or older.

Notes to Appendix B

1. The years in parentheses indicate the year of appointment to committees.

Bibliographical Note

The material for *Improving Cattle by the Millions* comes from many sources. The most pertinent references are mentioned in the footnotes of each chapter. Use was made of published reports on cattle improvement and artificial insemination in many countries; personal correspondence; and articles in various farm magazines and scientific journals. By virtue of being a pioneer worker in the AI program and the first executive secretary of NAAB (1953–1972), I have drawn heavily upon my own firsthand knowledge of the initiation and development of the AI program.

For sources of information on genetics and cattle development as developed prior to and following the introduction of AI and modern methods of sire evaluation, the reader will find the following texts of interest. Jay L. Lush, *Animal Breeding Plans*, 3d ed. (Ames: Iowa State University Press, 1945); I. M. Lerner and H. P. Donald, *Modern Developments in Animal Breeding* (New York, N.Y.: Academic Press, 1966); Everett J. Warwick and James E. Legates, *Breeding and Improvement of Farm Animals*, 6th ed. (New York, N.Y.: McGraw-Hill, 1942–1979); John L. Lasley, *Genetics of Livestock Improvement* (New Jersey: Prentice-Hall, 1979); and Richard C. Foley, Donald L. Bath, Frank N. Dickinson, and H. Allen Tucker, *Dairy Cattle: Principles, Practices, Problems, Profits,* 2d ed. (Philadelphia, Pa.: Lea and Febiger, 1972) 3d ed. (Philadelphia: Lea and Febiger, 1979).

The history and development of the AI program in the United States was documented in part from reports carried in the *Dairy Herd Improvement Letter* (USDA, Washington, D.C.) from the years 1936 to 1979. Generous use was also made of reports and summaries of the program from the NAAB files from 1946 to date. Included were reports appearing in the *NAAB News,* the *A.I. Digest,* and beginning in May 1977 the *Advanced Animal Breeder,* all official publications of NAAB; proceedings of the annual NAAB conventions (1947 to date); and proceedings of various conferences. These publications are found in many agriculture college libraries and likewise at the headquarters of NAAB, in Columbia, Missouri.

The history in development of the program by states and organizations was compiled by use of the above related sources; by a survey conducted from 1972 to 1977 by the author involving the extension dairymen, and agriculture colleges in each state, and the existing AI organizations located in each state. Use was also made of the early history of the NAAB by Samuel B. Rose, Findlay, Ohio, who was the first NAAB historian. Rose collected information (unpublished) on the AI program in each state through 1957, when the responsibility for historical records was delegated to the NAAB office.

The information on the worldwide use of AI resulted from a survey conducted by the writer during 1972–1978 of the principal livestock countries of the world. Results of surveys made by W. Belanski, of Poland, in 1963 (*A.I. Digest* 11:1), and Telesforo Bonadonna, of Italy, during 1974–1976 (*Proceeding VIII Congress on Animal Reproduction and Artificial Insemination,* Krakow, Poland) were also utilized. Information was further augmented by journal reports and correspondence from workers in overseas countries. The author on two occasions visited AI organizations throughout Europe and assisted leaders of the program in Central America and South America, thus gaining much firsthand information utilized in this writing.

The history of NAAB and the AI program in the United States is based upon reports of the annual meetings and conferences, minutes of directors' meetings, reports from NAAB members, and correspondence. The developments from 1946 to date are fairly well documented, and many pictures are maintained in the NAAB files.

Other sources of historical material are the reports and articles published in farm publications, particularly *Hoard's Dairyman* (from 1937 to date), the journal of each dairy breed registry association, and the publications of the beef recording associations. These publications are found in most libraries or can be obtained by contacting the organization concerned.

Other very important information can be found in hundreds of articles that have appeared in scientific publications from 1936 to date, which report the results of research and the developments regarding cattle improvement methods. These include reproductive efficiency, semen preservation, modern methods for sire evaluation, progeny testing, and herd management practices. The reader is referred to the *Journal of Dairy Science,* which is published by the American Dairy Science Association (Champaign, Illinois) and the *Journal of Animal Science,* which is published by the American Society of Animal Science (Champaign, Illinois). In addition, there are numerous articles on the technical phases of the AI program carried in the journals for agriculture and for veterinary medicine from countries around the world. Reference to these journals can be found in most libraries of agricultural colleges and universities.

Index